Beta-Glucan in Foods and Health Benefits

Beta-Glucan in Foods and Health Benefits

Editors

Seiichiro Aoe
Tatsuya Morita
Naohito Ohno

MDPI • Basel • Beijing • Wuhan • Barcelona • Belgrade • Manchester • Tokyo • Cluj • Tianjin

Editors

Seiichiro Aoe
Department of Food Science,
Otsuma Women's University,
Tokyo, Japan

Tatsuya Morita
Department of Applied
Life Sciences,
Shizuoka University, Japan

Naohito Ohno
School of Pharmacy, Tokyo
University of Pharmacy and
Life Science School of Pharmacy,
Tokyo 192-0392, Japan

Editorial Office
MDPI
St. Alban-Anlage 66
4052 Basel, Switzerland

This is a reprint of articles from the Special Issue published online in the open access journal
Nutrients (ISSN 2072-6643) (available at: https://www.mdpi.com/journal/nutrients/special_issues/
Beta_Glucan_in_Foods_and_Health_Benefits).

For citation purposes, cite each article independently as indicated on the article page online and as
indicated below:

LastName, A.A.; LastName, B.B.; LastName, C.C. Article Title. *Journal Name* **Year**, *Volume Number*,
Page Range.

ISBN 978-3-0365-5107-4 (Hbk)
ISBN 978-3-0365-5108-1 (PDF)

Contents

Preface to "Beta-Glucan in Foods and Health Benefits"

Many articles and manuscripts focusing on the structure, function, mechanism of action, and effects of β-glucan have been published recently. Beta-glucan is a general term for polysaccharides that consist of β-bonds. Structural studies report that combinations of β-1,3 and β-1,6 bonds form long linear β-glucans, and these structures can be detected by specific intestinal receptors, such as dectin-1, which then stimulate the immunological system. Cereal β-glucans that have been derived from barley and oats have also been widely researched in both animal and human studies. They are water soluble, viscous polysaccharides with a linear structure in which glucose is bound through β-1,4 and β-1,3 linkages. Many physiological functions, such as anti-obesity effects, reductions in postprandial glucose increases, and the normalization of serum cholesterol levels have been reported. The recent interest in barley and oat β-glucans has been sparked by reports discussing their prebiotic action, which is dependent on molecular weight. A marketing report discussing the health benefits of the β-glucans in oats and barley products has also been published. Another β-glucan that has been recently reported on is paramylon, a linear β-1,3-glucan in which glucose is β-1,3 bound and that is abundant in Euglena gracilis. It is an insoluble and unfermentable polysaccharide which is reported to have various physiological functions, including anti-obesity effects and anti-diabetic effects, and has been shown to stimulate immune function.

This Special Issue entitled "β-glucan in foods and health benefits" reports on the health benefits of indigestible carbohydrates with respect to metabolic diseases and immune functions. The effects of β-glucan have been investigated through the use isolated preparations or natural dietary fibers from whole grain cereals and brans, yeasts, or Euglena. This Special Issue includes original research articles that are based on human intervention studies that address the effects of β-glucan on metabolic diseases and immune function-related markers as well as in vitro and in vivo studies. It also reviews the health benefits of β-glucans in humans.

Seiichiro Aoe, Tatsuya Morita, and Naohito Ohno
Editors

 nutrients

Editorial

Beta-Glucan in Foods and Health Benefits

Seiichiro Aoe

The Institute of Human Culture Studies, Otsuma Women's University, Chiyoda-ku, Tokyo 102-8357, Japan; s-aoe@otsuma.ac.jp

Many articles and manuscripts focusing on the structure, function, mechanism of action, and effects of β-glucan have been published recently. Beta-glucan is a general term for polysaccharides that consist of β-bonds. Structural studies report that combinations of β-1,3 and β-1,6 bonds form long linear β-glucans, and these structures can be detected by specific intestinal receptors, such as dectin-1, which then stimulate the immunological system [1]. Cereal β-glucans that have been derived from barley and oats have also been widely researched in both animal and human studies [2–5]. They are water soluble, viscous polysaccharides with a linear structure in which glucose is bound through β-1,4 and β-1,3 linkages. Many physiological functions, such as anti-obesity effects, reductions in postprandial glucose increases, and the normalization of serum cholesterol levels have been reported [6,7]. The recent interest in barley and oat β-glucans has been sparked by reports discussing their prebiotic action, which is dependent on molecular weight [8,9]. A marketing report discussing the health benefits of the β-glucans in oats and barley products has also been published [10]. Another β-glucan that has been recently reported on is paramylon, a linear β-1,3-glucan in which glucose is β-1,3 bound and that is abundant in Euglena gracilis. It is an insoluble and unfermentable polysaccharide which is reported to have various physiological functions, including anti-obesity effects and anti-diabetic effects, and has been shown to stimulate immune function [11,12].

This Special Issue entitled "β-glucan in foods and health benefits" reports on the health benefits of indigestible carbohydrates with respect to metabolic diseases and immune functions. The effects of β-glucan have been investigated through the use isolated preparations or natural dietary fibers from whole grain cereals and brans, yeasts, or Euglena. This Special Issue includes original research articles that are based on human intervention studies that address the effects of β-glucan on metabolic diseases and immune function-related markers as well as in vitro and in vivo studies. It also reviews the health benefits of β-glucans in humans.

Funding: This research received no external funding.

Conflicts of Interest: The author declares no conflict of interest.

Citation: Aoe, S. Beta-Glucan in Foods and Health Benefits. *Nutrients* **2022**, *14*, 96. https://doi.org/10.3390/nu14010096

Received: 30 November 2021
Accepted: 16 December 2021
Published: 27 December 2021

Publisher's Note: MDPI stays neutral with regard to jurisdictional claims in published maps and institutional affiliations.

References

1. Suzuki, T.; Kusano, K.; Kondo, N.; Nishikawa, K.; Kuge, T.; Ohno, N. Biological Activity of High-Purity β-1,3-1,6-Glucan Derived from the Black Yeast Aureobasidium Pullulans: A Literature Review. *Nutrients* **2021**, *13*, 242. [CrossRef] [PubMed]
2. Wolever, T.M.S.; Mattila, N.; Rosa-Sibakov, N.; Tosh, S.M.; Jenkins, A.L.; Ezatagha, A.; Duss, R.; Steinert, R.E. Effect of Varying Molecular Weight of Oat β-Glucan Taken just before Eating on Postprandial Glycemic Response in Healthy Huma. *Nutrients* **2020**, *12*, 2275. [CrossRef] [PubMed]
3. Kopiasz, Ł.; Dziendzikowska, K.; Gajewska, M.; Oczkowski, M.; Majchrzak-Kuligowska, K.; Królikowski, T.; Gromadzka-Ostrowska, J. Effects of Dietary Oat Beta-Glucans on Colon Apoptosis and Autophagy through TLRs and Dectin-1 Signaling Pathways—Crohn's Disease Model Study. *Nutrients* **2021**, *13*, 321. [CrossRef] [PubMed]
4. Wolever, T.M.S.; Rahn, M.; Dioum, E.H.; Jenkins, A.L.; Ezatagha, A.; Campbell, J.E.; Chu, Y. Effect of Oat β-Glucan on Affective and Physical Feeling States in Healthy Adults: Evidence for Reduced Headache, Fatigue, Anxiety and Limb/Joint Pains. *Nutrients* **2021**, *13*, 1534. [CrossRef] [PubMed]
5. Gudej, S.; Filip, R.; Harasym, J.; Wilczak, J.; Dziendzikowska, K.; Oczkowski, M.; Jałosińska, M.; Juszczak, M.; Lange, E.; Gromadzka-Ostrowska, J. Clinical Outcomes after Oat Beta-Glucans Dietary Treatment in Gastritis Patients. *Nutrients* **2021**, *13*, 2791. [CrossRef] [PubMed]
6. Mio, K.; Yamanaka, C.; Matsuoka, T.; Kobayashi, T.; Aoe, S. Effects of Beta-Glucan Rich Barley Flour on Glucose and Lipid Metabolism in the Ileum, Liver, and Adipose Tissues of High-Fat Diet Induced-Obesity Model Male Mice Analyzed by DNA Microarray. *Nutrients* **2020**, *12*, 3546. [CrossRef] [PubMed]
7. Suzuki, S.; Aoe, S. High β-Glucan Barley Supplementation Improves Glucose Tolerance by Increasing GLP-1 Secretion in Diet-Induced Obesity Mice. *Nutrients* **2021**, *13*, 527. [CrossRef] [PubMed]
8. Mio, K.; Ootake, N.; Nakashima, S.; Matsuoka, T.; Aoe, S. Ingestion of High β-Glucan Barley Flour Enhances the Intestinal Immune System of Diet-Induced Obese Mice by Prebiotic Effects. *Nutrients* **2021**, *13*, 907. [CrossRef] [PubMed]
9. Aoe, S.; Mio, K.; Yamanaka, C.; Kuge, T. Low Molecular Weight Barley β-Glucan Affects Glucose and Lipid Metabolism by Prebiotic Effects. *Nutrients* **2021**, *13*, 130. [CrossRef] [PubMed]
10. Hughes, J.; Grafenauer, S. Oat and Barley in the Food Supply and Use of Beta Glucan Health Claims. *Nutrients* **2021**, *13*, 2556. [CrossRef] [PubMed]
11. Yasuda, K.; Nakashima, A.; Murata, A.; Suzuki, K.; Adachi, T. Euglena Gracilis and β-Glucan Paramylon Induce Ca^{2+} Signaling in Intestinal Tract Epithelial, Immune, and Neural Cells. *Nutrients* **2020**, *12*, 2293. [CrossRef] [PubMed]
12. Aoe, S.; Yamanaka, C.; Mio, K. Microarray Analysis of Paramylon, Isolated from Euglena Gracilis EOD-1, and Its Effects on Lipid Metabolism in the Ileum and Liver in Diet-Induced Obese Mice. *Nutrients* **2021**, *13*, 3406. [CrossRef] [PubMed]

Review

Biological Activity of High-Purity β-1,3-1,6-Glucan Derived from the Black Yeast *Aureobasidium pullulans*: A Literature Review

Toshio Suzuki [1,*], Kisato Kusano [2], Nobuhiro Kondo [3], Kouji Nishikawa [4], Takao Kuge [5,*] and Naohito Ohno [6]

[1] Research and Development Laboratories, Fujicco, Co., Ltd., 6-13-4 Minatojima-Nakamachi, Chuo-ku, Kobe, Hyogo 650-8558, Japan

[2] Aureo Co., Ltd., 54-1, Kazusa Koito, Kimitsu-shi, Chiba 292-1149, Japan; kusano-kisato@aureo.co.jp

[3] Research and Development Division, Itochu Sugar Co., Ltd., 3, Tamatsuura, Hekinan, Aichi 447-8506, Japan; nobuhiro.kondo@itochu-sugar.co.jp

[4] Innovation Center, Osaka Soda Co., Ltd., 9, Otakasu-cho, Amagasaki, Hyogo 660-0842, Japan; knishika@osaka-soda.co.jp

[5] Life Science Materials Laboratory, ADEKA Corporation., 7-2-34, Higashi-Ogu, Arakawa-ku, Tokyo 116-8553, Japan

[6] Tokyo University of Pharmacy and Life Sciences, 1432-1, Horinouchi, Hachioji, Tokyo 192-0392, Japan; ohnonao@toyaku.ac.jp

* Correspondence: t-suzuki@fujicco.co.jp (T.S.); kuge@adeka.co.jp (T.K.); Tel.: +81-78-303-5385 (T.S.); +81-3-4455-2829 (T.K.)

Citation: Suzuki, T.; Kusano, K.; Kondo, N.; Nishikawa, K.; Kuge, T.; Ohno, N. Biological Activity of High-Purity β-1,3-1,6-Glucan Derived from the Black Yeast *Aureobasidium pullulans*: A Literature Review. *Nutrients* **2021**, *13*, 242. https://doi.org/10.3390/nu13010242

Received: 14 December 2020
Accepted: 13 January 2021
Published: 16 January 2021

Abstract: The black yeast *Aureobasidium pullulans* produces abundant soluble β-1,3-1,6-glucan—a functional food ingredient with known health benefits. For use as a food material, soluble β-1,3-1,6-glucan is produced via fermentation using sucrose as the carbon source. Various functionalities of β-1,3-1,6-glucan have been reported, including its immunomodulatory effect, particularly in the intestine. It also exhibits antitumor and antimetastatic effects, alleviates influenza and food allergies, and relieves stress. Moreover, it reduces the risk of lifestyle-related diseases by protecting the intestinal mucosa, reducing fat, lowering postprandial blood glucose, promoting bone health, and healing gastric ulcers. Furthermore, it induces heat shock protein 70. Clinical studies have reported the antiallergic and triglyceride-reducing effects of β-1,3-1,6-glucan, which are indicators of improvement in lifestyle-related diseases. The primary and higher-order structures of β-1,3-1,6-glucan have been elucidated. Specifically, it comprises a single highly-branched glucose residue with the β-1,6 bond (70% or more) on a backbone of glucose with 1,3-β bonds. β-Glucan shows a triple helical structure, and studies on its use as a drug delivery system have been actively conducted. β-Glucan in combination with anti-inflammatory substances or fullerenes can be used to target macrophages. Based on its health functionality, β-1,3-1,6-glucan is an interesting material as both food and medicine.

Keywords: *Autreobasidium pullulans*; β-1,3-1,6-glucan; physiological function

1. Introduction

β-1,3-Glucan, a well-known functional food ingredient derived from mushrooms, has been reported to exhibit antitumor activity [1–5]. The Reishi mushrooms (*Ganoderma lucidum*) and other traditional medicines have been used for generations. For instance, schizophyllan derived from the Suehirotake mushroom (*Schizophyllum commune*) and lentinan derived from the Shiitake mushroom (*Lentinus edodes*) are manufactured and traded as injectable anticancer drugs. These compounds comprise β-1,3-glucan with β-1,6-branched structures. In addition to mushrooms, β-1,3-glucans are derived from ascomycetes and other fungi, such as baker's yeast and black yeast, which also compromise β-1,6-branched structures. Furthermore, β-glucans have been derived from oat and barley, which possess repeating structures with β-1,3 and β-1,4 bonds in their unbranched primary chains. β-Glucans derived from seaweeds, such as laminaran, comprise repeating structures with β-1,3 and β-1,6 bonds (Table 1).

Table 1. Some types of β-glucan with different structures derived from different natural sources.

Type of β-glucan Structure	Natural Source and Trivial Name of β-glucan
1,3-β-glucan (linear unbranched, homogeneous)	bacterium *Alcaligenes faecalis*, curdlan algae *Euglena gracilis*, paramylon fungus *Poria cocos*, pachyman grape *Vitis vinifera*, callose
1,3-1,6-β-glucan (linear with 1,6-linked β-glucosyl side branches)	algae *Laminaria* sp. laminarin (unbranched) algae *Eisenia bicyclis*, laminarin (some branched) fungus *Claviceps purpuria*, cell wall glucan fungus *Sclerotinia sclerotiorum*, cell wall or extracelluar glucan fungus (black yeast) *Aureobasidium pullulas*, extracellular glucan fungus/mushroom *schizophyllum commune*,extracelluar or cell wall glucan mushroom *Grifola frondosa*, cell wall glucan mushroom *Lentinula edodes*, cell wall glucan
1,3-1,6-β-glucan (branch on branch structure)	yeast *Saccharomyces cerevisiae*, cell wall glucan yeast *Candida albicans*, cell wall glucan yeast *Candia utilis*, cell wall glucan
1,3-1,4-β-glucan (linear)	cereal β-glucan, such as barley, oat, wheat, and rye Iceland moss *Centraria islandia*, lichenin
1,3-1,4-β-glucan (linear with 1,4-lincked β-glucosyl side branches)	oyster mushroom *Pleurotus ostreatus*, cell wall glucan

In particular, many studies in Europe and North America have investigated the effects of barley-derived β-glucans on lifestyle-related diseases, such as coronary heart disease and diabetes. In May 2006, the U.S. Food and Drug Administration acknowledged a health claim label stating that β-glucan "helps reduce the risk of coronary heart disease." In October 2009, Europe also approved a health claim that β-glucan "helps maintain normal blood cholesterol." In Japan, a Foods with Function Claim under the label of "cholesterol reduction effect and intestinal control effect" has been submitted. Recently, a study was conducted to clarify the physiological role underlying the metabolic benefits of barley β-glucan using a mouse model of high-fat diet (HFD)-induced obesity. Then, metabolic parameters, gut microbial composition, and the production of fecal short-chain fatty acids (SCFAs) were analyzed. As a result, the beneficial metabolic effects of barley β-glucan were found to be primarily due to the suppression of appetite and improvement of insulin sensitivity, which are induced by gut hormone secretion promoted via the gut microbiota, and subsequently induced the production of fecal SCFAs [6].

Furthermore, an important source of β-glucans is the cell wall of yeast, such as *Saccharomyces cerevisiae*, of which 55–60% is β-glucans. These β-glucans are built of a glucose backbone with β-1,3 linkages, from which short sidechains branch off linked by β-1,6 bonds. Approximately 30–35% of insoluble β-glucans are present in the inner layer of the yeast cell wall, 20–22% of soluble β-glucans in the middle layer, and 30% of glycoproteins in the outer layer. Therefore, β-glucans in the yeast cell wall are present in two forms such as insoluble in bases and soluble, and yeast-derived β-glucans are basically insoluble in water without sufficient purification. An important and traditional β-glucan derived from yeast is zymosan, an insoluble polymer of glucose with a nanoparticle size of 1–2 μm in diameter, demonstrating strong antibacterial properties, activation of macrophages, and induction of cytokine secretion that enhances the immune system [7]. Moreover, β-glucan from *Candida utilis* has also been introduced as a novel source of yeast β-glucan [8,9].

However, the molecular mechanisms underlying the immunostimulatory effects of β-glucan remain to be elucidated. Recently, macrophages have been reported to harbor receptors for β-1,3-1,6-glucans. Consequently, β-glucan has attracted much attention as a potent antigenic agent that stimulates innate immunity [10]. Yadomae (2000) [2] and Ohno (2000) [3] reported that if all β-1,6-branch side chains were removed from β-1,3-1,6-glucan, such as yeast β-glucan (zymosan) by a chemically oxidized reaction, little anticancer activity is observed compared with that before the reaction, suggesting that β-1,3-glucan with β-1,6-branch enhances immunoactivity.

The black yeast *Aureobasidium pullulans* produces abundant soluble β-1,3-1,6-glucan, which possesses strong immunostimulatory activity. *A. pullulans* (formerly known as *Pullularia pullulans, Hormonema dematioides,* or *Dematium pullulans*) is an imperfect fungus, which is common in nature and typically found growing in soil and water, as well as on weathered wood and many other plants. Recently, *A. pullulans* was taxonomically classified as an anamorph of the teleomorph *Sydowia polyspora*. In another taxonomic study, de novo genome sequencing of four *A. pullulans* strains was performed [11,12].

The recent taxonomic classification of *A. pullulans* is as follows:

- Kingdom: Fungi
- Phylum: Ascomycota
- Class: Dothideomycetes
- Subclass: Dothideomycetidae
- Order: Dothideales
- Family: Dothioraceae
- Genus: *Aureobasidium*
- Species: *A. pullulans*

In 1969, Arkadjeva first reported the presence of functional polysaccharides in *A. pullulans*. In addition [13], Han et al. (1976) used this fungus as a highly nutritious single-cell proteinaceous food produced via fermentation using ryegrass hydrolysate as the substrate [14]. β-glucan has also been produced via fermentation using soybean as the substrate. Furthermore, Anastassiadis et al. (2007) have reported gluconic acid production [15], and Bharti et al. (2013) have reported fructo-oligosaccharide production [16]. At present, functional food products containing *A. pullulans*-derived, high-purity, solubilized β-1,3-1,6-glucan produced by several Japanese companies are commercially available [5]. Moreover, the structural details and the three-dimensional structure of β-glucan have also been clarified, and studies on its use as a drug delivery system have also been conducted [4,17].

In this review, we discuss the production method, safety, and various physiological functions of β-glucan derived from the black yeast *A. pullulans* based on previous studies in MEDLINE database. We searched for articles in MEDLINE using the keywords "black yeasts," "functionality," and "clinical tests." We obtained 155 studies, including functional food cases in animals and clinical trials. Finally, a total of 47 relevant studies to biological activities, including those published by the authors of this review, were included and their results were summarized as a review of *A. pullulans*-derived β-glucan (APβG) functional food products.

2. Results

2.1. Production of APβG via Fermentation

During APβG production, food ingredients are used as the medium and sucrose as the sole carbon source for fermentation [18]. The obtained culture solution has a viscosity of several thousand cP (mPa·s). In addition, a unique manufacturing method by adjusting the metal salt concentration, pH, and temperature and reducing the viscosity of the culture solution has proven industrially useful. Accordingly, a stable fermentative production method for low-viscosity and high-purity APβG has been established.

In Japan, APβG is approved as a thickener under the label of "Existing Food Additives." In addition, the culture solution itself is also sold as functional food, and following recovery and purification, it is also used as a high-purity β-glucan material.

2.1.1. Fermentation

During fermentation, *A. pullulans* (AP) degrades sucrose as the carbon source into glucose and fructose under aerobic conditions and utilizes these sugars in that order. This is accompanied by β-glucan production. Consequently, the viscosity of the culture solution increases to several thousand cP. AP is an imperfect fungus and typically grows in the mycelial form; however, it assumes a yeast-like form at the APβG production stage—the reason it has acquired the moniker "black yeast." AP also produces black melanin in the late phase of culture, but ascorbic acid can be used to effectively suppress melanin production [18,19].

2.1.2. β-Glucan Recovery and Purification

Since the culture solution containing β-glucan has a high viscosity of several thousand cP (mPa.s) or higher, it can be directly used as a functional food material following sterilization. Upon purification, it is difficult to remove microorganisms because of its high viscosity. Therefore, various unique methods have been developed.

The black yeast culture solution is subjected to homogenization by physical stirring or a similar method, and then the yeast cells are separated through a filter press to obtain an aqueous polysaccharide solution. If necessary, APβG can be recovered as a precipitate using ethanol precipitation or a similar method and dried to obtain a powdered product of high purity [20].

β-glucan has a rigid triple helical structure under a neutral pH; however, under alkaline conditions, the structure changes to a random coil and the solution viscosity is reduced [1]. Consequently, microorganisms can be easily removed using a filter press and further dialysis steps can be performed. Finally, high-purity β-glucan can be mass produced using an alcohol (e.g., ethanol) precipitation method. Interestingly, the solubilized β-glucan thus obtained can form ultrafine particles (<200 nm) in an aqueous solution, which may be useful for stimulating intestinal immunity [18].

APβG can be precipitated by adding alum to the black yeast culture solution, following separation using a filter press to recover APβG. Thereafter, APβG is appropriately diluted with water and slurried, and then the pH is adjusted to a neutral range. Finally, the solution is subjected to hydrothermal treatment from 170 to 180 °C under 2 MPa for approximately 10 min. The purity of the solution is increased via filtration by adding diatomaceous earth or activated carbon and ultrafiltration following hydrothermal treatment; the final product is powdered using freeze drying. Of note, the hydrothermal treatment reduces molecular weight of APβG [21].

The concentration of high-purity APβG recovered by ethanol precipitation or a similar method is measured using the phenol–sulfuric acid method using glucose as a reference material; this method conforms to the one used for measuring reduced sugars. Furthermore, NMR spectroscopy can also be used for qualitative analysis. Structural analysis is performed using ^{13}C-^{1}H two-dimensional NMR spectroscopy. The C(1)-H signal integration values of β-1,3 and β-1,6 bonds are 4.8 and 4.5 ppm, respectively. With respect to the results, β-1,3 bonds are present in the primary chains and β-1,6 bonds in the side chain, confirming a high degree of branching. In addition, based on the analysis of enzymatic degradation by exotype β-1,3-glucanase, glucose, and gentiobiose have been confirmed as the products, and their structures have been demonstrated to harbor β-1,3 bonds in the primary chain and a single branched glucose with a β-1,6-bond showing over 70% branching. Gel chromatography using pullulan as a standard and sodium hydroxide as a developing solvent revealed that the MW of β-glucan thus produced is over 1,000,000 [18,19].

2.2. Structure and Bioactivity of APβG

APβG exhibits an array of physiological activities. The biological activity of β-glucan depends on its primary structure, conformation, and molecular weight. Therefore, the primary structure of APβG was clarified using NMR spectroscopy, and its physiological activity was assessed. It comprised a primary chain of β-1,3-D-glucan and side branches of β-1,6-β-D-glucopyranosyl units at every two residues as the major structure or a primary chain of β-1,3-D-glucan and a side chain of β-1,6-β-D-glucopyranosyl units every three residues as the minor structure [20]. Moreover, the higher-order conformation of APβG had a delicate triple helical structure [17].

In addition, APβG exhibits immunostimulatory effects, such as immune cell accumulation, as well as priming effects on the intestinal bacteria. According to many studies reviewed here, β-1,3-glucans with many β-1,6 glucopyranosyl branches, such as APβG, have unique structures, even though they are isolated from different organisms, including bacteria and plants, among others [5,20].

Interestingly, a unique hydrothermal process for the preparation of APβG may produce an active reagent [21]. Reprocessing APβG increased low molecular weight fractions, suppressive activities were markedly enhanced, and the resulting APβG was estimated to have a low MW of approximately 10,000. Lipopolysaccharide (LPS)-induced nitrogen oxide (NO) synthesis and tumor necrosis factor (TNF)-α production in RAW264.7 cells were suppressed by the resulting low molecular weight APβG in a dose-dependent manner. Therefore, low molecular weight fractions obtained by the hydrothermal processing of APβG may result in potential reagents that control inflammation induced by various pathogens [22].

2.3. Safety

The black yeast culture solution is a safe material for human consumption and has been registered as a thickener and stabilizer under the label "Existing Food Additives." Accordingly, it is also utilized in food products in Japan. According to the "Survey and Study on Review of Safety of Existing Food Additives" by the Japan Food Chemistry Research Foundation of the Ministry of Health, Labor and Welfare Foundation (June 2004), there were no problems associated with the administration of repeated doses of this solution for 90 days or later and the results of mutagenicity test were negative, confirming the safety of the black yeast culture solution as a food ingredient [23]. On the other hand, pullulan, which is an α-1,4-1,6-glucan, produced by *A. pullulans*, is admitted as a food additive in both the United States (NO.GRN000099, generally recognized as safe (GRAS) status) and Europe (EFSA-Q-2003-138); thus, *A. pullulans* are considered safe for consumption of food use.

In addition, β-1,3-1,6-glucan is safe for use in food products and cosmetics as a high-purity commercial ingredient as we describe below. In studies of APβG-containing food products in human participants, three times the recommended dose was administered for four months or two years and no obviously abnormal findings were noted in clinical tests, indicating its safety as a food material [24]. In addition, a Japanese company has been selling the APβG-containing food product named "*Aureobasidium* cultured solution" as a dietary supplement since 1999, and the estimated amount sold exceeds several hundred tons. Of note, there have been no reports of complaints related to adverse health effects from the customers (data not shown).

2.4. Physiological Function of β-1,3-1,6-glucan

Various physiological functions of β-1,3-1,6-glucan have been confirmed, and its novel functions have been revealed using oral administration experiments in animal models. Clinical trials have also been conducted. Table 2 summarizes the results of such studies, and the key physiological functions of β-1,3-1,6-glucan are described below.

Table 2. Various bioactive functions of β-1,3-1,6-glucan from black yeast *Aureobasidium pullulans* (APβG).

No.	Author, Year, Reference [1]	Item of Chracterization or Healthy Function	Outline of Functionality and Efficacy
1	Arkadjeva 1969 [13]	Single cell protein	The first information on bioactive polysaccharide from *A. pullulans*.
2	Han et al., 1976 [14]	Intestinal immunity	Production of single-cell protein from cellulosic wastes using *A. pullulans*. *A. pullulans* cells were not toxic, and the values of their feed intake, weight gain, and protein efficiency ratio were superior to those of other cells.
3	T. Suzuki et al., 2004 [25]	Intestinal immunity	Intestinal immunostimulatory and modulatory effects. Cellular-level in vitro experiments using mouse lymphocytes from Peyer's patch showed that immunoglobulin A was produced at APβG concentrations ranging from 0 to 200 µg/mL in a dose-dependent manner.
4	Kimura et al., 2006 [26]	Antitumor	Antitumor and antimetastatic actions in mice. The antitumor and antimetastatic actions of APβG may be involve in the enhancement of intestinal immune functions through the increases in NK- and IFN-gamma-positive cell numbers.
5	Kim et al., 2007 [27]	Reduction of inflamamatory	Reduction of the acute inflammatory responses induced by xylene application in mice. Xylene-induced acute inflammatory changes were significantly and dose-dependently decreased by beta-glucan treatment (up to 250 mg/kg).
6	Kimura et al., 2007 [28]	Anti food allergy	The anti-food allergic action of beta-glucan may be caused by the induction of IFN-g production in the small intestine and splenocytes. APβG diets (0.5–1%) significantly inhibited not only the OVA-specific IgE elevation but also reduced the production of IFN-g and the number of CD8- and IFN-gamma-positive cells from the splenocytes and in the small intestine, respectively.
7	Kimura et al., 2007 [29]	Anti stress	Inhibitory actions of APβG (100 mg/kg) on the increase in corticosterone level and reduction of NK activity induced by restraint stress. These effects may be associated with the abrogation of interleukin-6 (IL-6) and IL-12.
8	Shin et al., 2007 [30]	Reduction of osteoporosis	APβG exhibited favorable effects on ovariectomy-induced osteoporosis. It significantly and dose-dependently suppressed the decrease in bone weight, bone mineral content, failure load, bone mineral density, and serum calcium and phosphorus levels and the increase in serum osteocalcin levels.
9	Ikewaki et al., 2007 [31]	Immunomodulatory mechanism	APβG may have unique immune regulatory or enhancing properties. It stimulated the production of interleukin-8 (IL-8) or soluble Fas (sFas); however, it did not stimulate that of IL-1beta, IL-2, IL-6, IL-12 (p70 + 40), IFN-g, TNF-a, or soluble Fas ligand (sFasL), in either cultured PBMCs or cells of the human monocyte-like cell line U937.
10	Tada et al., 2008 [20]	Structure information	The primary structure of APβG and its biological activities were determined and evaluated, respectively using NMR spectroscopy. The structure comprises a mixture of a 1-3-beta-D-glucan backbone with single 1-6-beta-D-glucopyranosyl side-branching units in every two residues (major structure) and a 1-3-beta-D-glucan backbone with single 1-6-beta-D-glucopyranosyl side-branching units in every three residues (minor structure).

Table 2. *Cont.*

No.	Author, Year, Reference [1]	Item of Chracterization or Healthy Function	Outline of Functionality and Efficacy
11	Tada et al., 2009 [32]	Immunomodulation mechanism	Immunomodulatory effect of APβG on DBA/2 mouse-derived splenocytes *in vitro*. APβG strongly induced the production of various cytokines, especially Th1 cytokines (e.g., IFN-g and IL-12p70) and Th17 cytokines (e.g., IL-17A), but did not induce the production of IL-4, IL-10, and TNF-a on splenocytes in vitro.
12	Tada et al., 2009 [33]	Immunomodulatory mechanism	This is the first study in which the branched chains at position 6 of beta-D-glucan strongly contribute to its recognition by antibodies in human sera. APβG reacted to IgG antibodies in human sera and the IgGs recognize branched chains at position 6.
13	Kimura et al., 2009 [34]	Protection of intestine	Protective effects of APβG (50 and 100 mg/kg twice daily) against the toxicity of UFT (combination of tegafur (1-(2-tetrahydrofuryl)-5-fluorouracil) and uracil) in mice bearing colon 26 tumors. Histological analysis showed that the damage found in the villi of the small-intestine by UFT was inhibited by the orally administered beta-glucan.
14	Sumiyoshi et al., 2009 [35]	Reduction and control of blood glucose	Reduction and control of blood glucose level in mice. In the 100 mg/kg and 200 mg/kg APβG dose groups, the increase in blood glucose from 15 to 30 min after glucose administration was minimal, after which the blood glucose level decreased, and significantly decreased 60 min after administration.
15	Yoon et al., 2010 [36]	Immunomodulatory mechanism	Immunomodulatory effects of exopolymers of *A. pullulans* containing APβG, which were orally administered at 10 mL/kg, were evaluated on cyclophosphamide (CPA)-treated mice. APβG can be effectively used to prevent an immunosuppress mediated (at least partially) and the recruitment of T cells and TNF-a-positive cells or enhancement of their activity.
16	Tada et al., 2011 [37]	Immunomodulatory mechanism	Immunomodulatory production of various cytokines in DBA/2 mouse-derived splenocytes in vitro was found via dectin-1-independent pathways. The production of IFN-γ in DBA/2 mouse-derived splenocytes by APβG was not inhibited following a treatment with an anti-dectin-1 neutralizing antibody.
17	Tada et al., 2011 [38]	Immunomodulatory mechanism	The induction of cytokines by APβG was dependent on the existence of a granulocyte macrophage colony-stimulating factor (GM-CSF). GM-CSF is indispensable for the induction of cytokines by APβG in mouse-derived splenocytes, similar to a typical 1,3-β-D-glucan from *Sparassis crispa* (SCG).
18	Tanaka et al., 2011 [39]	Reduction of ulcer	Oral administration of APβG (>100 mg/kg) ameliorated the gastric lesions induced by ethanol (EtOH) or HCl in mice. The administration of APβG also suppressed EtOH-induced inflammatory responses through the protection of the gastric mucosa from the formation of irritant-induced lesions by increasing the levels of defensive factors, such as HSP70 and mucin.
19	Muramatsu et al., 2012 [40]	Anti virus	Oral administration of *A. pullulans*-cultured fluid enriched with APβG exhibits efficacy in protecting mice infected with a lethal titer of the A/Puerto Rico/8/34 (PR8; H1N1) strain of the influenza virus.

Table 2. *Cont.*

No.	Author, Year, Reference [1]	Item of Chracterization or Healthy Function	Outline of Functionality and Efficacy
20	Ku et al., 2012 [41]	Anti allergy	The effect of APβG, orally administered at 125 mg/kg, on ovalbumin (OVA)-induced allergic asthma was found in OVA-inducing asthmatic mice. The increase in body weight after OVA aerosol challenge, lung weight, total leukocytes and eosinophils in peripheral blood, total cell numbers, and neutrophil and eosinophils in BALF were detected in the OVA control compared to sham control (non-OVA).
21	Uchiyama et al., 2012 [42]	Microbiome	The effects of oral administration on bacterial flora in the intestines of domestic animals, using Holstein cows and newborn Japanese Black calves, were observed. The expressions of TNF-α and interleukin (IL)-6 in all cows became slightly lower and the bacterial flora were tendentiously changed.
22	Kim et al., 2012 [43]	Reduction of osteoarthritis	Osteoarthritis (OA) was effectively induced by anterior cruciate ligament transection and partial medial meniscectomy (ACLT&PMM) by APβG (42.5mg/kg).
23	Sato et al., 2012 [44]	Anti allergy	Effective therapeutic treatment of allergic diseases, inhibition of mast cell degranulation, and passive cutaneous anaphylaxis (PCA) were shown. APβG (100 to 150 mg/kg) dose-dependently inhibited the degranulation of both rat basophilic leukemia (RBL-2H3) and cultured mast cells (CMCs) activated by calcium ionophore A23187 or IgE.
24	Tanioka et al., 2012 [45]	Anti microorganism	Positive effect of oral administration of APβG on *Candida albicans* or methicillin-resistant *Staphylococcus aureus* (MRSA) infection in immunosuppressed mice fed 2.5% APβG diet for 14–30 days.
25	Tanioka et al., 2013 [46]	Immunomodulatory mechanism	APβG had effects on intestinal immune systems by Peyer's patch (PP) cells and interleukin (IL)-5, IL-6, and IgA production in culture media. The production of IL-6 and IgA by PP cells and that of IL-6 by PP dendritic cells (PPDCs) in APβG-fed and cyclophosphamide (CY)-treated mice also increased.
26	Tamegai et al., 2013 [47]	Immunomodulatory mechanism	Activation of several distinct innate immune receptor signaling pathways enhances the immune response induced by R-848, indicating non-influenza antiviral efficacy. The expression of TNF-a and IL-12p40 was significantly enhanced when co-stimulated with culture supernatants of R-848 and APβG compared with the culture supernatant of R-848 alone.
27	Iwai et al., 2013 [48]	Anti virus	Antiviral effects of the expression of interferon-inducible genes, through the induction of interferon, and the enhancement of the transcriptional activity of STAT1 were observed.
28	Kim et al., 2014 [49]	Anti oxidant	Several antioxidants may serve as a functional ingredient in cosmetic products by reducing hyaluronidase, elastase, collagenase, and MMP-1 activities and inhibiting melanin production and tyrosinase activities.
29	Muramatsu et al., 2014 [50]	Immunomodulatory mechanism	Stimulation with APβG effectively induces the interferon (IFN) stimulated genes (ISGs) in macrophage-like cell lines through the induction of IFN and the enhancement of STAT1-mediated transcriptional activation.

Table 2. *Cont.*

No.	Author, Year, Reference [1]	Item of Chracterization or Healthy Function	Outline of Functionality and Efficacy
30	Ganesh et al., 2014 [51]	Reduction of triglyceride and cholesterol, in human	In a clinical study, one male was orally administered 1.5 mg of APβG for two months, and his triglyceride, VLDL, and HDL cholesterol levels decreased from 523 mg/dL to 175 mg/dL, 104.6 mg/dL to 35 mg/dL, and 27 mg/dL to 38 mg/dL, respectively.
31	Oboshi et al., 2014 [52]	Immunomodulatory mechanism	When AGβP-containing foods were orally administered to mice (BALB/c six-week old females), an increase in the titer of antibodies in the blood and the phagocytic capacity of blood macrophages were observed.
32	Aoki et al., 2015 [53]	Reduction of triglyceride and cholesterol	Oral administration of APβG is effective in preventing the development of high-fat diet (HFD)-induced fatty liver in mice. After 16 weeks of oral administration of APβG, serological analysis showed that HFD-induced high blood cholesterol and triglyceride levels were reduced by the oral administration of APβG. HFD induced-fatty liver was also significantly reduced.
33	Kim et al., 2015 [54]	Reduction osteoporosis	The effects of purified exopolymers from APβG were evaluated in UVB-induced hairless mice. E-AP-SM2001 consists of 1.7% β-1,3/1,6-glucan, fibrous polysaccharides, and other organic materials, such as amino acids and mono- and di-unsaturated fatty acids (linoleic and linolenic acids), and shows anti-osteoporotic and immunomodulatory effects through antioxidant and anti-inflammatory mechanisms.
34	Kawata et al., 2015 [55]	Anti-tumor	Mechanism of anti-tumor activities in mice have been demonstrated. Stimulation with APβG induces TRAIL expression in mouse and human macrophage-like cell lines. TRAIL is known to be a cytokine that specifically induces apoptosis in transformed cells, but not in untransformed cells.
35	Lai et al., 2015 [56]	Improvement of intestine morphology	The effects of co-fermented *Pleurotus eryngii* stalk residues (PESR) and soybean hulls with *A. pullulans* on performance and intestinal morphology of broiler chickens significantly increased the ratio of lactic acid bacteria to *Clostridium perfringens* in ceca, ileum villus height, and jejunum villus height/crypt depth ratio of 35-day old birds.
36	Li et al., 2015 [57]	Anti inflammatory effect	Anti-inflammatory effect of water solubilized 1′-Acetoxychavicol acetate (ACA) on contact dermatitis by complexation with β-1,3-glucan isolated form *A. pullulans* black yeast was reported.
37	Aoki et al., 2015 [58]	Reduction of atherosclerosis	The effect of oral administration of APβG on high-fat diet (HFD)-induced atherosclerosis was evaluated using apolipoprotein E deficient mice, a widely used mouse model for atherosclerosis. HFD-induced atherosclerosis was significantly reduced in the APβG-treated mice.
38	Jippo et al., 2015 [59]	Anti allergy in human	A clinical study, comprising a randomized, single blind, placebo-controlled, and parallel group, was performed in 65 subjects (aged 22 to 62). All subjects consumed one bottle of placebo or beta glucan (150 mg) daily and their allergic symptoms were recorded in a diary. APβG group had a significantly lower prevalence of sneezing, nose blowing, tears, and hindrance to the activities of daily living.

Table 2. *Cont.*

No.	Author, Year, Reference [1]	Item of Chracterization or Healthy Function	Outline of Functionality and Efficacy
39	Iinuma et al., 2016 [60]	Anti colongenomic cancer	APβG was orally administered in combination with anticancer drugs, such as Avastin, Elplat, Levofolinate and fluorouracil, to treat for stage III colon cancer patients, on which two sites of cologenomic cancer were transferred, and remarkable cancer elimination was reported in the all cases.
40	Hirabayashi et al., 2016 [21]	Structural characterization of by hydrothermal treatment	The chemical structure of hydrothermally treated APβG was characterized using techniques such as gas chromatography/mass spectrometry (GC/MS) and nuclear magnetic resonance (NMR). It became water soluble, with an average molecular weight of 128,000 and was completely hydrolyzed to glucose by enzymatic reaction. Gentiobiose and glucose were released as products during the reaction with the maximum yield approximately 70% (w/w) of gentiobiose.
41	Yamamoto et al., 2017 [61]	Wound healimg	Muscle transplantation was avoided, and rapid healing was observed when there was external application of APβG in the case of third degree deep low temperature burn.
42	Ikeda et al., 2017 [62]	Inclusion of [63] Fullerene	[63] Fullerene was dissolved in water via complexation with β-1,3-glucan using a mechanochemical highspeed vibration milling apparatus. The photodynamic activity of APβG-complexed C70 was highly dependent on the expression level of dectin-1 on the cell surfaces of macrophages. The photodynamic activity increased owing to the synergistic effect between β-1,3-glucan-complexed 1'-acetoxychavicol acetate and the C70 complex.
43	Cho et al., 2018 [64]	Anti osteoporotic	The synergistic anti-osteoporotic potential of mixtures containing different proportions of APβG and TM compared with that of single formulations of each herbal extract using bilateral ovariectomized (OVX) mice. The EAP:TM (3:1) formulation synergistically enhanced the anti-osteoporotic potential of individual EAP or TM formulations, possibly owing to the enhanced variety of active ingredients.
44	Fujikura et al., 2018 [65]	Anti virus	APβG exhibits adjuvant activity and renders mice to be resistant to influenza A virus infection. Intraperitoneal administration of APβG increased the serum level of IL-18 and the number of splenic IFN-γ producing CD4+ cells during influenza A virus infection. The adjuvant effect of APβG was distinct from that of alum. In addition, AP-CF injection barely increased the number of peritoneal neutrophils and inflammatory macrophages.
45	Lim et al., 2018 [66]	Muscle preserving effect	The beneficial skeletal muscle-preserving effects of extracellular polysaccharides from APβG on dexamethasone (DEXA)-induced catabolic muscle atrophy in mice. APβG at 400 mg/kg exhibited favorable muscle protective effects against DEXA-induced catabolic muscle atrophy; the effects are comparable with those of oxymetholone (50 mg/kg).
46	Hayashi et al., 2019 [22]	Anti inflammatory effect	Inflammatory immunostimulatory action of nitrogen oxide (NO) synthesis and TNF a production in RAW264.7 cells induced by Lipopolysaccharide (LPS) were suppressed by APβG with lower molecular weight less than 10,000.

Table 2. *Cont.*

No.	Author, Year, Reference [1]	Item of Chracterization or Healthy Function	Outline of Functionality and Efficacy
47	Hino et al., 2019 [67]	Inclusion of porphyrin	Fluorescence intensities of water-soluble APβG-complexed porphyrin derivatives were very weak owing to self-quenching. However, APβG-complexed tetra (aminophenyl) porphyrin exhibited 'off-state' to 'on-state' fluorescence switching activity via intracellular uptake. Furthermore, the internalized complex showed a high level of photodynamic activity toward HeLa cells under photoirradiation at long wavelengths.

[1] Reference numbers in "Author, Year and Reference" column is shown in reference numbers in the text.

2.4.1. Intestinal Immunostimulatory Effect

Organisms absorb nutrients in the intestine from food, assimilate them for growth and other physiological functions, and generate energy to survive. However, food contains pathogens and, therefore, an efficient immune system is required for protection against such pathogens. A prominent organ related to the immune function is the gut-associated lymphoid tissue (GALT) [68].

The gut microbes, such as lactobacilli and bifidobacteria, are important for the activation of intestinal immunity. Recent studies have revealed that polysaccharides or nucleic acid components of the cell surface, which constitutes the cellular body, are the essential ligands for the activation of innate immunity and induction of acquired immunity in the intestine. By recognizing the components of various microbial cells, innate immune activation and signaling enhance the responsiveness of acquired immunity. In other words, through exposure to food (ingredients), GALT matures, gains immune responsiveness, and accumulates and exerts its functions. In 2011, the discovery of the interactions between the innate and adaptive immunity was awarded the Nobel Prize in Physiology or Medicine.

The Peyer's patch plays pivotal roles in the intestinal immunity, specifically the GALT immunity. At the Peyer's patch, useful immunostimulatory components of food are taken up into the lymph node, where antigen-presenting cells, such as macrophages and dendritic cells, are present. These cells are taken in by the M cells in the Peyer's patches, which are tens of microns in size. These antigen-presenting cells express innate immune signal receptors, such as TLRs and C-type lectins, and transmit various signals to the systemic immune system [10].

The intestinal immunostimulatory and modulatory effects of APβG have been evaluated in animal models such as mice. Cellular-level in vitro experiments using lymphocytes derived from the Peyer's patches in mice showed that immunoglobulin (Ig)A was produced at APβG concentrations ranging from 0 to 200 μg·mL^{-1} in a dose-dependent manner, and APβG was more active than yeast-derived zymosan and bacterial inflammatory substances (e.g., LPS). Simultaneously, interleukin (IL)-5 and IL-6 production was also increased In in vivo experiments, APβG was orally administered to each mouse at a concentration of 10–20 mg·kg^{-1} body weight per day for seven7 days; on day 8, lymphocytes from the Peyer's patches were extracted and their IgA level were measured. APβG promoted the production of IgA in the intestine, and activated lymphocytes and production of IL5 and IL6 in the Peyer's patch, which is the immune tissue of the intestine, by triggering the intestinal immune system [25].

Furthermore, when APβG-containing food products were orally administered to six-week-old BALB/c female mice, antibody titers in the blood were elevated and the phagocytic capacity of blood macrophages was enhanced, suggesting an immunostimulatory effect via intestinal immune activation [52].

In addition, the Peyer's patch cells were isolated, and IL-5, IL-6, and IgA levels in the medium were measured following APβG administration. The levels of both cytokines and IgA were increased, and the level of IL-6 secreted by the Peyer's patch dendritic cells was also elevated. In another study, APβG was orally administered for two weeks, and IgA levels were measured; APβG tended to promote small intestinal IgA production. Interestingly, following treatment with the immunosuppressant cyclophosphamide (CY), mice receiving an APβG-supplemented diet showed significantly increased IgA production compared with mice receiving the control diet. Moreover, the levels of IL-6 and IgA secreted by the Peyer's patch lymphocytes as well as of IL-6 secreted by the Peyer's patch dendritic cells were increased. Overall, these effects of oral administration indicate the potential use of APβG as a functional food with immunomodulatory activity [46].

2.4.2. Splenic Immunomodulation

Although the biological action of β-glucan depends on its structure, the effect of highly branched 1,3-β-D-glucan on cytokine production in mouse leukocytes remains poorly understood. In an in vitro study, APβG strongly induced the production of various cytokines in DBA/2 murine splenocytes. Specifically, APβG induced interferon-γ (IFN-γ), IL-12p70 (a Th1-type cytokine), and IL-17A (a Th17-type cytokine), but not IL-4, IL-10, and tumor necrosis factor-α (TNF-α) [32]. Furthermore, anti-dectin-1 neutralizing antibodies could not inhibit this APβG-induced IFN-γ production in DBA/2 murine solenocytes. This result indicates that APβG induces IFN-γ activity through signaling pathways other than those involving dectin-1, which is a major β-1,3-d-glucan receptor [37].

APβG has been reported to induce cytokine production in the presence of granulocyte–macrophage colony-stimulating factor (GM-CSF). This APβG-induced production of cytokines in DBA/2 murine splenocytes was completely blocked by anti-GM-CSF neutralizing antibodies. Moreover, the addition of GM-CSF to C57BL/6 murine splenocytes that were less responsive to APβG showed APβG-induced cytokine production. These findings suggest that GM-CSF is essential for APβG-induced cytokine production in murine splenocytes. This discovery is expected to offer novel insights into the effects of β-1,3-1,6-glucan as well as to help design and develop highly efficient β-glucan formulations [38].

2.4.3. Antitumor and Antimetastatic Activities

The antitumor and antimetastatic activities of APβG have been investigated. In a murine oral administration study, 10,000 colon-26 cancer cells were transplanted into murine spleens to investigate the antitumor effects of APβG on primary tumors and its inhibitory effects on cancer metastasis to the liver; significant ($p < 0.05$) reduction in primary tumor incidence and inhibition of tumor metastasis to the liver were observed even at an APβG dose as low as 50 mg·kg^{-1}. Furthermore, to clarify its mechanism of action in the intestine, the immune function of the small intestine was examined. The number of IFN-γ–positive cells was remarkably increased. Based on this result, the antitumor and antimetastatic effects of APβG were assumed to have been exerted through the enhancement of IFN-γ production from the small intestinal mucosal cells. Natural killer (NK) cell activity was also enhanced. That was the first study to demonstrate that the oral administration of high-purity β-glucan exhibits antitumor and antimetastatic activities by enhancing IFN-γ production and NK cell activity by triggering the intestinal immunity [26].

Furthermore, the mechanism underlying the antitumor effects of APβG has been explored. APβG induced the expression of TNF-related apoptosis-inducing ligand (TRAIL) in murine and human macrophage-like cells, suggesting that TRAIL expression induces tumor cell apoptosis [55].

In addition, APβG has a β-1,3-1,6 structure, similar to lentinan, which is an antitumor agent approved for clinical use in Japan. However, lentinan is intravenously injected. In three clinical cases, APβG was orally administered in combination with anticancer drugs, such as Avastin, Elplat, Levofolinate, and fluorouracil, to treat for stage III colon cancer patients, on which two sites of cologenomic cancer were transferred, and remarkable cancer elimination was reported in the all cases [60].

2.4.4. Antimicrobial Activities

The efficacy of APβG was tested against *Candida* in intravenously infected mice and methicillin-resistant *Staphylococcus aureus* (MRSA) in intestinally infected mice. Mice sensitive to CY were intravenously infected with 6×10^4 cells of *C. albicans*, and APβG was administered intraperitoneally at 1 mg per day for four days, which significantly prolonged survival. In CY-treated mice exhibiting a mild infection (intravenous administration of 2×10^4 cells of *C. albicans*), oral administration of 2.5% APβG significantly prolonged survival and reduced renal microbial viability at 30 days of infection. In mice intestinally infected with MRSA, oral administration of APβG did not reduce fecal MRSA, although it inhibited CY-induced weight loss. Prophylactic oral administration of APβG improved the resistance of CY-treated mice to *C. albicans* infection [45].

2.4.5. Effect of Alleviating Influenza Symptoms

Oral administration of APβG has been reported to protect mice from lethal influenza [caused by A/Puerto Rico/8/34 (PR8; H1N1) strain], which is an infectious disease. The survival of mice infected with a sublethal dose of the influenza A virus was significantly increased following the oral administration of APβG. Furthermore, viral titers in mouse lungs were significantly reduced following the oral administration of APβG. Therefore, APβG administration likely enhanced the expression of viral sensor molecules to exert protective effects against influenza through the inhibition of viral replication [40].

Previous studies have shown that APβG exhibited adjuvant activity and induced resistance against influenza; however, further investigation into the underlying mechanisms revealed that the intraperitoneal administration of APβG increased the serum level of IL-18 and the number of splenic IFN-γ-producing CD4-positive cells after influenza A viral infection. In addition, APβG induced IL-18 production in DC2.4 cells, a dendritic cell line, as well as in peritoneal exudate cells, including peritoneal macrophages. Thus, APβG acts as an adjuvant inducing the Th1-type response during influenza A viral infection [65].

To test its non-influenza antiviral efficacy, THP-1 macrophages were treated with APβG along with R-848, which is an anti-herpesvirus agent; the expression of TNF-α and IL-12p40 was significantly enhanced when the cells were co-stimulated with the M2 cell culture supernatants of R-848 and APβG compared to when stimulated with the M2 cell culture supernatant of R-848 alone. Furthermore, co-stimulation with R-848 and APβG significantly enhanced the phagocytosis-promoting ability of apoptotic Jurkat cells. These findings suggest that the APβG-induced activation of several distinct innate immune receptor signaling pathways enhances the overall immune response induced by R-848 [47].

As noted above, the antiviral effects of APβG may also be realized through the expression of interferon-stimulated genes (ISGs) in macrophage-like cell lines. These findings suggest that APβG stimulation effectively promotes the expression of ISGs through inducing IFNs and enhancing STAT1-mediated transcriptional activity [50].

2.4.6. Effects on Improving Lifestyle-Related Diseases, Including Obesity

Oral administration of APβG modulated the development of fatty liver caused by an HFD. Increased blood cholesterol and triglyceride levels triggered by HFD intake were suppressed by the oral administration of APβG. In addition, triglyceride accumulation in the liver was significantly suppressed by the oral administration of APβG. In HFD-fed mice, the elevated serum alanine aminotransferase levels associated with hepatic injury were lowered by the oral administration of APβG. These findings indicate that the oral administration of APβG may be effective in preventing the development of nonalcoholic fatty liver disease. In this case, the concentration of oral administration was less than 1% aqueous solution. APβG led to a significant upregulation of cholesterol 7 alpha-hydroxylase (CYP7A1) and IL-6 and normalized lipid metabolism. There is a possibility that this is due to the immunity-mediated cytokine network rather than the black yeast β-glucan physically binding to bile acids. [53].

Moreover, the effect of APβG on atherosclerosis induced by HFD intake was also confirmed in apolipoprotein E-deficient mice, which is a common animal model of atherosclerosis. Atherosclerosis induced by HFD intake was significantly suppressed in APβG-treated mice compared to that in control mice. Serological analysis showed that the blood levels of oxidized low-density lipoprotein cholesterol, a well-known risk factor for atherosclerosis, were significantly reduced in APβG-treated mice. In addition, APβG reduced macrophage accumulation in the blood vessels. These data suggest that oral administration of APβG is effective in preventing the development of atherosclerosis induced by HFD intake [58].

In a clinical study, a male subject was orally administered 1.5 mg APβG for 2 months. As a result, his triglyceride levels decreased from 523 (at the start of the study) to 175 mg·dL^{-1} (after two months). Dyslipidemia is a major risk factor for the development of cardiovascular diseases, and statins are the routine drugs used to manage dyslipidemia. In this case report, the subject was a patient with dyslipidemia but without diabetes, who was treated with Rosuvastatin. Upon APβG administration, his very low-density lipoprotein level decreased from 104.6 (at the start of the study) to 35 mg·dL^{-1} after two months, whereas the high-density lipoprotein cholesterol level increased from 27 to 38 mg·dL^{-1}. That was the first report on the effects of APβG on dyslipidemia not associated with diabetes. Thus, APβG supplementation along with routine medication is beneficial to treat dyslipidemia, although a large-scale prospective trial is warranted to confirm these effects [51].

2.4.7. Postprandial Blood Glucose Reduction

The effects of the oral administration of APβG on increased postprandial blood glucose levels were examined. APβG was orally administered to mice at doses of 50, 100, 200, and 500 mg·kg^{-1} per day in the morning and evening for seven days. On day 8, after fasting for 4 h or more, an aqueous solution containing APβG and an aqueous glucose solution (100 mg, 0.2 mL per mouse) were orally administered. In the control group, the blood glucose levels increased up to 30 min after administration. In the APβG-treated group, the blood glucose levels decreased 30 min after administration, and the blood glucose levels at 60 min after administration were significantly decreased compared with those in the control group. In the groups administered 100 and 200 mg·kg^{-1} APβG, blood glucose level slightly increased from 15 to 30 min after glucose administration but decreased thereafter, and the blood glucose level at 60 min after administration was significantly decreased compared with the control level.

In addition to the single-dose study described above, a seven-day free-intake study using an aqueous solution containing APβG (0.1–1.0%) was also conducted. When insulin levels were measured (at 15 min after glucose administration) during the glucose tolerance test on day 8, the increase in postprandial blood glucose tended to be suppressed, and blood insulin levels were significantly decreased. Based on these results, APβG intake with a meal might be effective in controlling postprandial blood glucose and insulin levels, thereby reducing the risk of metabolic syndromes [35].

2.4.8. Anti-Type I Allergic and Anti-Inflammatory Effects

The effect of APβG on type-I allergies was examined. Mice were provided an APβG-containing diet (0%, 0.25%, 0.5%, and 1.0%) ad libitum for 37 days. Food allergy was induced by intraperitoneal administration of 3 mg ovalbumin (OVA) saline on days 16 and 30 and oral administration of 15 mg OVA on day 37. As a result, the increased levels of OVA-specific IgE due to food allergy were lowered by APβG intake, and these levels were significantly decreased in mice receiving 1% APβG.

Th1-dominant response was observed due to increased IFN-γ and IL-12 production in splenic lymphocytes, and the numbers of CD8-positive, IFN-γ-positive, and IgA-positive cells were increased in the small intestinal mucosa. These results indicate that APβG produces an anti-type-I allergic effect through a mechanism suppressing OVA-specific IgE production. OVA ingestion mediates IgE production, which induces an allergic food reaction, leading to a Th1-dominant state. Moreover, APβG-mediated activation of the intestinal immune system under OVA-induced food allergies may protect against bacterial and viral invasion by increasing the numbers of CD8-positive T-cells, IgA-positive cells, and INF-γ-positive cells [28]. Furthermore, the anti-inflammatory effect of APβG may be strengthened by reducing its molecular weight [17].

APβG showed an allergy-alleviating effect through mast cells. The precise role of APβG in type-1 allergic reactions remains to be fully investigated. The inhibitory effects of low-molecular-weight β-glucan on mast cell degranulation and passive cutaneous anaphylaxis (PCA) were investigated. APβG inhibited the degranulation of both rat basophilic leukemia (RBL-2H3) and cultured mast cells (CMCs) activated by the calcium ionophore A23187 or IgE in a dose-dependent manner. Furthermore, oral administration of APβG inhibited IgE-induced PCA in mice. Specifically, a single dose of APβG ($100\text{--}150$ mg·kg^{-1} per mouse) significantly reduced PCA. Of note, tranilast (active ingredient of Rizaben, Kyorin Pharmaceutical) has been shown to be as effective as APβG [44]. Based on this result, a single-blind two-group parallel-monitoring study on pollinosis in humans (60 participants) was conducted. Consumption of drinks containing 150 mg APβG every day significantly reduced end-points such as sneezing and nasal discharge associated with pollinosis. The incidence of allergic diseases, such as allergic rhinitis, atopic dermatitis, asthma, and food allergies, has increased in several countries. Mast cells play critical roles in various biological processes related to allergic diseases. These cells express a high-affinity receptor for IgE on their surface, and the interaction of multivalent antigens with surface-bound IgE leads to the secretion of granule-stored mediators as well as the de novo synthesis of cytokines. These mediators and cytokines promote the development of allergic diseases [59].

The symptoms of mice with OVA-induced allergic asthma were alleviated following the administration of 125 mg·kg^{-1} APβG. In another study, the effects of APβG, derived from a UV-induced mutant of *A. pullulans*, on OVA-induced allergic asthma were examined and compared with the effects of intraperitoneally administered dexamethasone (DEXA) (3 mg·kg^{-1}) in mice. Following OVA aerosol challenge, body weight, lung weight, total leukocyte and eosinophil count in the peripheral blood, total cell number, neutrophil count, and eosinophil count in bronchoalveolar lavage fluid were increased in the OVA-control group compared with values in the sham-control (non-OVA) group. Therefore, APβG produces favorable effects against OVA-induced allergic asthma. The efficacy of 125 mg·kg^{-1} APβG was similar to or slightly lower than that of 3 mg·kg^{-1} DEXA [41].

2.4.9. Anti-stress and Immunomodulatory Effects

The effects of APβG administration on mice subjected to forced fasting and restraint stress were examined. The mice were orally administered APβG (25, 50, or 100 mg·kg^{-1} mice) every morning for seven consecutive days after one week of acclimatization. Forced fasting and restraint stress were induced using a 50 mL plastic tube with vents for 12 h from night to morning on days 3, 5, and 7 of administration. The restraint group, forced fasting plus restraint group (food and water were not provided plus forced restraint was performed), forced fasting group (food or water were not provided but no forced restraint was performed) were compared with the normal group (control; food and water provided ad libitum, and no forced restraint was performed).

The blood levels of corticosterone, which a stress marker, were significantly increased in the restraint and forced fasting plus restraint groups compared with those in the control group, and these levels tended to increase in the forced fasting group. In contrast, APβG treatment suppressed the increase in corticosterone levels. Specifically, the increase in blood corticosterone level was significantly suppressed in the 50 mg·kg^{-1} APβG administration group compared with that in the forced fasting plus restraint and forced fasting groups, indicating a stress-relieving effect of APβG. Therefore, the oral administration of APβG significantly alleviated restraint stress in mice.

IL-12 secretion from splenocytes was significantly reduced after forced fasting plus restraint stress. Moreover, restraint stress significantly suppressed immune function. Spleen weight was significantly lower in the forced fasting plus restraint and forced fasting groups than in the control group. In contrast, the decrease in IL-12 due to forced fasting and restraint stress was improved following APβG treatment. In particular, the IL-12 secretory volume was significantly increased in the 100 mg·kg^{-1} administration group compared with that in the forced fasting group. The efficacy of APβG in improving the immune function by restoring IL-6 secretion was also observed. The reduction in spleen weight due to restraint stress was not significantly different among the APβG-treated groups. In addition, the NK activity of splenic lymphocytes was markedly decreased in the forced fasting and forced fasting plus restraint stress groups, but APβG significantly ameliorated the decrease in NK activity caused by forced fasting and restraint stress.

Thus, repeated restraint stress increased blood corticosterone level but decreased splenic weight, splenic cytokine (IL-6 and IL-12) production, and NK activity. In contrast, the oral administration of APβG prevented the increase in blood corticosterone levels and the decrease in cytokine production and NK activity caused by restraint stress. Although restraint stress reduced immune functions, such as cytokine production and NK activity, the oral administration of APβG ameliorated this reduction in immunocompetence [29].

Along with the aforementioned anti-stress results, the authors planned and performed a clinical trial with a randomized crossover test (unpublished data). The duration of the clinical trial was two weeks and involved 28 participants (aged 22–55 years). In this trial, all participants consumed a bottle of placebo or APβG (100 mg) daily. After one and two weeks, corticosterone concentration in the saliva was measured as a stress marker. The anti-stress effect was physiologically estimated based on the corticosterone concentration after a stress test, such as the calculation and Visual Analogue Scale test (VAS) evaluation conducted on participants. The results suggested that, after two weeks, both physiological and mental stresses were reduced significantly. [5].

2.4.10. Protective Effect on the Intestinal and Gastrointestinal Mucosa

The redacting effects of APβG on the side effect of oral anticancer drug cocktail tegafur/uracil (UFT) were examined in mice. In the group treated with UFT alone, diarrhea, which is one of the side effects, was observed, whereas in the group treated with both APβG (oral; 25–100 mg·kg^{-1}) and UFT, diarrhea was suppressed. Further examination of the tissue condition confirmed that the oral administration of APβG markedly inhibited damage to the small intestinal mucosa [34].

In mice treated with the anticancer drug CY, the weight of the spleen and thymus was reduced and the level of the corresponding cytokines, such as IL10, was also decreased. In mice receiving APβG, this immunosuppressive effect of CY was reduced. The immunomodulatory effects of APβG were evaluated in CY-treated mice. APβG could effectively prevent CY-mediated immunosuppression, at least partially, by recruiting T-cells and TNF-α-positive cells or enhancing their activity. Therefore, APβG could effectively prevent the immunosuppressive effects of treatment regimens for diseases such as cancer, sepsis, and high-dose chemotherapy or radiotherapy [36].

The effects of APβG on gastric ulcers were also investigated. APβG was orally administered to 6–8-week-old male mice at a dose of 0, 5, 20, 50, 100, and 200 mg·kg^{-1} body weight. After 1 h, 100% ethanol was administered orally at 5 mL·kg^{-1} body weight to induce gastric injury (gastric ulcer). The stomach was then excised by laparotomy 4 h after the induction of the gastric injury and scored for hemorrhagic injury (hemorrhagic damage). The hemorrhagic injury score was calculated in 1 mm squares in the wounded areas and summed up to yield the gastric injury index. The results showed that the oral administration of 100% ethanol caused significant gastric injury and that APβG ameliorated this injury in a dose-dependent manner.

Therefore, the oral administration of APβG reduced ethanol-induced gastric injury (gastric ulcer), and this protective effect on the intestinal mucosa was mediated through the induction of heat shock protein 70 (HSP70) and increased production of mucin [39].

2.4.11. Effects on the Microbiome

The effects of APβG on the microbiome have also been demonstrated in domestic animals. APβG was orally administered for three months to Holstein cows fed raw milk with a somatic cell count of 3×10^5 mL^{-1} or less in milk, and the concentrations of milk and blood cytokines were analyzed. The number of somatic cells in milk did not change significantly but the concentration of solid non-fat in milk tended to increase following the oral administration of APβG. Analysis of blood cytokine levels using enzyme-linked immunosorbent assay revealed that TNF-α and IL-6 expression was slightly lower in the APβG-administered cows than in the control cows after two months of oral administration. Moreover, the expression of IL-8 tended to be moderately higher in the APβG-administered cows than in the control cows after three months of oral administration. APβG was orally administered to peripartum Japanese Black beef cows and their newborn calves, and the intestinal flora of the calves was analyzed using terminal restriction fragment length polymorphism. The intestinal flora was altered following the oral administration of APβG. For instance, the population of the genus *Prevotella* (317 bp operational taxonomic units) tended to increase in the APβG-administered calves compared with that in the control calves. Therefore, the oral administration of APβG may affect the blood cytokine levels in Holstein cows and the intestinal flora in Japanese Black calves [42].

The effects of APβG produced via fermentation using soybean as the substrate on the intestinal morphology of broiler chickens were examined. Improved gut microbiota (increased ratio of lactobacilli and *Clostridium perfringens*) and weight gain were observed. Soybean hulls (SBH) are a by-product of soybean processing for oil and meal production. *Pleurotus eryngii* stalk residues (PESR) are a by-product of the edible portion of the fruiting body enriched in bioactive metabolites. The effects of APβG produced via the co-fermentation of PESR and SBH on the performance and intestinal morphology of broiler chickens were evaluated [56]. A total of 400 broiler chicken (Ross 308) were randomly assigned into four groups receiving the basal diet (control) or the basal diet supplemented with 0.5% fermented SBH (0.5% FSBH), 0.5% co-fermented SBH plus PESR (0.5% FSHP), and 1.0% co-fermented SBH plus PESR (1.0% FSHP) until 35 days of age. Compared with the control group, the 0.5% FSHP group showed significantly increased ratio of lactobacilli to *Clostridium perfringens* in the ceca, ileum villus height, and jejunum villus height-to-crypt depth ratio at 35 days of age. In conclusion, dietary supplementation of 0.5% FSHP in broiler chickens improved the body weight gain and intestinal morphology through the effects of its bioactive metabolites.

Recently it was found that suppression of IL-17F provides protection against colitis by inducing Treg cells through modification of the intestinal microbiota with increasing growth of *Lactobacillus murinus* by intereaction with low morecular β-glucan, such as laminaran and dectin 1 receptor [69].

2.4.12. Application to the Skin

External application of APβG avoided muscle transplantation and promoted healing in cases with third-degree low-temperature burns [61]. APβG has been reported to serve cosmetic functions when applied to the skin. APβG has been reported to show anti-aging effects, whitening effects, and inhibitory effects on melanin production and tyrosinase activity.

Moreover, APβG reduced the activity of hyaluronidase, elastase, collagenase, MMP-1, and tyrosinase, as well as inhibited the production of melanin. Owing to its antioxidative and whitening effects, APβG may be used as a functional cosmetic material [49].

In UVB-induced hairless mice, APβG application to the skin reduced reactive oxygen species-mediated inflammation through its antioxidative and anti-inflammatory effects. APβG may be used as a raw material for cosmetics. Since antioxidants from natural sources may be an effective approach to prevent and treat UV-induced skin damage, the effects of purified APβG were evaluated in UVB-induced hairless mice. In that study, APβG containing 1.7% β-1,3/1,6-glucan, fibrous polysaccharides, and other organic materials, such as amino acids, and mono- and di-unsaturated fatty acids (linoleic and linolenic acids), showed anti-osteoporotic and immunomodulatory effects through antioxidative and anti-inflammatory mechanisms [54].

In traditional data, the effects of APβG on acute xylene-induced inflammation were observed. APβG at a dose of 62.5, 125, and 250 mg·kg^{-1} was orally administered once to xylene-treated mice (0.03 mL xylene was applied to the anterior surface of the right ear to induce inflammation). APβG showed a somewhat favorable effect in reducing the xylene-induced acute inflammatory response in mice [27].

2.4.13. Osteoporosis Improvement

The anti-osteoporotic effects of APβG at doses of 31.25, 62.5, and 125 mg·kg^{-1} were investigated in ovariectomized mice. APβG was orally administered once daily for 28 days to bilateral ovariectomized mice, beginning four weeks after surgery. Changes in body weight, bone weight, bone mineral content, bone mineral density, failure load, histologic profiles, and histomorphometric parameters were evaluated, and serum osteocalcin, calcium, and phosphorus levels were measured. Alendronate was used as the reference drug. APβG significantly suppressed the decrease in bone weight, bone mineral content, failure load, bone mineral density, and serum calcium and phosphorus levels and increased serum osteocalcin levels in a dose-dependent manner. In addition, APβG significantly suppressed the decrease in histomorphometric parameters, such as volume, length, and thickness of the trabecular bone, the thickness of the cortical bone, and increased the number of osteoclasts in the femur and tibia. Although the effects of APβG were generally moderate and inferior to those of alendronate, its effects on the cortical bone thickness were superior. In addition, APβG exhibited favorable effects on ovariectomy-induced osteoporosis. However, further long-term studies are warranted to confirm the effects of APβG on osteoporosis [30].

Another study examined the synergistic anti-osteoporotic potential of mixtures containing different proportions of APβG, 13% β-glucan-containing extract, and *Textoria morbifera* Nakai (TM) extract compared with that of the single formulations of each herbal extract in bilateral ovariectomized mice, a well-known rodent model for studying human osteoporosis. The APβG:TM (3:1) formulation synergistically enhanced the anti-osteoporotic potential of single APβG or TM formulation, possibly due to the enhanced array of active ingredients. Furthermore, the effects of APβG:TM were comparable to those of RES (2.5 mg·kg^{-1}). These results suggest that the APβG:TM (3:1) combination can serve as a novel pharmaceutical agent and/or functional food for managing osteoporosis in menopausal women [64].

The additional studies also assessed the beneficial skeletal muscle-preserving effects of extracellular polysaccharides from APβG on dexamethasone (DEXA)-induced catabolic muscle atrophy and osteoarthritis in mice [43,66].

2.4.14. Use as a Material Inclusion Agent

APβG can solubilize pharmacologically active hydrophobic substances through complex formation without compromising their pharmacological activity. 1′-Acetoxychavicol acetate (ACA) is isolated from the roots of *Alpinia galanga* and shows various physiological activities such as antineoplastic and immunosuppressive properties. As a complex with APβGs, ACA can serve as a pharmaceutic agent, functional food material, and cosmetic material. The clathrate complex between ACA and APβG was successfully produced using a high-speed vibrational grinding method. The ACA–APβG complex was intraperitoneally administered to treat dinitrofluorobenzene-induced contact dermatitis in a mouse model, and auricle thickening, immune cell infiltration, and blood inflammatory cytokine levels were evaluated. Consequently, treatment with the ACA–APβG complex suppressed auricle thickening and immune cell infiltration and reduced the abundance of inflammatory cytokines in the blood, demonstrating anti-inflammatory effects [57].

Fullerenes are selected as target drugs, and their application in photodynamic therapy (PDT) is anticipated because of their high photosensitization ability; in this context, APβG was considered as a drug carrier. In addition, β-1,3-glucans are specifically recognized by dectin-1, which is a specific Toll-like receptor (TLR) expressed on macrophage surfaces. Therefore, the application of APβG is expected to increase the selectivity of PDT. Various fullerene derivatives were water-solubilized by APβG, and the efficacy of PDT using APβG as a drug carrier was evaluated in mouse peritoneal macrophages [62,67]. Upon complexing with APβG using a high-speed vibration pulverization method, fullerene, which is poorly water-soluble, could be efficiently solubilized. In atomic force microscopic images, the thickness of the APβG–C70 complex was about twice that of the APβG-only complex. The APβG–C70 complex functioned as an excellent photosensitizer both in vitro and in vivo and could specifically target macrophages expressing dectin-1 [62].

3. Discussion

3.1. Physiological Functions and Polysaccharides

As described earlier, β-1,3-glucan has traditionally been recognized as a functional food ingredient obtained from mushrooms and has been reported to possess antitumor activity. β-glucans are homopolysaccharides composed solely of glucose and have been reported to be functional in different binding modes. In addition to mushrooms, β-glucans have been derived from oat and barley, which contain repeating structures with β-1,3 and β-1,4 bonds in their primary chains without branching. Meanwhile, β-glucans derived from seaweeds, such as laminaran, contain repeating structures with β-1,3 and β-1,6 bonds. β-glucans with multiple linkages in their straight chains have been characterized to be highly water soluble [1–5].

As described above, the immunostimulatory mechanisms of β-glucans have recently been elucidated, and receptors on macrophages and dendritic cells, which are antigen-presenting cells, have been revealed to be involved in these mechanisms. The use of β-glucans as a functional food material to boost the intestinal immunity has been proposed, as discussed below. The mechanism underlying the immunostimulatory actions of APβG have also been clarified. Interestingly, the β-glucan polysaccharide, which is a macromolecule with a MW of a few million has been reported to show a triple helical structure. Recent studies have demonstrated that specific molecular weight is required for its action, and studies have accumulated the evidence of the association between APβG conformation and function, such as the recognition of β-1,6 bonds by antibodies, such as several kinds of IgG [31,33]. In addition, the higher-order structure of β-glucan has attracted much attention in host–guest chemistry based on its ability to solubilize poorly water-soluble hydrophobic substances, such as pharmaceutical compounds, vitamins, and physically or chemically treated nanoparticle materials. Therefore, the application of APβG as a drug delivery system is anticipated in vaccine adjuvant research.

3.2. Recognition Mechanism of β-Glucans

The immunomodulatory functions of β-glucans are realized through the interaction between the activation of innate immunity and the acquisition of adaptive immunity. Discoveries and evidence linking their roles are essential to elucidate the presence and function of receptors expressed on the surface of cells, such as macrophages, dendritic cells, and neutrophils. Among them, TLRs are specific receptors with some well-known ligands, such as nucleic acid molecules (DNA and RNA) and polysaccharides on the surface of flagella and cells. The specific receptors for β-glucan, in particular, include the β2-integrin-type complement receptor CR3 [70], lactosylceramide [71], and the C-type lectin dectin-1 [63], and there have been numerous recent studies focusing on these molecules. In the early 21st century, dectin-1 was recognized as a ligand for β-glucan. Using knockout mice, Brown et al. (2001) demonstrated that dectin-1, a C-type lectin, is expressed on the plasma membrane of mammalian leukocytes and is deeply involved in biodefense mechanisms, including phagocytosis, reactive oxygen species generation, and cytokine production; many studies on the roles of dectin-1 are currently underway [72]. Ohno et al. (2009) reported that different activation pathways have also been identified in the spleen [23,24,68]. The recognition structures include β-1,3-1,6-glucans with a β-1,3 primary chain with β-1,6 branches in fungi, linear β-1,3 and β-1,4-glucans in plants, and β-1,3-oligosaccharides with more than 16 sugars [4,73].

Anti-β-glucan IgG have been detected in vivo and reported to recognize β-1,6 branches [33]. Recently, the unique structure of APβG was determined, and it was revealed to be human serum antibody-reactive. The reactivity of APβG was stronger than that of *Grifola frondosa*-derived β-1,3-glucans but weaker than that of *Candida*-derived β-1,3-glucans. Specifically, APβG is reactive to human serum IgG antibodies, such as IgG2, IgG1, and IgG3.

APβG can bind to R-848, a low-molecular-weight ligand that activates TLR7. Moreover, APβG induced TNF-α and IL-12p40 and enhanced the phagocytotic ability of macrophages. In combination with R-848, β-glucan, which binds to the cell membrane-localized TLRs and the C-type lectin receptor dectin-1, could target THP-1 macrophages. Compared to R-848 treatment alone, co-treatment of R-848 and APβG significantly augmented TNF-α and IL-12p40 cytokine expression. These results suggest that APβG activates an array of innate immune receptor pathways to enhance the immune response of R-848 [47].

Stimulation with APβG effectively induced the interferon stimulated genes (ISGs) in macrophage-like cell lines, perhaps through the induction of IFNs and enhancement of STAT1-mediated transcriptional activation. The mRNA expression of ISGs was significantly increased in RAW264.7 cells following stimulation with APβG. APβG stimulation induced transient mRNA expression of IFN-β. Additionally, in IFN-α receptor-knockdown RAW264.7 cells, APβG stimulation induced the expression of viperin mRNA more efficiently than did IFN-α stimulation. Phosphorylation of Ser727, which is involved in promoting STAT1 activation, was rapidly increased following APβG stimulation. In addition, the expression of viperin mRNA induced by IFN-α stimulation was significantly enhanced by co-stimulation with APβG. These findings suggest that APβG stimulation effectively induces the expression of ISGs through the induction of IFNs and enhancement of STAT1-mediated transcriptional activation [48,50].

Based on these results, APβG has been reported to serve numerous health functions. Specifically, the antitumor and antimetastatic activities, antimicrobial effects, inflammation-reducing effects, food allergy-alleviating effects, and stress-relieving effects of APβG are known. Moreover, APβG improves lifestyle-related diseases by protecting the intestinal mucosa, reducing fat, lowering postprandial blood sugar, promoting bone and joint health, and alleviating gastric ulcers. APβG has been reported to induce HSP70. Interestingly, a randomized, single-blind, placebo-controlled, parallel-group clinical trial involving 65 participants (age, 22–62 years) with pollinosis was performed for three weeks from before and until the end of the cedar pollen season. In this trial, all participants consumed a bottle of placebo or APβG (150 mg) daily and recorded allergic symptoms in a diary. The APβG-administered group had a significantly reduced prevalence of sneezing, nose blowing, tears, and hindrance in the activities of daily life compared to the placebo group. These results suggest that APβG can serve as an effective treatment for allergic diseases [59].

4. Conclusions

The term "functional food" was coined in the 1980s in Japan [74]. A new and more widely established term used in Japan is "Food with Health Claims (FHC)." There are also a few definitions of functional foods. One of them is included in the final document of the research program Functional Food Science in Europe (FUFOSE), financed by the European Commission. According to this definition, food can be recognized as functional if its beneficial influence on one or more organism functions has been proven [75]. The need to continue the current research to systematize and verify the reports is warranted. This research should mainly focus on the practical application of β-glucan. There is still no precise information on the amount of β-glucan that should be consumed to obtain the desired health effect. Most observations were made using either very high doses of β-glucan or in combination with other biologically active ingredients. On this basis, it is difficult to unequivocally recommend the concentration of this ingredient that should be supplied daily to achieve the expected result. It is also unknown whether a potentially effective dose could be obtained by the consumption of products rich in β-glucan, or that the usage of the enrichment process of selected food products in this ingredient would be necessary. Therefore, consumers must be educated about β-glucan along with the benefits associated with their usual diet.

In the future, studies unveiling the precise mechanisms underlying the immunostimulatory functions of β-glucan as well as its effects on the autonomic nervous system and endocrine system are warranted. In addition, solubilized APβG is anticipated to serve as a functional food for modern and aging societies that experience high stress and as a potent pharmaceutical agent in the field of preventive medicine. Recently, owing to its unique triple helical structure, APβG has attracted much attention as a nanomaterial, hemostatic agent, and drug delivery system in the medical field.

Author Contributions: Original draft and conceptualization: T.S. and T.K. Data extraction and appraisal of evidence: T.S., T.K., K.K., N.K., N.O. Manuscript writing and approval of the final draft: T.S., T.K., K.K., N.K., K.N., N.O. All authors have read and agreed to the published version of the manuscript.

Funding: This research received no external funding.

Conflicts of Interest: The authors declare no conflict of interest.

References

1. Ohno, N. Immunopharmacological Activity of Polysaccharides. *Fragr. J.* **1995**, *5*, 70–75. [CrossRef]
2. Yadomae, T. Structure and Biological Activity of Fungus β-1,3-Glucans. *J. Pharm. Soc. Jpn.* **2000**, *120*, 413–431. [CrossRef]
3. Ohno, N. Modulation of Host Defense Systems by Beta-Glucans. *Jpn. J. Bacteriol.* **2000**, *55*, 527–537. [CrossRef]
4. Ohno, N. Structural Diversity and Physiological Functions of β-Glucans. *Int. J. Med. Mushrooms* **2005**, *7*, 167–174. [CrossRef]
5. Suzuki, T.; Nishikawa, K.; Nakamura, S.; Suzuki, T. [Review: Prize-Awarded Article] Research and Development of β-1,3–1,6-Glucan From Black Yeast for a Functional Food Ingredient. *Bull. Appl. Glycosci.* **2012**, *2*, 51–60. [CrossRef]
6. Miyamoto, J.; Watanabe, K.; Taira, S.; Kasubuchi, M.; Li, X.; Irie, J.; Itoh, H.; Kimura, I. Barley β-Glucan Improves Metabolic Condition via Short-Chain Fatty Acids Produced by Gut Microbial Fermentation in High Fat Diet Fed Mice. *PLoS ONE* **2018**, *13*, e0196579. [CrossRef]
7. Sobieralski, K.; Siwulski, M.; Lisiecka, J.; Jedryczka, M.; Sas-Golak, I.; Fruzynska-Józwiak, D. Fungi-Derived β-Glucans as a Component of Fungal *Food*. *Acta Sci. Pol. Hortorum Cultus* **2012**, *11*, 111–128.
8. Bzducha, A.-W.; Pobiega, K.; Błażejak, S.; Kieliszek, M. The Scale-Up Cultivation of *Candida utilis* in Waste Potato Juice Water With Glycerol Affects Biomass and β (1, 3)/(1, 6)–Glucan Characteristic and Yield. *Appl. Microbiol. Biotechnol.* **2018**, *102*, 9131–9145. [CrossRef]
9. Bzducha, A.-W.; Koczoń, P.; Błażejak, S.; Kozera, J.; Kieliszek, M. Valorization of Deproteinated Potato Juice Water Into β-Glucan Preparation of *C. utilis* Origin: Comparative Study of Preparations Obtained by Two Isolation Methods. *Waste Biomass Valori.* **2020**, *11*, 3257–3271. [CrossRef]
10. Gantner, B.N.; Simmons, R.M.; Canavera, S.J.; Akira, S.; Underhill, D.M. Collaborative Induction of Inflammatory Responses by dectin-1 and Toll-Like receptor 2. *J. Exp. Med.* **2003**, *197*, 1107–1117. [CrossRef]
11. Nagano, Y.; Elborn, J.S.; Millar, B.C.; Goldsmith, C.E.; Rendall, J.; Moore, J.E. Development of a Novel PCR Assay for the Identification of the Black Yeast, Exophiala (Wangiella) Dermatitidis from Adult Patients with Cystic Fibrosis (CF). *J. Cyst. Fibros.* **2008**, *7*, 576–580. [CrossRef] [PubMed]
12. Gostinčar, C.; Ohm, R.A.; Kogej, T.; Sonjak, S.; Turk, M.; Zajc, J.; Zalar, P.; Grube, M.; Sun, H.; Han, J.; et al. Genome Sequencing of Four *Aureobasidium pullulans* Varieties: Biotechnological Potential, Stress Tolerance, and Description of New species. *BMC Genom.* **2014**, *15*, 549–576. [CrossRef] [PubMed]
13. Arkadjeva, G.F. The Biological Activity of Polysaccharides Isolated From *Rhodotorula glutinis* and *Aureobasidium pullulans*. *Antonie Leeuwenhoek* **1969**, *35*, E9–E10.
14. Han, Y.W.; Cheeke, P.R.; Anderson, A.W.; Lekprayoon, C. Growth of *Aureobasidium pullulans* on Straw Hydrolysate. *Appl. Environ. Microbiol.* **1976**, *32*, 799–802. [CrossRef] [PubMed]
15. Anastassiadis, S.; Morgunov, I.G. Gluconic Acid Production. *Recent Pat. Biotechnol.* **2007**, *1*, 167–180. [CrossRef] [PubMed]
16. Bharti, S.K.; Krishnan, S.; Kumar, A.; Rajak, K.K.; Murari, K.; Bharti, B.K.; Gupta, A.K. Antidiabetic Activity and Molecular Docking of Fructooligosaccharides Produced by *Aureobasidium pullulans* in poloxamer-407-induced T2DM Rats. *Food Chem.* **2013**, *136*, 813–821. [CrossRef]
17. Tamano, K.; Nakasha, K.; Iwamoto, M.; Numata, M.; Suzuki, T.; Uyama, H.; Fukuhara, G. Chiroptical Properties of Reporter-Modified or Reporter-Complexed Highly 1,6-Glucose-Branched β-1,3-Glucan. *Polym. J.* **2019**, *51*, 1063–1071. [CrossRef]

18. Suzuki, T.; Nakamura, S.; Nishikawa, K.; Nakayama, S.; Suzuki, T. Production of Highly Purified β-Glucan From *Aureobasidium pullulans* and Its Characteristics. *Food Funct.* **2006**, *2*, 45–50.

19. Hamada, N.; Tsujisaka, Y. The Structure of the Carbohydrate Moiety of an Acidic Polysaccharide Produced by *Aureobasidium* sp. K-1. *Agric. Biol. Chem.* **1983**, *47*, 1167–1172. [CrossRef]

20. Tada, R.; Tanioka, A.; Iwasawa, H.; Hatashima, K.; Shoji, Y.; Ishibashi, K.I.; Adachi, Y.; Yamazaki, M.; Tsubaki, K.; Ohno, N. Structural Characterisation and Biological Activities of a Unique Type Beta-D-Glucan Obtained From *Aureobasidium pullulans*. *Glycoconj. J.* **2008**, *25*, 851–861. [CrossRef]

21. Hirabayashi, K.; Kondo, N.; Hayashi, S. Characterization and Enzymatic Hydrolysis of Hydrothermally Treated β-1,3–1,6-Glucan From *Aureobasidium pullulans*. *World J. Microbiol. Biotechnol.* **2016**, *32*, 206. [CrossRef] [PubMed]

22. Hayashi, N.; Shoubayashi, Y.; Kondo, N.; Fukudome, K. Hydrothermal Processing of β-Glucan From Aureobasidium pullulans Produces a Low Molecular Weight Reagent That Regulates Inflammatory Responses Induced by TLR Ligands. *Biochem. Biophys. Res. Commun.* **2019**, *511*, 318–322. [CrossRef] [PubMed]

23. Survey and Research on the Review of the Safety of Existing Additives. In Proceedings of the Pharmaceutical Affairs and Food Sanitation Council Additives Sub-Committee, Tokyo, Japan, 24 June 2004; Japan Food Chemical Research Foundation: Osaka, Japan, 2004.

24. Amakawa, M.; Nakano, T.; Kusaka, M.; Ichihara, N.; Kameko, F.; Kato, R. Evaluation of the Clinical Laboratory Data Depends on Long Oral Taking of Aureobasidium Cultured Solution. *Jpn. J. PharmSci* **2016**, *73*, 477–487.

25. Suzuki, T.; Hosono, A.; Hachimura, S.; Suzuki, T.; Kaminogawa, S. Modulation of Cytokine and Immunoglobulin A Release by Beta-(1,3–1,6)–Glucan From *Aureobasidium pullulans* strain 1A1. In *Animal Cell Technology: Basic & Applied Aspects*; Iijima, S., Nishijima, K., Eds.; Springer: Dordrecht, The Netherlands, 2004; Volume 14, pp. 369–375.

26. Kimura, Y.; Sumiyoshi, M.; Suzuki, T.; Sakanaka, M. Antitumor and Antimetastatic Activity of a Novel Water-Soluble Low Molecular Weight β-1,3-D-glucan (branch β-1,6) isolated from AP 1A1 strain black yeast. *Anticancer Res.* **2006**, *26*, 4131–4142.

27. Kim, H.D.; Cho, H.R.; Moon, S.B.; Shin, H.D.; Yang, K.J.; Park, B.R.; Jang, H.J.; Kim, L.S.; Lee, H.S.; Ku, S.K. Effects of Beta-Glucan From *Aureobasidium pullulans* on Acute Inflammation in Mice. *Arch. Pharm. Res.* **2007**, *30*, 323–328. [CrossRef] [PubMed]

28. Kimura, Y.; Sumiyoshi, M.; Suzuki, T.; Suzuki, T.; Sakanaka, M. Inhibitory Effects of Water-Soluble Low-Molecular-Weight Beta-(1,3–1,6) d-Glucan Purified From *Aureobasidium pullulans* GM-NH-1A1 Strain on Food Allergic Reactions in Mice. *Int. Immunopharmacol.* **2007**, *7*, 963–972. [CrossRef] [PubMed]

29. Kimura, Y.; Sumiyoshi, M.; Suzuki, T.; Suzuki, T.; Sakanaka, M. Effects of Water-Soluble Low-Molecular-Weight β-1,3-D-Glucan(Branch β-1,6) Isolated From *Aureobasidium pullulans* 1A1 Strain Black Yeast on Restraint Stress in Mice. *J. Pharm. Pharmacol.* **2007**, *59*, 1137–1144. [CrossRef]

30. Shin, H.D.; Yang, K.J.; Park, B.R.; Son, C.W.; Jang, H.J.; Ku, S.K. Antiosteoporotic Effect of Polycan, Beta-Glucan From *Aureobasidium*, in Ovariectomized Osteoporotic Mice. *Nutrition* **2007**, *23*, 853–860. [CrossRef]

31. Ikewaki, N.; Fujii, N.; Onaka, T.; Ikewaki, S.; Inoko, H. Immunological Actions of Sophy Beta-Glucan (Beta-1,3–1,6 Glucan), Currently Available Commercially as a Health Food Supplement. *Microbiol. Immunol.* **2007**, *51*, 861–873. [CrossRef]

32. Tada, R.; Yoshikawa, M.; Kuge, T.; Tanioka, A.; Ishibashi, K.I.; Adachi, Y.; Tsubaki, K.; Ohno, N. A Highly Branched 1,3-beta-D-glucan Extracted From *Aureobasidium pullulans* Induces Cytokine Production in DBA/2 Mouse-Derived Splenocytes. *Int. Immunopharmacol.* **2009**, *9*, 1431–1436. [CrossRef]

33. Tada, R.; Tanioka, A.; Ishibashi, K.I.; Adachi, Y.; Tsubaki, K.; Ohno, N. Involvement of Branched Units at position 6 in the Reactivity of a Unique Variety of Beta-D-Glucan From *Aureobasidium pullulans* to Antibodies in Human Sera. *Biosci. Biotechnol. Biochem.* **2009**, *73*, 908–911. [CrossRef]

34. Sumiyoshi, M.; Suzuki, T.; Kimura, Y. Protective Effects of Water-Soluble Low-Molecular-Weight β-(1,3–1,6)D-Glucan Purified From *Aureobasidium pullulans* GM-NH-1A1 Against UFT Toxicity in Mice. *J. Pharm. Pharmacol.* **2009**, *61*, 795–800. [CrossRef]

35. Kimura, Y.; Sumiyoshi, M.; Suzuki, T.; Suzuki, T.; Sakanaka, M. Effects of Water-Soluble Low-Molecular Weight b-1,3-D-glucan (Branch b-1,6) Isolated From *Aureobasidium pullulans* 1A1 Strain (Black Yeast) on Blood Glucose and Insulin Concentrations in Oral Glucose Tolerance Test of Mice. *J. Pharm. Pharmacol.* **2009**, *61*, 115–119. [CrossRef]

36. Yoon, H.S.; Kim, J.W.; Cho, H.R.; Moon, S.B.; Shin, H.D.; Yang, K.J.; Lee, H.S.; Kwon, Y.S.; Ku, S.K. Immunomodulatory Effects of *Aureobasidium pullulans* SM-2001 Exopolymers on the Cyclophosphamide-Treated Mice. *J. Microbiol. Biotechnol.* **2010**, *20*, 438–445. [CrossRef] [PubMed]

37. Tada, R.; Yoshikawa, M.; Ikeda, F.; Adachi, Y.; Kato, Y.; Kuge, T.; Tanioka, A.; Ishibashi, K.I.; Tsubaki, K.; Ohno, N. Induction of IFN-γ by a Highly Branched 1,3-β-d-glucan From *Aureobasidium pullulans* in Mouse-Derived Splenocytes via dectin-1-independent Pathways. *Biochem. Biophys. Res. Commun.* **2011**, *404*, 1105–1110. [CrossRef] [PubMed]

38. Tada, R.; Yoshikawa, M.; Kuge, T.; Tanioka, A.; Ishibashi, K.I.; Adachi, Y.; Tsubaki, K.; Ohno, N. Granulocyte Macrophage Colony-Stimulating Factor Is Required for Cytokine Induction by a Highly 6-Branched 1,3-β-D-glucan From *Aureobasidium pullulans* in Mouse-Derived Splenocytes. *Immunopharmacol. Immunotoxicol.* **2011**, *33*, 302–308. [CrossRef] [PubMed]

39. Tanaka, K.; Tanaka, Y.; Suzuki, T.; Mizushima, T. Protective Effect β-(1, 3→ 1, 6)-D-Glucan Against Irritant-Induced Gastric Lesions. *Br. J. Nutr.* **2011**, *106*, 475–485. [CrossRef] [PubMed]

40. Muramatsu, D.; Iwai, A.; Aoki, S.; Uchiyama, H.; Kawata, K.; Nakayama, Y.; Nikawa, Y.; Kusano, K.; Okabe, M.; Miyazaki, T. β-Glucan Derived From *Aureobasidium pullulans* Is Effective for the Prevention of Influenza in Mice. *PLoS ONE* **2012**, *7*, e41399. [CrossRef]

41. Ku, S.K.; Kim, J.W.; Cho, H.R.; Kim, K.Y.; Min, Y.H.; Park, J.H.; Kim, J.S.; Park, J.H.; Seo, B.I.; Roh, S.S. Effect of β-Glucan Originated From *Aureobasidium pullulans* on Asthma Induced by Ovalbumin in Mouse. *Arch. Pharm. Res.* **2012**, *35*, 1073–1081. [CrossRef]

42. Uchiyama, H.; Iwai, A.; Asada, Y.; Muramatsu, D.; Aoki, S.; Kawata, K.; Kusano, K.; Nagashima, K.; Yasokawa, D.; Okabe, M.; et al. A Small Scale Study on the Effects of Oral Administration of the β-Glucan Produced by *Aureobasidium pullulans* on Milk Quality and Cytokine Expressions of Holstein Cows, and on Bacterial Flora in the Intestines of Japanese Black Calves. *BMC Res. Notes* **2012**, *5*, 189. [CrossRef]

43. Kim, J.W.; Cho, H.R.; Ku, S.K. Efficacy Test of Polycan, a Beta-Glucan Originated From *Aureobasidium pullulans* SM-2001, on Anterior Cruciate Ligament Transection and Partial Medial Meniscectomy-Induced-Osteoarthritis Rats. *J. Microbiol. Biotechnol.* **2012**, *22*, 274–282. [CrossRef] [PubMed]

44. Sato, H.; Kobayashi, Y.; Hattori, A.; Suzuki, T.; Shigekawa, M.; Jippo, T. Inhibitory Effects of Water-Soluble Low-Molecular-Weight β-(1,3–1,6) D-Glucan Isolated From *Aureobasidium pullulans* 1A1 Strain Black Yeast on Mast Cell Degranulation and Passive Cutaneous Anaphylaxis. *Biosci. Biotechnol. Biochem.* **2012**, *76*, 84–88. [CrossRef] [PubMed]

45. Tanioka, A.; Hayama, K.; Mitsuya, M.; Tansho, S.; Ono, Y.; Tsubaki, K.; Abe, S. Effect of Oral Administration of β-D-Glucan From *Aureobasidium pullulans* ADK-34 on Candida and MRSA Infections in Immunosuppressed Mice. *Med. Mycol. J.* **2012**, *53*, 41–48. [CrossRef] [PubMed]

46. Tanioka, A.; Tanabe, K.; Hosono, A.; Kawakami, H.; Kaminogawa, S.; Tsubaki, K.; Hachimura, S. Enhancement of Intestinal Immune Function in Mice by β-D-Glucan From *Aureobasidium pullulans* ADK-34. *Scand. J. Immunol.* **2013**, *78*, 61–68. [CrossRef] [PubMed]

47. Tamegai, H.; Takada, Y.; Okabe, M.; Asada, Y.; Kusano, K.; Katagiri, Y.U.; Nagahara, Y. *Aureobasidium pullulans* Culture Supernatant Significantly Stimulates R-848-activated Phagocytosis of PMA-Induced THP-1 Macrophages. *Immunopharmacol. Immunotoxicol.* **2013**, *35*, 455–461. [CrossRef]

48. Iwai, A.; Shiozaki, T.; Miyazaki, T. Relevance of Signaling Molecules for Apoptosis Induction on Influenza A Virus Replication. *Biochem. Biophys. Res. Commun.* **2013**, *441*, 531–537. [CrossRef]

49. Kim, K.H.; Park, S.J.; Lee, J.E.; Lee, Y.J.; Song, C.H.; Choi, S.H.; Ku, S.K.; Kang, S.J. Anti-Skin-Aging Benefits of Exopolymers From *Aureobasidium pullulans* SM2001. *J. Cosmet. Sci.* **2014**, *65*, 285–298.

50. Muramatsu, D.; Kawata, K.; Aoki, S.; Uchiyama, H.; Okabe, M.; Miyazaki, T.; Kida, H.; Iwai, A. Stimulation With the *Aureobasidium pullulans*-Produced β-Glucan Effectively Induces Interferon Stimulated Genes in Macrophage-Like Cell Lines. *Sci. Rep.* **2014**, *4*, 4777. [CrossRef]

51. Aoki, S.; Iwai, A.; Kawata, K.; Muramatsu, D.; Uchiyama, H.; Okabe, M.; Ikesue, M.; Maeda, N.; Uede, T. Oral Administration of the β-Glucan Produced by *Aureobasidium pullulans* Ameliorates Development of Atherosclerosis in Apolipoprotein E Deficient Mice. *J. Funct. Foods* **2015**, *18*, 22–27. [CrossRef]

52. Oboshi, W.; Amakawa, M.; Kato, R. Effects of β-Glucan and Lactic Acid Bacteria on Gut Immune System. *Jpn. J. Med. Technol.* **2014**, *63*, 673–679.

53. Ganesh, J.S.; Rao, Y.Y.; Ravikumar, R.; Jayakrishnan, G.A.; Iwasaki, M.; Preethy, S.; Abraham, S.J. Beneficial Effects of Black Yeast Derived 1–3, 1–6 Beta Glucan-Nichi Glucan in a Dyslipidemic Individual of Indian Origin—A Case Report. *J. Diet. Suppl.* **2014**, *11*, 1–6. [CrossRef] [PubMed]

54. Kim, K.H.; Park, S.J.; Lee, Y.J.; Lee, J.E.; Song, C.H.; Choi, S.H.; Ku, S.K.; Kang, S.J. Inhibition of UVB-Induced Skin Damage by Exopolymers From *Aureobasidium pullulans* SM-2001 in Hairless Mice. *Basic Clin. Pharmacol. Toxicol.* **2015**, *116*, 73–86. [CrossRef] [PubMed]

55. Kawata, K.; Iwai, A.; Muramatsu, D.; Aoki, S.; Uchiyama, H.; Okabe, M.; Hayakawa, S.; Takaoka, A.; Miyazaki, T. Stimulation of Macrophages With the β-Glucan Produced by *Aureobasidium pullulans* Promotes the Secretion of Tumor Necrosis Factor-Related Apoptosis Inducing Ligand (TRAIL). *PLoS ONE* **2015**, *10*, e0124809. [CrossRef] [PubMed]

56. Lai, L.P.; Lee, M.T.; Chen, C.S.; Yu, B.; Lee, T.T. Effects of Co-Fermented *Pleurotus eryngii* Stalk Residues and Soybean Hulls by *Aureobasidium pullulans* on Performance and Intestinal Morphology in Broiler Chickens. *Poult. Sci.* **2015**, *94*, 2959–2969. [CrossRef] [PubMed]

57. Li, J.; Aizawa, Y.; Hiramoto, K.; Kasahara, E.; Tsuruta, D.; Suzuki, T.; Ikeda, A.; Azuma, H.; Nagasaki, T. Anti-Inflammatory Effect of Water-Soluble Complex of 1′-Acetoxychavicol Acetate With Highly Branched β-1,3-Glucan on Contact Dermatitis. *Biomed. Pharmacother.* **2015**, *69*, 201–207. [CrossRef] [PubMed]

58. Aoki, S.; Iwai, A.; Kawata, K.; Muramatsu, D.; Uchiyama, H.; Okabe, M.; Ikesue, M.; Maeda, N.; Uede, T. Oral Administration of the *Aureobasidium pullulans*-Derived β-Glucan Effectively Prevents the Development of High Fat Diet-Induced Fatty Liver in Mice. *Sci. Rep.* **2015**, *5*, 10457. [CrossRef]

59. Jippo, T.; Suzuki, T.; Sato, H.; Kobayashi, Y.; Shigekawa, M. Water Soluble Low Molecular Weight β -(1, 3 1, 6)D-Glucan Inhibit Cedar Pollinosis. *Funct. Foods Health Dis.* **2015**, *5*, 80–88. [CrossRef]

60. Iinuma, K. Case Report: A Patient Was Completely Recovered From stage 3 of Colorectal Cancer and 2 Sites of Liver Metastatic Cancer by Anti-Cancer Agents and β-Glucan EX After Resection of Colorectal Cancer. *New Food Indust.* **2016**, *58*, 29–34.

61. Yamamoto, K.; Yamauchi, Y.; Kusano, K.; Iinuma, K. Healing of Third-Degree Moderate-Temperature Burn Promoted Containing β -Glucan: A Case Report. *Jpn. J. PharmSci* **2017**, *74*, 697–703.

62. Ikeda, A.; Akiyama, M.; Sugikawa, K.; Koumoto, K.; Kagoshima, Y.; Li, J.; Suzuki, T.; Nagasaki, T. Formation of β-(1,3–1,6)-D-Glucan-Complexed Fullerene and Its Photodynamic Activity Towards Macrophages. *Org. Biomol. Chem.* **2017**, *15*, 1990–1997. [CrossRef]

63. Brown, G.D.; Herre, J.; Williams, D.L.; Willment, J.A.; Marshall, A.S.J.; Gordon, S. Dectin-1 Mediates the Biological Effects of β-Glucans. *J. Exp. Med.* **2003**, *197*, 1119–1124. [CrossRef] [PubMed]

64. Cho, C.S.; Jeong, H.S.; Kim, I.Y.; Jung, G.W.; Ku, B.H.; Park, D.C.; Moon, S.B.; Cho, H.R.; Bashir, K.M.I.; Ku, S.K.; et al. Anti-Osteoporotic Effects of Mixed Compositions of Extracellular Polymers Isolated From *Aureobasidium pullulans* and *Textoria Morbifera* in Ovariectomized Mice. *BMC Complement. Altern. Med.* **2018**, *18*, 295. [CrossRef] [PubMed]

65. Fujikura, D.; Muramatsu, D.; Toyomane, K.; Chiba, S.; Daito, T.; Iwai, A.; Kouwaki, T.; Okamoto, M.; Higashi, H.; Kida, H.; et al. *Aureobasidium pullulans*-Cultured Fluid Induces IL-18 Production, Leading to Th1-Polarization During Influenza A Virus Infection. *J. Biochem.* **2018**, *163*, 31–38. [CrossRef]

66. Lim, J.M.; Lee, Y.J.; Cho, H.R.; Park, D.C.; Jung, G.W.; Ku, S.K.; Choi, J.S. Extracellular Polysaccharides Purified from Aureobasidium Pullulans SM-2001 (Polycan) Inhibit Dexamethasone-Induced Muscle Atrophy in Mice. *Int. J. Mol. Med.* **2018**, *41*, 1245–1264. [CrossRef] [PubMed]

67. Hino, S.; Funada, R.; Sugikawa, K.; Koumoto, K.; Suzuki, T.; Nagasaki, T.; Ikeda, A. Turn-On Fluorescence and Photodynamic Activity of β-(1,3–1,6)-D-Glucan-Complexed Porphyrin Derivatives Inside HeLa Cells. *Photochem. Photobiol. Sci.* **2019**, *18*, 2854–2858. [CrossRef]

68. Tsuji, N.M.; Kosaka, A. Oral Tolerance: Intestinal Homeostasis and Antigen-Specific Regulatory T Cells. *Trends Immunol.* **2008**, *29*, 532–540. [CrossRef]

69. Tang, C.; Kakuta, S.; Shimizu, K.; Kadoki, M.; Kamiya, T.; Shimazu, T.; Kubo, S.; Saijo, S.; Ishigame, H.; Nakae, S.; et al. Suppression of IL-17F, but not of IL-17A, provides protection against colitis by inducing Treg cells through modification of the intestinal microbiota. *Nat. Immunol* **2018**, *19*, 755–765. [CrossRef]

70. Ross, G.D.; Cain, J.A.; Lachmann, P.J. Membrane Complement Receptor Type Three (CR3) Has Lectin-Like Properties Analogous to Bovine Conglutinin as Functions as a Receptor for Zymosan and Rabbit Erythrocytes as Well as a Receptor for iC3b. *J. Immunol.* **1985**, *134*, 3307–3315.

71. Vetvicka, V.; Thornton, B.P.; Ross, G.D. Soluble Beta-Glucan Polysaccharide Binding to the Lectin Site of Neutrophil or Natural Killer Cell Complement Receptor type 3 (CD11b/CD18) Generates a Primed State of the Receptor Capable of Mediating Cytotoxicity of iC3b- Opsonized Target Cells. *J. Clin. Investig.* **1996**, *98*, 50–61. [CrossRef]

72. Brown, G.D.; Gordon, S. Immune Recognition. A New Receptor for Beta-Glucans. *Nature* **2001**, *413*, 36–37. [CrossRef]

73. Ohno, N. Analysis of Glycans; Polysaccharide Functional Properties. *Compr. Glycosci.* **2007**, *2*, 559–577.

74. Hardy, G. Nutraceuticals and Functional Foods: Introduction and *Meaning*. *Nutrition* **2000**, *16*, 688–689. [CrossRef]

75. Roberfroid, M.B. Global View on Functional Foods: European Perspectives. *Br. J. Nutr.* **2002**, *88* (Suppl. 2), S133–S138. [CrossRef] [PubMed]

 *nutrients*

Article

Effect of Varying Molecular Weight of Oat β-Glucan Taken just before Eating on Postprandial Glycemic Response in Healthy Humans

Thomas M. S. Wolever [1], Outi Mattila [2], Natalia Rosa-Sibakov [2], Susan M. Tosh [3], Alexandra L. Jenkins [1], Adish Ezatagha [1], Ruedi Duss [4] and Robert E. Steinert [4,5,*

[1]   INQUIS Clinical Research, Ltd. (formerly GI Labs), Toronto, ON M5C 2N8, Canada; twolever@inquis.com (T.M.S.W.); alexandrajenkins@inquis.com (A.L.J.); aezatagha@inquis.com (A.E.)
[2]   VTT Technical Research Centre of Finland Ltd., 1000 Espoo, Finland; Outi.Mattila@vtt.fi (O.M.); Natalia.Rosa-Sibakov@vtt.fi (N.R.-S.)
[3]   School of Nutrition Sciences, University of Ottawa, Ottawa, ON K1N 6N5, Canada; Susan.Tosh@uottawa.ca
[4]   DSM Nutritional Products Ltd., R&D Human Nutrition and Health, 4002 Basel, Switzerland; Ruedi.Duss@dsm.com
[5]   Department of Surgery, Division of Visceral and Transplantation Surgery, University Hospital Zürich, 8091 Zürich, Switzerland
*   Correspondence: robert.steinert@dsm.com

Received: 24 June 2020; Accepted: 27 July 2020; Published: 29 July 2020

Abstract: To see if the molecular weight (MW) and viscosity of oat β-glucan (OBG) when taken before eating determine its effect on postprandial glycemic responses (PPRG), healthy overnight-fasted subjects ($n = 16$) were studied on eight separate occasions. Subjects consumed 200 mL water alone (Control) or with 4 g OBG varying in MW and viscosity followed, 2–3 min later, by 113 g white-bread. Blood was taken fasting and at 15, 30, 45, 60, 90, and 120 min after starting to eat. None of the OBG treatments differed significantly from the Control for the a-priori primary endpoint of glucose peak-rise or secondary endpoint of incremental area-under-the-curve (iAUC) over 0–120 min. However, significant differences from the Control were seen for glucose iAUC over 0–45 min and time to peak (TTP) glucose. Lower log(MW) and log(viscosity) were associated with higher iAUC 0–45 ($p < 0.001$) and shorter TTP ($p < 0.001$). We conclude that when 4 g OBG is taken as a preload, reducing MW does not affect glucose peak rise or iAUC0-120, but rather accelerates the rise in blood glucose and reduces the time it takes glucose to reach the peak. However, this is based on post-hoc calculation of iAUC0-45 and TTP and needs to be confirmed in a subsequent study.

Keywords: humans; oat β-glucan; acute glycemic response; dietary fiber; preload; carbohydrates

1. Introduction

Reducing the postprandial glycemic response (PPGR) is generally considered to be a beneficial effect [1,2]. Oats and oat products elicit lower glycemic responses than most other types of ready-to-eat and cooked breakfast cereals when comparing equivalent amounts of available carbohydrate (avCHO) [3–7]. Oatmeal is rich in β-glucan, a highly viscous soluble dietary fiber found predominantly in the endospermic cell wall of oats and barley [8] thought to be responsible, at least in part, for the low glycemic impact of oats. Oat β-glucan (OBG) is thought to reduce PPGR by increasing the viscosity of the contents of the upper gut [8,9], which, in turn, may slow gastric emptying, reduce the rate of starch digestion, and increase the thickness of the so-called unstirred water layer in the small intestine, thus delaying the absorption of carbohydrates [9]. The addition of OBG to test meals results in a dose-dependent reduction in PPGR, an effect which is evident with as little as 0.04 g OBG per gram avCHO in the test meal [10].

Previous studies suggest that taking viscous fibers separate from a test-meal may not reduce the PPGR [11,12]. Thus, in order to reduce PPGR, it may be necessary to mix viscous fibers with the food; however, this alters the taste and texture of foods in a way that may reduce palatability. A potential solution to this problem may be to consume a fiber 'preload' which develops viscosity slowly so that it can be consumed in a palatable form just before eating and remains dispersed in the stomach long enough to be able to mix effectively with the main meal, but becomes viscous by the time the stomach starts to empty. To this end, we showed that taking commercial oat bran containing 0.9, 2.6, or 5.3 g OBG mixed in water a few minutes before consuming 50 g avCHO as white bread reduced postprandial glycemic responses in a dose-dependent fashion [13]. Since the different doses of OBG were mixed into a fixed amount of water, the viscosity of the preload increased as the dose of OBG increased. Thus, it is not possible to determine from that experiment whether it was the amount of OBG or the viscosity of the preload which affected the glycemic response. We hypothesized that the extent to which an OBG preload reduces the glycemic response depends on its viscosity. Since the viscosity of solutions containing a fixed amount of OBG can be changed by altering the molecular weight (MW) of the OBG, the purpose of this study was to determine the effect of varying the MW, and hence viscosity, of 4 g OBG preloads on the PPGR elicited by a 51.5 g avCHO portion of white bread.

2. Material and Methods

2.1. Subjects and Study Design

We conducted an open-label study with a randomized cross-over design at a contract research organization; participant visits occurred between 3 January and 6 February 2018. We recruited 16 healthy subjects (9 males and 7 non-pregnant females) of mixed ethnic background (7 Caucasian and 9 non-Caucasian as follows: 4 East- or South-East Asian, 2 South Asian, 2 Hispanic and 1 mixed Aboriginal/Caucasian), aged (mean ± SD) 41 ± 11 y with body mass index 25.6 ± 2.7 kg/m^2, and all 16 completed the study according to protocol. Compared to the 7 Caucasian participants, the 9 non-Caucasian participants were younger (mean ± SD, 36.4 ± 10.8 vs. 45.7 ± 8.6 y, p = 0.084) and had a lower BMI (23.9 ± 1.6 vs. 27.8 ± 2.1 kg/m^2, pp = 0.001) (Supplemental Table S1). Twelve (12) participants took no prescription medications or supplements, 3 took daily supplements (1 took 750 mg omega-3 fatty acids, 125 mg calcium, and 125 mg magnesium daily; 1 took 2 capsules of the hair growth supplement Priorin daily; and 1 took 1000 mg omega-3 fatty acids and 1000 IU vitamin D daily), and 1 took a birth control pill. The procedures used were in accordance with the protocol used at INQUIS for determining the GI of foods, which was approved by the Western Institutional Review Board®. The work described in this manuscript was carried out in accordance with The Code of Ethics of the World Medical Association (Declaration of Helsinki) for experiments involving human subjects. The protocol used was approved by the Western Institutional Review Board® (WIRB) which meets all requirements of the US Food and Drug Administration (FDA), the Department of Health and Human Services (DHHS), the Canadian Health Protection Branch (HPB), Canadian Institutes for Health Research (CIHR) and the European Community Guidelines. Prior to their participation, all subjects provided written informed consent by signing the approved consent form (WIRB protocol number: 971199).

2.2. Procedures

Each subject underwent 8 treatments on separate days, with each subject performing up to 3 tests per week separated by at least one day. On each test day, subjects came to the facility (20 Victoria Street, 3rd floor, Toronto, ON, Canada) in the morning after a 10–14 h overnight fast. After being weighed, subjects gave 2 fasting blood samples obtained by finger-prick ~5 min apart. After the first fasting blood sample, subjects consumed a preload consisting of 200 mL water either alone or mixed (by hand, avoiding clumps) with one of 6 OBG sources (described below) each of which contained 4 g OBG of varying MW (Table 1). Subjects consumed water alone (Control) on 2 separate occasions and each of the OBG treatments once in randomized order. After the preload, a second fasting blood sample was obtained. Subjects then consumed a portion of white bread containing 51.5 g avCHO within 12 min. Subjects could choose to have, with each test meal, a drink of coffee, tea, or water, to which 30 mL 2% milk and/or artificial sweetener could be added; the drink chosen remained the same for all 8 treatments. After consuming the test meal, subjects rated the palatability of the test meal using a visual analogue scale (VAS) consisting of a 100 mm line anchored at the left end by "very unpalatable" and at the right end by "very palatable". Subjects make a vertical mark along the line to indicate their perceived palatability. The distance from the left end of the line to the mark made by the subject was the palatability rating; the higher the value, the higher the perceived palatability. Further blood samples were obtained at 15, 30, 45, 60, 90, and 120 min after starting to eat. Subjects remained seated quietly during the 2 h of the test. After the last blood sample had been obtained, subjects were offered a snack and then permitted to leave.

Blood samples (2–3 drops each) were collected into 5 mL tubes containing sodium fluoride/potassium oxalate, mixed by rotating the tube vigorously, placed into a refrigerator until the end of the test session, then stored at −20 °C prior to glucose analysis using a YSI (Yellow Springs Instruments, Yellow Springs, OH, USA) analyser within 5 days.

Table 1. Composition of the Control oat β-glucan preloads.

Preload	Amount (g)	Protein (g)	Fat (g)	Total Carb. (g)	Total Dietary Fiber (g)	Oat β-Glucan			
						Amount (g)	MW (kDa)	Viscosity (cP)	
								Initial	PD *
Control	0	0	0	0	0	0	-	-	-
MW1	7.48	na	na	na	5.0	4.0	52 ± 5	14 ± 4	10 ± 1
MW2	7.41	na	na	na	5.1	4.0	76 ± 8	17 ± 3	14 ± 1
MW3	8.28	na	na	na	5.7	4.0	153 ± 5	28 ± 2	31 ± 1
MW4	10.42	na	na	na	5.1	4.0	393 ± 31	35 ± 5	40 ± 1
OP1	13.7	3.2	0.7	8.4	7.2	4.0	1980 ± 265	25 ± 1	112 ± 9
OP2	11.90	0.4	0.1	10.7	5.5	4.0	841 ± 110	117 ± 7	143 ± 7

The Control preload was 200 mL water; the oat β-glucan preloads (MW1, MW2, MW3, MW4, OP1, and OP2) were added to 200 mL water. Values for protein, fat and total carbohydrate (Carb.) were provided by the sponsors ("na" = not available). Total dietary fiber was measured by Eurofins CLF Specialised Nutrition Testing Services GmBH, Professor-Wagner-Straße 11, D-61381 Friedrichsdorf, Germany. Values for MW and viscosity are means ± SD of quadruplicate measures. * PD = post-digestion.

2.3. Preparation of Test Meals and OBG Treatments

White bread was baked in a bread maker in loaves containing 500 g avCHO. The ingredients for each loaf (520 mL warm water, 708 g all-purpose flour, 14 g sugar, 8 g salt, and 13 g yeast) were placed into an automatic bread maker according to instructions, and the machine turned on. After the loaf had been made, it was allowed to cool for an hour, and then weighed and after discarding the crust ends, the remainder was divided into portions each of which weighed ~113 g and contained 9 g protein, 0.7 g fat, 54.4 g total carbohydrate, and 2.9 g dietary fiber. These portions were frozen prior to use, and reheated in the microwave prior to consumption.

The sources of OBG in 2 of the treatments were commercial oat products (OatWell®, DSM Nutritional Products, Heerlen, The Netherlands, termed OP1; and PromOat®, Tate & Lyle, London, UK, termed OP2), while the source of OBG in 4 of the treatments (MW1, MW2, MW3, and MW4) consisted of OP1 which had been treated to reduce the MW of the OBG it contained using a method based on Zurbriggen et al. [14] and Aktas-Akyildiz et al. [15]. The details of this procedure are described in Supplemental Information.

2.4. Measurement of OBG Viscosity and MW

The amount of OBG in the test sources was measured using the American Organization of Analytical Chemists approved method 995.16 [16]. The viscosity of the test OBG sources was measured using a method adapted from the American Association of Cereal Chemists International approved method 32-24.01 [17]. Viscosity measured by this method has previously been shown to correlate with post-prandial blood glucose response [18]. The MW of β-glucan was determined using a Gilson HPLC with a Shimadzu RF-551 fluorescence detector (excitation: 415 nm; emission: 450 nm) using a method adapted from previous research [19]. Details of these methods are described in Supplemental Information.

The characteristics of the OBG preloads are shown in Table 1. The MW and post-digestion viscosities of each OBG treatment were found to be significantly different from every other treatment.

2.5. Sample Size and Randomization

The primary endpoint of the study was glucose peak rise and the secondary endpoint glucose iAUC over 0–120 min (iAUC0-120). Based on our previous study [13], we estimated that 4 g of highly viscous OBG could reduce glucose peak rise by about 25%. We studied 16 subjects because, using the t-distribution and assuming an average coefficient of variation (CV) of within-individual variation of 20%, $n = 16$ subjects has a 90% power to detect a 25% difference in glucose peak rise with two-tailed $p < 0.05$. The order of treatments was randomized using a balanced Latin-square design [20]. The 16 orders created were assigned to subjects in the order they attended for the first visit after being recruited.

2.6. Data Management and Statistical Analysis

Data were entered into a spreadsheet by 2 different individuals and the values were compared to assure accurate transcription. Values for the glucose were missing for 4 (0.4%) of the 1024 blood samples due to clotted blood; 2 values at –5 min were replaced by the value at 0 min; 2 values at 15 min were imputed using a procedure described by Snedecor and Cochran [21]. The mean ± SD glucose concentration in the 113 fasting (0 min) samples in which duplicates could be measured was 4.37 ± 0.0409 mmol/L (analytical coefficient of variation (CV = $100 \times$ SD/mean) of 0.9%). Glucose peak rise was the maximum glucose concentration minus the fasting concentration. Incremental areas under the curve (iAUC) were calculated using the trapezoid rule, ignoring area beneath the baseline [22], for the time interval of 0–120 min (iAUC0-120; secondary endpoint) and the time intervals of 0–45 min (iAUC0-45) and 45–120 min (iAUC45-120; additional endpoints). For calculation of iAUC and peak rise, fasting glucose was taken to be the mean of the first measurement of glucose in the 2 fasting samples. The time taken to reach the peak (time to peak) was taken to be the time of the maximum measured glucose concentration. The iAUC and peak rises for each OBG treatment were expressed relative to the mean iAUC or peak rise after the 2 tests of White Bread alone taken by the same subject. Relative response values >2 × SD greater than the mean were deemed outliers and replaced by the mean for the respective treatment.

For statistical analysis, the mean of the values for the 2 Control tests were considered as one treatment. Blood glucose concentrations and increments were subjected to a repeated-measures analysis of variance using the general linear model procedure (ANOVA) and examining for the main effects of time and treatment and the time × treatment interaction. After the demonstration of a significant time × treatment interaction, the difference between each OBG treatment and Control at each time was determined using Dunnett's test. Values for iAUC and peak rise were analysed using ANOVA, examining for the main effect of the treatment. After the demonstration of a significant main effect, the significance of the difference between each OBG treatment and the Control was determined using Dunnett's test. Differences were considered to be statistically significant if two-tailed $p < 0.05$. The significance of the regressions of log MW, log initial viscosity and log post-digestion viscosity of the 6 OBG treatments on the various outcomes were determined by ANOVA, with $p < 0.05$ taken to be the criterion for statistical significance.

3. Results

The test-meals were all consumed within the specified time and there were no adverse events or protocol deviations. All the OBG treatments were rated as being significantly less palatable than the Control (Table 2). Mean palatability was not related to MW ($p = 0.51$) or viscosity ($p = 0.52$) (not shown). Mean fasting glucose was similar for the 7 treatments (Table 2).

Table 2. Palatability, fasting glucose, and measures of glycemic response.

Preload	Palatability (mm)	Fasting Glucose (mmol/L)	Peak Rise (mmol/L)	Incremental Area under the Curve (mmol × min/L)			Time to Peak (min)
				0–45 min	45–120 min	0–120 min	
Control	68 ± 6	4.39 ± 0.08	3.10 ± 0.27	80 ± 6	118 ± 15	198 ± 20	37 ± 3
MW1	47 ± 7 *	4.48 ± 0.07	2.89 ± 0.23	75 ± 6	98 ± 11	173 ± 15	34 ± 2
MW2	46 ± 6 *	4.38 ± 0.08	2.89 ± 0.23	73 ± 6	85 ± 11	159 ± 14	36 ± 2
MW3	34 ± 7 *	4.34 ± 0.06	2.93 ± 0.30	63 ± 6 *	123 ± 16	186 ± 21	43 ± 2
MW4	34 ± 8 *	4.44 ± 0.08	2.70 ± 0.32	60 ± 7 *	122 ± 17	181 ± 22	52 ± 5 *
OP1	43 ± 7 *	4.47 ± 0.08	2.90 ± 0.31	52 ± 4 *	143 ± 18	195 ± 21	53 ± 5 *
OP2	45 ± 7 *	4.40 ± 0.07	2.55 ± 0.29	61 ± 5 *	122 ± 18	183 ± 21	48 ± 4
p **	<0.0001	0.095	0.19	<0.0001	0.002	0.20	<0.0001

Values are means ± SEM for $n = 16$ subjects. The Control preload was 200 mL water; the oat β-glucan preloads (MW1, MW2, MW3, MW4, OP1 and OP2) were added to 200 mL water. Preloads were taken 2–3 min before consuming ~113 g white bread. * Significantly different from the Control by Dunnett's test for two-tailed $p < 0.05$. ** Significance of main effect of treatment by ANOVA.

Compared to Caucasians, non-Caucasian subjects had higher iAUC45-120 ($p < 0.05$) and higher iAUC0-120 ($p = 0.06$), and subjects with BMI <25 kg/m^2 had higher iAUC0-45, iAUC45-120 and iAUC0-120 than those with BMI ≥25 kg/m^2. However, when expressed as a percentage of the Control, all subjects responded similarly regardless of their sex (not shown), age, ethnicity or BMI (Supplementary Table S2).

There were significant time×treatment interactions for blood glucose concentrations ($p < 0.0001$) and increments ($p < 0.0001$). The mean blood glucose increments after MW3, MW4, OP1, and OP2 were significantly less than after the Control at both 15 and 30 min, while increments after MW2 were significantly less than the Control at 90 min (Figure 1).

There was no significant difference among treatments for the primary endpoint of the glucose peak rise and no significant difference for the secondary endpoint of glucose iAUC0-120.

Additional analysis showed that the iAUC0-45 after MW3, MW4, OP1, and OP2 were significantly less than the Control, and, after MW4 and OP1, it took significantly longer for blood glucose to reach a peak than after the Control (Table 2). Glucose iAUC45-120 differed significantly among treatments, but none of the treatments differed from the Control (Table 2).

When expressed as a % of Control, iAUC0-45 was significantly associated with log MW ($p < 0.0001$), log initial viscosity ($p < 0.01$), and log post-digestion viscosity ($p = 0.006$) (Figure 2). Ten-fold (1 log unit) increases in MW, initial viscosity, and post-digestion viscosity, respectively, were associated with ~17, ~16 and ~20% reductions in iAUC0-45 (Figure 2).

When expressed as a % of Control, mean iAUC45-120 was significantly associated with log MW ($p = 0.002$), log initial viscosity ($p = 0.046$), and log post-digestion viscosity ($p = 0.0004$).

The time it took blood glucose to reach the peak was significantly associated with log MW ($p < 0.0001$), log initial viscosity ($p = 0.003$), and log post-digestion viscosity ($p < 0.0001$) (Figure 2). Ten-fold (1 log unit) increases in MW, initial viscosity, and post-digestion viscosity, respectively, were associated with ~12, ~14, and ~15 min delays in the time to reach peak blood glucose (Figure 2).

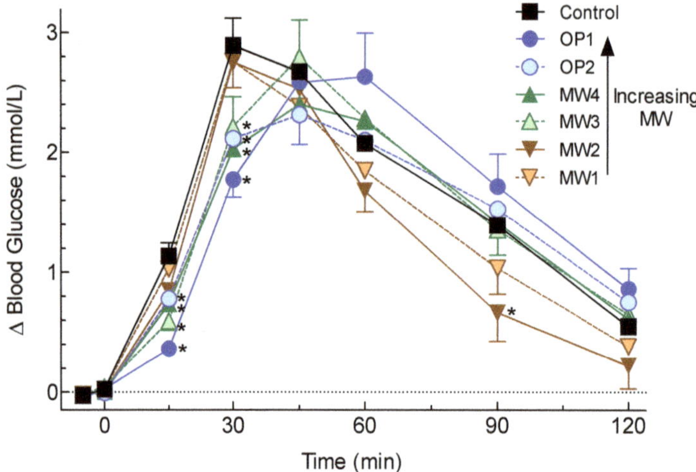

Figure 1. Glycemic responses elicited by the test meals. Values are means ± SEM for $n = 16$ subjects. The Control (200 mL water) and oat β-glucan preloads (MW1, MW2, MW3, MW4, OP1, and OP2 added to 200 mL water) were taken 2–3 min before consuming ~113 g white bread. * Significantly different from the Control by Dunnett's test for two-tailed $p < 0.05$.

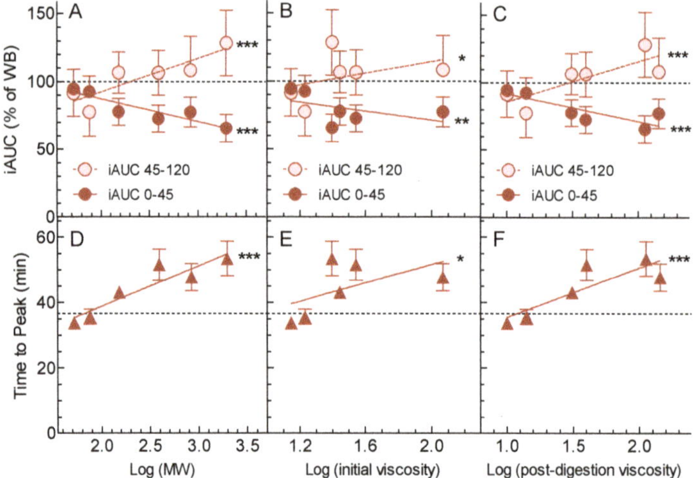

Figure 2. Relationships between glycemic response and β-glucan MW and viscosity. Values are means ± SEM for $n = 16$ subjects. Panels A, B, and C show incremental areas under the curve (iAUC) expressed as a percentage of that elicited by the Control; dark red circles are iAUC over 0–45 min, light red circles are iAUC over 60–120 min. Panels D, E, and F show the time taken to reach peak glucose in minutes. Panels A and D show the relationships between glycemic responses and log (MW); panels B and E between glycemic responses and log (initial viscosity); and panels C and F between glycemic responses and log (post-digestion viscosity). The black dashed lines show the value for the Control, and dark red solid and light red dashed lines are regression lines. Significance of the regressions by ANOVA: * $p < 0.05$; ** $p < 0.01$; *** $p < 0.001$.

4. Discussion

The results showed that there was no significant difference among treatments for the primary endpoint of glucose peak rise or the secondary endpoint of iAUC0-120 (Table 2). Thus, our a-priori hypothesis that the extent to which an OBG preload reduces the glycemic response (as measured by glucose peak rise) depends on its viscosity was not supported. However, post-hoc additional analysis showed significant differences among treatments for glucose iAUC0-45 and for the time taken to reach the peak (Table 2), with greater log MW and log viscosity being associated with lower iAUC0-45 min ($p < 0.001$) and a longer time to reach peak glucose ($p < 0.001$) (Figure 2).

The lack of any significant reduction in iAUC0-120 and peak rise after 4 g OBG preloads was surprising, given our previous demonstration that, when 0.9, 2.6 and 5.3 g doses of OBG were given as a preload before consuming a 50 g avCHO portion of white bread, OBG reduced both glucose peak rise and iAUC0-120 in a dose-dependent fashion [13]. However, since only the 5.3 g OBG dose elicited statistically significant reductions, it is possible that that the 4 g dose used here may not have been large enough to elicit significant reductions in peak rise of iAUC0-120. Furthermore, we did not report results for iAUC0-45, iAUC45-120, and time to peak in the previous publication [13]. One of the sources of OBG included in the present study, OP1, was also used as the source of OBG in our previous study [13]. When the results for 4 g OBG from OP1 from the present study are plotted on dose-response curves for iAUC0-45, iAUC45-120, time to peak, and peak rise from the previous study [13], the results from the present study are consistent with those from the previous study, since the 95% confidence intervals (95% CI) from the 2 studies overlap (Figure 3). Taken together, the results of these 2 studies suggest that, when high MW OBG is taken as a preload, increasing the dose of OBG without changing MW reduces glucose peak rise, iAUC0-45, and iAUC0-120, and increases the time to peak in a dose-dependent fashion, whereas reducing the MW of 4 g of OBG only affects glucose iAUC0-45 and time to reach the peak. However, the effects of reducing the MW of preloads containing <4 g or >4 g OBG may differ from those seen here.

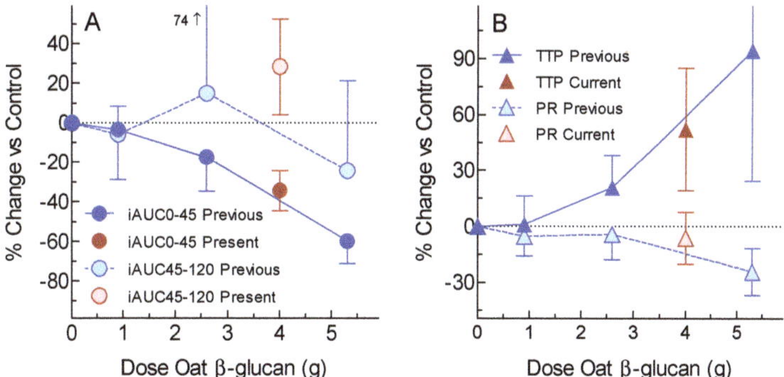

Figure 3. Current results compared to previous results for the same oat-bran treatment. Values are means ± 95% confidence intervals of the percentage change from control (0 g oat β-glucan). The previous study [13] tested the effects of a commercially available oat bran containing 0.9, 2.6 and 5.3 g of oat-β-glucan given as a preload on the glycemic response elicited by white bread in $n = 10$ subjects (blue symbols) using the same protocol as the present study. The present results for the same source of oat β-glucan used in the previous study (i.e., OP1) are shown with red symbols. Panel A shows the incremental area under the glucose curve (iAUC) from 0–45 min (dark circles) and from 45–120 min (light circles). Panel B shows the time to reach the peak (TTP; dark triangles) and peak rise (PR; light triangles).

It is generally considered that the effect of OBG on glycemic responses is determined by its ability to increase the viscosity of the contents of the gastrointestinal tract. The viscosity of OBG can be decreased either by decreasing the MW of OBG or by decreasing the amount of OBG consumed. When OBG is mixed with the food consumed, the glucose peak rise is reduced in proportion to log (MW × C), where C is the concentration of OBG in the intestine [23,24]. Nevertheless, when OBG is mixed into a glucose solution, the effects of viscosity on glycemic response are not fully understood. For example, reducing the viscosity of an OBG-containing glucose solution by reducing MW significantly increased the glucose peak rise and iAUC0-120, but the same reduction in glucose solution viscosity obtained by increasing the volume of water in the OBG-containing glucose solution had no effect on glucose peak rise or iAUC0-120 [25]. This result is consistent with observations that high viscosity meals are diluted in the stomach by oral and gastric secretions more than low viscosity meals [26]. Furthermore, the viscosity of OBG in the small intestine may differ from the viscosity of test-meals because the digestion process may release OBG from the food matrix. This is illustrated in the present results where the post-digestion viscosities of OP1 and OP2 were higher than their initial viscosities; this may explain why post-digestion viscosity was more closely correlated to the glycemic outcomes than initial viscosity (Figure 2). However, the precise mechanisms by which OBG reduces glycemic responses are not fully understood.

One weakness of this study is that the effect of altering MW using different doses of OBG was not determined, and thus our conclusions only apply to a dose of 4 g. Also, the study was powered to detect a 25% reduction in peak rise, but was underpowered to detect the 18% reduction observed for OP2 vs. the control. Post-hoc power analysis indicates that $n = 20$ subjects would have provided 80% power to detect the 18% difference, and $n = 25$ subjects would have provided 90% power. Finally, we did not measure serum insulin responses. When OBG is mixed with the test meal consumed, the insulinemic response is reduced in proportion to the reduction in glucose [27,28]. Here, however, as the MW of OBG in a preload increased, we found a delay in the rise and fall of postprandial glucose but no reduction in peak rise. We would expect the delayed rise in glucose to be accompanied by a delayed rise in insulin, but we do not know if the lack of effect on the glucose peak rise and delayed fall in glucose were associated with lower or higher insulin responses. Such information would be useful to better understand the mechanism of this effect.

Our results raise the question of whether delaying the rise in postprandial glucose (shown here by the reduced glucose iAUC0-45 and delayed time to peak), in the absence of a reduction in peak-rise or iAUC0-120, is beneficial for the prevention or treatment of diabetes. There is evidence that blood glucose 30 or 60 min after a 75 g oral glucose tolerance test is a similar or better predictor of future type 2 diabetes (T2D) than the traditional 120 min [29]. This suggests that a delayed rise in postprandial blood glucose may be associated with reduced risk for T2D. Consistent with this are the results of a study showing that when subjects with T2D took 5 g of guar granules prior to consuming mashed potatoes (23 g avCHO), the glucose peak rise was delayed by 30–45 min, but there was no significant effect on glucose peak rise or iAUC; however, 5 g of guar mixed into the mashed potato test-meal reduced glucose iAUC ~40% [12]. Nevertheless, 4 g of guar granules twice daily before meals for 6 wk reduced HbA1c to the same extent as 8 g of guar per day incorporated into bread, with both guar treatments eliciting a significantly greater reduction in HbA1c than the control treatment [30]. Thus, OBG preloads which delay the rise in blood glucose without reducing it may be beneficial in T2D.

A 10–15 min delay in the postprandial glucose rise may also be useful in people with type 1 diabetes using subcutaneously administered insulin. Modern rapid-acting insulins begin to act 5–15 min after insulin administration [31]. Delaying meal consumption for 15 min after rapid-acting insulin administration reduced the glycemic response compared to eating the same meal immediately before or after insulin administration [32,33]. Thus, the ability of an OBG preload to delay the rise in blood glucose after eating could be useful in the management of type 1 diabetes and may warrant investigation. In addition, long-term OBG consumption may have beneficial effects on glycemic control by virtue of its fermentation in the large intestine with the production of short-chain fatty acids [34] and potential alteration of the colonic microbiota [35].

The effects of the OBG treatments on glucose iAUC0-45 were nearly opposite to those on iAUC45-120 (Figure 2). This likely reflects the different kinetics of glucose absorption of the treatments along the upper GI tract depending on MW and viscosity. The Control, MW1, and MW2, with lower MWs ($\leq$76 kDa) and viscosities ($\leq$14 cP) than MW3, MW4, OP1, and OP2, may have emptied from the stomach more quickly and been absorbed from the small intestine more rapidly, resulting in a high early glucose response (iAUC0-45) and a low late glucose response (iAUC45-120). In contrast, glucose absorption from MW3, MW4, OP1, and OP2, with higher MWs (153–1980 kDa) and viscosities (31–143 cP), may have been delayed, resulting in a lower early glucose response, a delayed peak, and higher late glucose response. The inverse correlations of iAUC0-45 and iAUC45-120, respectively, explain the lack of effect of the OBG treatments on iAUC0-120.

The pre-determined primary endpoint for this study was glucose peak rise. We previously showed that incorporating 4 g OBG with MW 132 kDa into a muffin containing 48 g avCHO had no significant effect on glucose peak rise, whereas 4 g of native OBG (MW ~2200 kDa) reduced glucose peak rise by 40% [23]. Although a similar range of MWs was investigated here (52 to 1980 kDa), altering the MW of 4 g OBG ingested separately from the avCHO had no significant effect on glucose peak rise, possibly because the viscous preloads were quickly diluted by oral and gastric secretions as the bread was being eaten. However, we used a preload containing 4 g OBG with a test meal of white bread containing ~50 g avCHO. The results might differ with different doses of OBG or with different types or sizes of the test meal.

5. Conclusions

We conclude that, when 4 g OBG is taken as a preload, reducing MW does not affect the peak rise or iAUC0-120, but rather accelerates the rise in blood glucose and reduces the time it takes for blood glucose to reach the peak. However, this conclusion is based on additional post-hoc comparisons of iAUC0-45 and the time to peak, and need to be confirmed in a subsequent study.

Supplementary Materials: The following are available online at http://www.mdpi.com/2072-6643/12/8/2275/s1, Table S1: Demographic details of Caucasian and non-Caucasian participants, Table S2: Glycemic responses in Caucasian and non-Caucasian participants.

Author Contributions: R.E.S. and R.D. conceived and designed the experiments; T.M.S.W., A.L.J. and A.E. performed the experiments; T.M.S.W. analyzed the data; O.M., N.R.-S. and S.M.T. contributed materials and analyses; T.M.S.W. wrote the paper. All authors have read and agreed to the published version of the manuscript.

Funding: Financial and in-kind support for this study was provided by DSM Nutritional Products Ltd. and VTT Technical Research Centre of Finland Ltd.

Conflicts of Interest: T.M.S.W., his wife, and A.J. are owners and employees of INQUIS Clinical Research, Ltd., a contract research organization; neither they nor INQUIS has any financial interest in any company involved in producing, distributing, or marketing foods or food ingredients and, in particular, no financial interest in DSM or VTT. O.M. and N.R.S. are employees of VTT. A.H. is an employee of INQUIS. R.D. and R.E.S. are employees of DSM Nutritional Products, Basel, Switzerland; they do not have any conflict of interest regarding the publication of this article. S.M.T. has no conflicts of interest to declare.

References

1. European Food Safety Authority (EFSA) Panel on dietetic Products, Nutrition and Allergies (NDA). Scientific Opinion on the substantiation of health claims related to beta-glucans from oats and barley and maintenance of normal blood LDL-cholesterol concentrations (ID1236, 1299), increase in satiety leading to a reduction in energy intake (ID 851, 852), reduction in post-prandial glycaemic response (ID 821, 824), and "digestive function" (ID 850) pursuant to Article 13(1) of Regulation (EC) No 1924/2006. *EFSA J.* **2011**, *9*, 2207.

2. Bureau of Nutritional Sciences, Food Directorate, Health Products and Food Branch, Health Canada. Draft guidance document on food health claims related to the reduction in post-prandial glycaemic response. Available online: https://www.canada.ca/en/health-canada/services/food-nutrition/public-involvement-partnerships/technical-consultation-draft-guidance-document-food-health-claims-related-post-prandial-glycaemia.html (accessed on 2 May 2020).

3. Jenkins, D.J.A.; Wolever, T.M.S.; Taylor, R.H.; Barker, H.M.; Fielden, H.; Baldwin, J.M.; Bowling, A.C.; Newman, H.C.; Jenkins, A.L.; Goff, D.V. Glycemic index of foods: A physiological basis for carbohydrate exchange. *Am. J. Clin. Nutr.* **1981**, *34*, 362–366. [CrossRef]

4. Jenkins, D.J.A.; Wolever, T.M.S.; Jenkins, A.L.; Thorne, M.J.; Lee, R.; Kalmusky, J.; Reichert, R.; Wong, G.S. The glycaemic index of foods tested in diabetic patients: A new basis for carbohydrate exchange favouring the use of legumes. *Diabetologia* **1983**, *24*, 257–264. [CrossRef] [PubMed]

5. Tosh, S.M.; Chu, Y. Systematic review of the effect of processing of whole-grain oat cereals on glycaemic responses. *Br. J. Nutr.* **2015**, *114*, 1256–1262. [CrossRef] [PubMed]

6. Wolever, T.M.S.; Katzman-Relle, L.; Jenkins, A.L.; Vuksan, V.; Josse, R.G.; Jenkins, D.J.A. Glycaemic index of 102 complex carbohydrate foods in patients with diabetes. *Nutr. Res.* **1994**, *14*, 651–669. [CrossRef]

7. Wolever, T.M.S.; van Klinken, B.J.W.; Spruill, S.E.; Jenkins, A.L.; Chu, Y.; Harkness, L. Effect of serving size and addition of sugar on the glycemic response elicited by oatmeal: A randomized, cross-over study. *Clin. Nutr. ESPEN* **2016**, *16*, 48–54. [CrossRef]

8. Wood, P.J. Oat β-glucan: Structure, location and properties. In *Oats: Chemistry and Technology*; Webstger, F.H., Ed.; AACC Inc.: St Paul, MN, USA, 1986; pp. 121–152.

9. Jenkins, D.J.A.; Wolever, T.M.S.; Leeds, A.R.; Gassull, M.A.; Haisman, P.; Dilawari, J.; Goff, D.V.; Metz, G.L.; Alberti, K.G.M. Dietary fibres, fibre analogues, and glucose tolerance: Importance of viscosity. *Br. Med. J.* **1978**, *1*, 1392–1394. [CrossRef]

10. Wolever, T.M.S.; Jenkins, A.L.; Prudence, K.; Johnson, J.; Duss, R.; Chu, Y.; Steinert, R.E. Effect of adding oat bran to instant oatmeal on glycaemic response in humans—A study to establish the minimum effective dose of oat β-glucan. *Food Funct.* **2018**, *9*, 1692–1700. [CrossRef]

11. Jenkins, D.J.; Nineham, R.; Craddock, C.; Craig-McFeely, P.; Donaldson, K.; Leigh, T.; Snook, J. Fibre in diabetes. *Lancet* **1979**, *1*, 434–435. [CrossRef]

12. Fuessl, S.; Adrian, T.E.; Bacarese-Hamilton, A.J.; Bloom, S.R. Guar in NIDD: Effect of different modes of administration on plasma glucose and insulin responses to a starch meal. *Pract. Diabetes Int.* **1986**, *3*, 258–260. [CrossRef]

13. Steinert, R.E.; Raederstorff, D.; Wolever, T.M.S. Effect of consuming oat bran mixed in water before a meal on glycemic responses in healthy humans—A pilot study. *Nutrients* **2016**, *8*, 524. [CrossRef]

14. Zurbriggen, B.Z.; Bailey, M.J.; Penttilä, M.E.; Poutanen, K.; Linko, M. Pilot scale production of a heterologous *Trichoderma reesei* cellulase by *Saccharomyces cerevisiae*. *J. Biotechnol.* **1990**, *13*, 267–278. [CrossRef]

15. Aktas-Akyildiz, E.; Sibakov, J.; Nappa, M.; Hytönen, E.; Koksel, H.; Poutanen, K. Extraction of soluble b-glucan from oat and barley fractions: Process efficiency ad dispersion stability. *J. Cereal Sci.* **2018**, *81*, 60–68. [CrossRef]

16. McCleary, B.V.; Mugford, D.C. Determination of β-glucan in barley and oats by streamlined enzymic method: Summary of collaborative study. *J. AOAC Int.* **1997**, *80*, 580–583. [CrossRef]

17. Gamel, T.H.; Abdel-Aal, E.S.M.; Ames, N.P.; Henderson, K.; Prothon, F.; Kongraksawech, T.; Tosh, S.M. AACCI approved methods technical committee report: A new AACCI approved method (32-24.01) for measuring viscosity of β-glucan in cereal products using the rapid visco analyzer. *Cereal Foods World* **2015**, *60*, 279–283. [CrossRef]

18. Gamel, T.H.; Abdel-Aal, E.S.M.; Ames, N.P.; Duss, R.; Tosh, S.M. Enzymatic extraction of beta-glucan from oat bran cereals and oat crackers and optimization of viscosity measurement. *J. Cereal Sci.* **2014**, *59*, 33–40. [CrossRef]

19. Tosh, S.M.; Brummer, Y.; Miller, S.S.; Regand, A.; Defelice, C.; Duss, R.; Wolever, T.M.; Wood, P.J. Processing affects the physicochemical properties of β-glucan in oat bran cereal. *J. Agric. Food Chem.* **2010**, *58*, 7723–7730. [CrossRef]

20. Brown, J.K.M. Experimental Design Generator and Randomizer. Available online: http://www.edgarweb.org.uk/ (accessed on 20 June 2019).

21. Snedecor, G.W.; Cochran, W.G. *Statistical Methods*, 7th ed.; Iowa State University Press: Ames, IA, USA, 1980; p. 277.

22. Wolever, T.M.S. Effect of blood sampling schedule and method calculating the area under the curve on validity and precision of glycaemic index value. *Br. J. Nutr.* **2004**, *91*, 295–300. [CrossRef]

23. Wood, P.J.; Beer, M.U.; Butler, G. Evaluation of the role of concentration and molecular weight of oat β-glucan in determining effect of viscosity on plasma glucose and insulin following an oral glucose load. *Br. J. Nutr.* **2000**, *84*, 19–23. [CrossRef]

24. Tosh, S.M.; Brummer, Y.; Wolever, T.M.S.; Wood, P.J. Glycemic response to oat bran muffins treated to vary molecular weight of β-glucan. *Cereal Chem.* **2008**, *85*, 211–217. [CrossRef]

25. Kwong, M.G.; Wolever, T.M.S.; Brummer, Y.; Tosh, S.M. Increasing the viscosity of oat β-glucan beverages by reducing solution volume does not reduce glycaemic responses. *Br. J. Nutr.* **2013**, *110*, 1465–1471. [CrossRef]

26. Marciani, L.; Gowland, P.A.; Spiller, R.C.; Manoj, P.; Moore, R.J.; Young, P.; Fillery-Travis, A.J. Effect of meal viscosity and nutrients on satiety, intragastric dilution, and emptying assessed by MRI. *Am. J. Physiol. Gastrointest. Liver Physiol.* **2001**, *280*, G1227–G1233. [CrossRef] [PubMed]

27. Wood, P.J.; Braaten, J.T.; Scott, F.W.; Riedel, K.D.; Wolynetz, M.S.; Collins, M.W. Effect of dose and modification of viscous properties of oat gum on plasma glucose and insulin following an oral glucose load. *Br. J. Nutr.* **1994**, *72*, 731–743. [CrossRef] [PubMed]

28. Biörklund, M.; van Rees, A.; Mensink, R.P.; Önning, G. Changes in serum lipids and postprandial glucose and insulin concentrations after consumption of beverages with β-glucans from oats or barley: A randomized dose-controlled trial. *Eur. J. Clin. Nutr.* **2005**, *59*, 1272–1281. [CrossRef]

29. Abdul-Ghani, M.A.; Williams, K.; DeFronzo, R.A.; Stern, M. What is the best predictor of future type 2 diabetes? *Diabetes Care* **2007**, *30*, 1544–1548. [CrossRef] [PubMed]

30. Peterson, D.B.; Ellis, P.R.; Baylis, J.M.; Fielden, P.; Ajodhia, J.; Leeds, A.R.; Jepson, E.M. Low dose guar in a novel food product: Improved metabolic control in non-insulin-dependent diabetes. *Diabet. Med.* **1987**, *4*, 111–115. [CrossRef] [PubMed]

31. McGibbon, A.; Adams, L.; Ingersoll, K.; Kader, T.; Tugwell, B. Diabetes Canada 2018 Clinical Practice Guidelines for the Prevention and Management of Diabetes in Canada: Glycemic Management of Adults with Type 1 Diabetes in Adults. *Can. J. Diabetes* **2018**, *42* (Suppl. 1), S80–S87. [CrossRef]

32. Luijf, Y.M.; van Bon, A.C.; Hoekstra, J.B.; DeVries, J.H. Premeal injection of rapid-acting insulin reduces postprandial glycemic excursions in type 1 diabetes. *Diabetes Care* **2010**, *33*, 2152–2155. [CrossRef]

33. Thuillier, P.; Sonnet, E.; Alavi, Z.; Roudaut, N.; Nowak, E.; Dion, A.; Kerlan, V. Comparison between preprandial vs. postprandial insulin aspart in patients with type 1 diabetes in insulin pump and real-time continuous glucose monitoring. *Diabetes Metab. Res. Rev.* **2018**, *34*, e3019.

34. Kim, Y.A.; Keogh, J.B.; Clifton, P.M. Probiotics, prebiotics, synbiotics and insulin sensitivity. *Nutr. Res. Rev.* **2018**, *31*, 35–51. [CrossRef]

35. McPhee, J.B.; Schertzer, J.D. Immunometabolism of obesity and diabetes: Microbiota link compart-mentalized immunity in the gut to metabolic tissue inflammation. *Clin. Sci.* **2015**, *129*, 1083–1096. [CrossRef]

 *nutrients*

Article

Effects of Dietary Oat Beta-Glucans on Colon Apoptosis and Autophagy through TLRs and Dectin-1 Signaling Pathways—Crohn's Disease Model Study

Łukasz Kopiasz [1], Katarzyna Dziendzikowska [1,*], Małgorzata Gajewska [2], Michał Oczkowski [1], Kinga Majchrzak-Kuligowska [2], Tomasz Królikowski [1] and Joanna Gromadzka-Ostrowska [1]

[1] Department of Dietetics, Institute of Human Nutrition Sciences, Warsaw University of Life Sciences, Nowoursynowska 159c, 02-776 Warsaw, Poland; lukasz_kopiasz@sggw.edu.pl (Ł.K.); michal_oczkowski@sggw.edu.pl (M.O.); tomasz_krolikowski@sggw.edu.pl (T.K.); joanna_gromadzka_ostrowska@sggw.edu.pl (J.G.-O.)

[2] Department of Physiological Sciences, Institute of Veterinary Medicine, Warsaw University of Life Sciences, Nowoursynowska 159, 02-776 Warsaw, Poland; malgorzata_gajewska@sggw.edu.pl (M.G.); kinga_majchrzak@sggw.edu.pl (K.M.-K.)

* Correspondence: katarzyna_dziendzikowska@sggw.edu.pl; Tel.: +48-2259-37-033

Citation: Kopiasz, Ł.; Dziendzikowska, K.; Gajewska, M.; Oczkowski, M.; Majchrzak-Kuligowska, K.; Królikowski, T.; Gromadzka-Ostrowska, J. Effects of Dietary Oat Beta-Glucans on Colon Apoptosis and Autophagy through TLRs and Dectin-1 Signaling Pathways—Crohn's Disease Model Study. *Nutrients* **2021**, *13*, 321. https://doi.org/10.3390/nu13020321

Academic Editor: Seiichiro Aoe

Received: 23 December 2020
Accepted: 19 January 2021
Published: 22 January 2021

Publisher's Note: MDPI stays neutral with regard to jurisdictional claims in published maps and institutional affiliations.

Abstract: Background: Crohn's disease (CD) is characterized by chronic inflammation of the gastrointestinal tract with alternating periods of exacerbation and remission. The aim of this study was to determine the time-dependent effects of dietary oat beta-glucans on colon apoptosis and autophagy in the CD rat model. Methods: A total of 150 Sprague–Dawley rats were divided into two main groups: healthy control (H) and a TNBS (2,4,6-trinitrobenzosulfonic acid)-induced *colitis* (C) group, both including subgroups fed with feed without beta-glucans (βG−) or feed supplemented with low- (βGl) or high-molar-mass oat beta-glucans (βGh) for 3, 7, or 21 days. The expression of autophagy (LC3B) and apoptosis (Caspase-3) markers, as well as Toll-like (TLRs) and Dectin-1 receptors, in the colon epithelial cells, was determined using immunohistochemistry and Western blot. Results: The results showed that in rats with *colitis*, after 3 days of induction of inflammation, the expression of Caspase-3 and LC3B in intestinal epithelial cells did not change, while that of TLR 4 and Dectin-1 decreased. Beta-glucan supplementation caused an increase in the expression of TLR 5 and Dectin-1 with no changes in the expression of Caspase-3 and LC3B. After 7 days, a high expression of Caspase-3 was observed in the colitis-induced animals without any changes in the expression of LC3B and TLRs, and simultaneously, a decrease in Dectin-1 expression was observed. The consumption of feed with βGl or βGh resulted in a decrease in Caspase-3 expression and an increase in TLR 5 expression in the CβGl group, with no change in the expression of LC3B and TLR 4. After 21 days, the expression of Caspase-3 and TLRs was not changed by colitis, while that of LC3B and Dectin-1 was decreased. Feed supplementation with βGh resulted in an increase in the expression of both Caspase-3 and LC3B, while the consumption of feed with βGh and βGl increased Dectin-1 expression. However, regardless of the type of nutritional intervention, the expression of TLRs did not change after 21 days. Conclusions: Dietary intake of βGl and βGh significantly reduced colitis by time-dependent modification of autophagy and apoptosis, with βGI exhibiting a stronger effect on apoptosis and βGh on autophagy. The mechanism of this action may be based on the activation of TLRs and Dectin-1 receptor and depends on the period of exacerbation or remission of CD.

Keywords: oat beta-glucan; *colitis*; Crohn's disease; apoptosis; autophagy; TLRs; Dectin-1; rats

1. Introduction

Inflammatory bowel disease (IBD) is becoming an increasingly common disease in the population of developed countries. One of the reasons for the increase in its incidence is changes in lifestyle, including eating habits, which have been observed over the past few decades. The transition from a diet based on natural or mildly processed products to a Western-type diet is a consequence of change from a rural lifestyle to an urban one, which took place on a large scale in the 20th century [1]. IBD is a disease characterized by chronic inflammation of the gastrointestinal tract, and includes ulcerative colitis (UC) and Crohn's disease (CD) [2]. The etiology of both these diseases is multifactorial, in which interactions between genetic and environmental factors as well as changes in the profile of the gut microbiota (dysbiosis) consequently contribute to disturbances in immune response mechanisms, leading to the development of inflammation [3]. UC first appears in the mucosa and submucosa of the large intestine, while in CD the chronic inflammatory process may occur in any part of the gastrointestinal tract, covering the entire thickness of its wall [4,5].

One of the main effects of chronic inflammation related to CD is damage to the intestinal epithelium and a significant increase in the programmed death of intestinal epithelial cells (IECs). In addition, disturbances occur in the mechanisms regulating the metabolic activity of cells, including autophagy [6]. Apoptosis contributes to the dynamic control of cells by removing damaged cells or that function abnormally without triggering inflammatory response or oxidative stress, thus ensuring appropriate tissue function. Simultaneously, autophagy plays an important role in maintaining the homeostasis of living organisms, as they regulate the basic metabolic functions and pathogenesis of many diseases, including malignant neoplasms, neurodegenerative and cardiovascular diseases, and IBD [6–8]. Both these processes regulate tissue homeostasis and response to extracellular factors, including the production of proinflammatory cytokines, and intensification of the apoptosis of IECs has been reported in people with IBD and in in vivo studies on animal models [9]. Caspase-3 is considered to be one of the key executive enzymes of apoptosis, the activation of which is common to both extrinsic and intrinsic pathways, as well as is dependent on granzyme B [10]. A protein specific to autophagy is LC3, which is involved in the final formation of the autophagosome [8]. In mammalian cells, this protein occurs in three isoforms—LC3A, LC3B, and LC3C—of which the LC3B form is the marker of autophagosomal activity [11,12].

In the light of the latest research, factors regulating the proliferation, autophagy, and apoptosis of intestinal cells include membrane Toll-like receptors (TLRs), such as TLR 4, TLR 5, and TLR 9, and Dectin-1 receptor. Activation of TLRs and Dectin-1 receptors stimulates the immune response. The main function of these receptors is the recognition of bacterial and fungal antigens. They also recognize a wide variety of bacterial and fungal ligands, of which the best known are lipopolysaccharide (LPS), flagellin, and triacyl lipopeptides. Plant-origin polysaccharides, including beta-glucans, are ligands that bind to TLRs and Dectin-1 receptors. Due to their specific structure, beta-glucans exhibit a number of health-promoting properties, and their biological activity results from their ability to bind with TLRs and Dectin-1 receptors, as well as with complement receptor 3 and type C lectin [13,14]. Beta-glucans are recognized by different cells expressing TLRs and Dectin-1 receptors, which in turn activate them [15–17]. It should be emphasized that the activation of Dectin-1 receptor by oat beta-glucans is influenced by the size of these polysaccharide molecules and is greater when they are subjected to preliminary digestion, i.e., when large molecules are broke down into smaller ones [17]. Dectin-1 receptors are involved, among others, in regulating the differentiation of macrophages and dendritic cells during ongoing inflammation, as well as modulating the immune response and controlling the recruitment of autophagy-related proteins, mainly LC3 [18–21]. The deficiency of TLRs inhibits the proliferation and differentiation of IECs and increases apoptosis, which affects the reconstruction of the intestinal epithelium [22,23]. This is very important with respect to the treatment of IBD, because due to chronic inflammation, the IECs are vulnerable to damage caused by high levels of reactive oxygen species (ROS). Moreover, as a consequence of the increased secretion of proinflammatory cytokines by immune cells, the permeability of the intestinal barrier increases [24,25].

The results of our research have shown the beneficial anti-inflammatory and indirect antioxidant effects of oat beta-glucans with low and high molar mass. These properties have been demonstrated in both the LPS-induced enteritis model [26,27] and the TNBS (2,4,6-trinitrobenzosulfonic acid)-induced colitis model [28,29], but the mechanisms underlying these effects have not yet been elucidated. Therefore, this study aimed to determine the effect of two oat beta-glucan forms with different molar mass on the activity of autophagy and apoptosis in the colonocytes of rats with colitis. Rectal administration of an ethanolic solution of TNBS causes strictly localized transmural inflammatory lesions in the large intestine, with dense lymphocyte infiltration and secretion of proinflammatory cytokines in the whole wall, which are characteristic of CD in humans. Moreover, this animal model is characterized by reproducibility, technical simplicity and low cost [30,31]. The molecular effects of both beta-glucan forms have been analyzed at different stages of colitis development.

2. Materials and Methods

2.1. Animals and Experimental Design

The experiment was performed on 150 adult male Sprague–Dawley rats (Charles River Laboratories, Sulzfeld, Germany) with an initial body weight of 414.8 ± 1.3 g. The rats were divided into two main groups—an experimental group (*colitis* group—C) with colitis induced by one-time rectal administration of 1 mL of 2,4,6-trinitrobenzenesulfonic acid (Sigma Aldrich, Darmstadt, Germany) solution (150 mg/kg bw) dissolved in 50% ethanol and a control group (H) without colitis, which was given 0.9% NaCl in the same way. Administration of TNBS was performed using the polyethylene catheter (FTP-18-75-50; Instech Laboratories, Inc., Plymouth Meeting, PA, USA) according to Parra et al. (2015) [32]. The animals from both groups were then divided into three nutritional subgroups, which received a diet with different types of supplementation for 3, 7, or 21 days—AIN-93M feed with 1% (w/w) low-molar-mass oat beta-glucans (βGl+ group), AIN-93M feed with 1% (w/w) high-molar-mass oat beta-glucans (βGh+ group), and AIN-93M feed without beta-glucans (βG− group) (Figure 1). A detailed description of the in vivo experiment and the method used for the extraction of beta-glucans from oats as well as the composition of the three types of rat feed used in this study is provided in our previous papers [28,29].

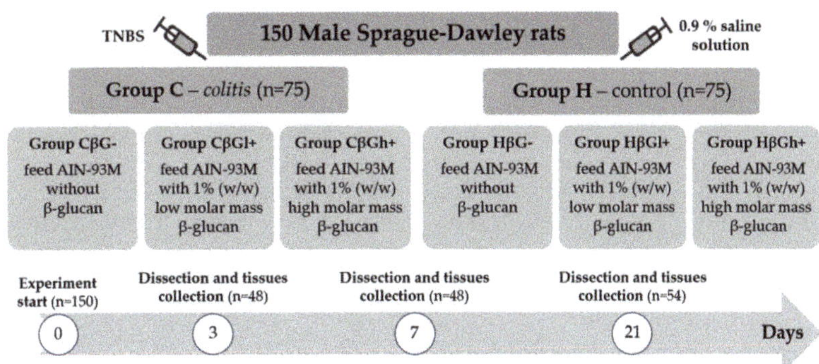

Figure 1. Scheme of experimental design of the study. TNBS: 2,4,6,-trinitrobenzenesulfonic acid alcohol solution.

The large intestine samples obtained for histological and immunohistochemical analysis were fixed in 10% buffered formaldehyde and routinely embedded in paraffin. Then, 5-μm sections were cut from each paraffin block and fixed on microscope glass slides. The slides prepared were used for further laboratory analyses.

The animal experiment was conducted after the approval of the II Local Animal Care and Use Committee in Warsaw (Resolution # 60/2015). All the procedures will be designed and conducted according to Polish and EU law regulations and with respect to 3R rules (Replacement, Reduction and Refinement).

2.2. Western Blot Analysis

Colon tissue samples were homogenized in RIPA buffer (50 mM Tris, pH 7.5, 150 mM NaCl, 1 mM EDTA, 1% NP-40, 0.25% Na-deoxycholate, and 1 mM PMSF) supplemented with protease inhibitor cocktail and phosphatase inhibitor cocktail (Sigma-Aldrich) with a tissue homogenizer (Bio-Gen PRO 200; PRO Scientific, Oxford, CT, USA). Following the mechanical disruption of tissue, the homogenates were incubated for 30 min at 4 °C. The lysates were cleared for 30 min at 14,000 rpm, and supernatants were collected. Protein concentration in the lysates was determined using Thermo Scientific™ Pierce™ BCA Protein Assay Kit according to the producer's instructions (Thermo Scientific, Waltham, MA, USA). Then, the proteins (50 µg) were resolved by sodium dodecyl sulfate-polyacrylamide gel electrophoresis and transferred onto PVDF membrane (Sigma-Aldrich). The membranes were blocked in 5% (w/v) nonfat dry milk in TBS (20 mM Tris-HCl, 500 mM NaCl) containing 0.5% Tween20. Next, they were incubated under gentle shaking at 4 °C overnight with primary antibodies, including anti-Cleaved Caspase-3 antibody (Cell Signaling Technology, Danvers, MA, USA), anti-LC3B antibody (Novus Biologicals, Centennial, CO, USA), anti-Beclin-1 antibody (Novus Biologicals, Centennial, CO, USA), and anti-β-actin (8H10D10) mouse monoclonal antibody (#3700; Cell Signaling Technology, Danvers, MA, USA). The membranes were incubated with chosen primary antibodies at a dilution range between 1:200 and 1:1000, depending on the antibody. Following incubation, the membranes were washed three times for 15 min and incubated with appropriate secondary antibodies conjugated with IR fluorophores: IRDye® 680 or IRDye® 800 CW (at 1:5000 dilution). Odyssey Infrared Imaging System (LI-COR Biosciences, Lincoln, NE, USA) was used to analyze the protein expression. Scan resolution of the instrument was set at 169 µm, and the intensity at 4. Quantification of the integrated optical density (IOD) was performed using the analysis software provided with the Odyssey scanner (LI-COR Biosciences). Immunoblot analysis was carried out on samples obtained from three independent experiments. For the purpose of publication, the color immunoblot images were converted into black and white images in the Odyssey software.

2.3. Immunohistochemical Staining

Five-micrometer-large intestine sections were deparaffinized in xylene and rehydrated in a series of decreasing concentrations of ethanol. To recover the antigen, the slides were placed in citrate buffer (pH 6.0) and boiled in a microwave two times for 5 min. After cooling and washing with distilled water, the samples were incubated in the Peroxidase Blocking Reagent (Dako, Denmark) for 13 min at room temperature. Then, these samples were incubated at room temperature in 2% bovine serum albumin (BSA) (Sigma Aldrich, Germany) for 30 min. Next, they were treated with the following primary antibodies (diluted in 2% BSA): mouse anti-TLR 4 antibody (1:100 dilution; Novus Biologicals, Centennial, CO, USA), rabbit anti-Dectin-1/CLEC7A antibody (1:400 dilution; Novus Biologicals, Centennial, CO, USA), rabbit anti-TLR 5 antibody (1:500 dilution; Novus Biologicals, Centennial, CO, USA), rabbit anti-TLR 6 antibody (1:250 dilution; Novus Biologicals, Centennial, CO, USA), rabbit anti-LC3B antibody (1:500 dilution; Novus Biologicals, Centennial, CO, USA), rabbit anti-Cleaved Caspase-3 antibody (1:400 dilution; Cell Signaling Technology, Danvers, MA, USA). The slides with antibodies were incubated overnight in a refrigerator at +4 °C.

EnVision kit (Dako, Glostrup Kommune, Denmark) was used for staining the slides. After incubating overnight and washing with phosphate-buffered saline, labeled polymers consisting of secondary antimouse or antirabbit antibodies conjugated with HRP enzyme complex was used (Dako). To obtain brown staining, 3,3'-diaminobenzidine was applied, which was washed with water after 25 s. Then, hematoxylin was used for nuclei counterstaining. Finally, the samples were dehydrated using a series of alcohols of increasing concentration and xylene, and preserved by sticking coverslips.

A negative control (staining without primary antibodies) was used for each immunohistochemical experiment.

2.4. Image Analysis

The immunohistochemically stained slides were observed under a $\times 20$ objective lens in a NIKON Eclipse Ti2 microscope. In the recorded photos, 20 areas of colonocytes were marked, and the colorimetric intensity (brown color reflecting antigen expression) and object area were measured by using NIS-Elements BR 5.01 program. In the next step, IOD was calculated from the following formula:

$$\text{Integrated Optical Density (IOD)} = \frac{Object\ area}{Measured\ area} \times Mean\ intensity$$

2.5. Statistical Analysis

The obtained data were analyzed using Statistica software (version 13.1 PL; StatSoft, Cracow, Poland). The normality of distribution and equality of variance were determined for all data. Statistical analysis was carried out in two stages. In the first stage, for each time point (3, 7, and 21 days), a two-way analysis of variance (ANOVA) (colon inflammation vs. dietary intervention) was performed, while in the second stage, a three-way ANOVA (colon inflammation vs. dietary intervention vs. period of its use) was carried out. The significance of differences between the groups was determined by the Tukey post hoc test. In addition, the results of all nutritional subgroups were compared with the control subgroup (HβG$-$) by the Dunnett post hoc test. The differences were considered statistically significant at $p < 0.05$. To assess the time-dependent effect between the experimental factors, Fisher's linear discriminant (FLD) analysis was performed using R statistics software (version 3.1.3; www.r-project.org).

3. Results

3.1. Histological Changes in the Large Intestine

Histological analysis revealed extensive inflammation in the colon tissue of rats in the CβG$-$ group (Figure 2). The lesions were of transwall nature, covering not only mucosa but also the deeper layers of the intestinal wall, which is a characteristic of CD. Inflammation in the colon (colitis) was confirmed in the same animals not only histologically but also biochemically, including analysis of colon level of pro-inflammatory cytokines (i.e., Il-1; Il-6 or TNF alfa) and CRP protein [28]. These microscopic and biochemical observations also confirmed the results of gene expression analysis of pro-inflammatory cytokines in colon tissue affected by colitis, including *Ifng* and *Tnf* after 3 days and *Il21* after 21 days (Figure S9). In the groups of animals fed with feed supplemented with oat beta-glucans during 21 days, these lesions were clearly reduced, as was presented in our previous study [28]. The expression of pro-inflammatory cytokine genes was also significantly reduced (Figure S9). Hematoxylin-eosin staining showed that the histopathology of the colon tissue in the control groups was within the normal limits.

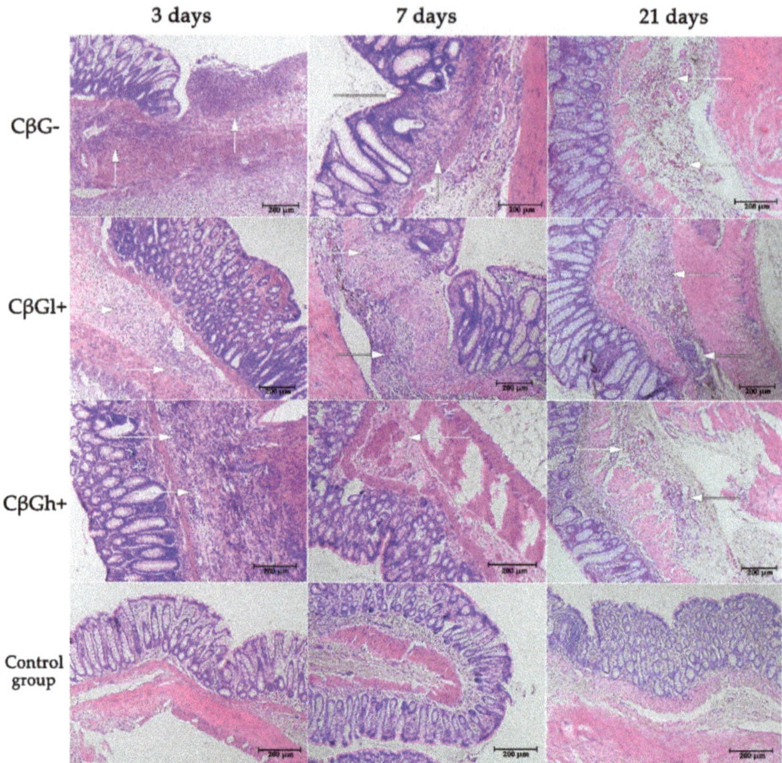

Figure 2. Microscopic changes in the colon caused by TNBS-induced inflammation. White arrows indicate diffused multifocal inflammation (lymphocyte infiltration) of the submucosa of varying severity.

3.2. Dectin-1 and Toll-like Receptors 4, 5, and 6 Expression in the Colonocytes

The expression of Dectin-1 beta-glucan receptor is presented in Figure 3. As shown by ANOVA, the expression of this receptor in colonocytes was lower in the rats with TNBS-induced *colitis* (ANOVA, $p < 0.001$; Figure S1A), whereas dietary intervention with oat beta-glucans resulted in a change in its expression after 3 and 21 days of feeding, which was confirmed by ANOVA. Furthermore, one-way ANOVA showed that the expression of Dectin-1 was lower after 21 days (ANOVA, $p < 0.001$; Figure S1B). After 7 days, induced *colitis* caused a significant decrease in Dectin-1 expression in all dietary subgroups compared with the control group (HβG−) (for each subgroup: Dunnett post hoc test, $p < 0.01$). In addition, analysis of variance showed the effect of interaction between time after TNBS/saline administration and the occurrence of inflammation on Dectin-1 expression (ANOVA, $p < 0.01$; Figure S1C). In the control groups, 3 and 7 days after saline solution administration, the expression of this receptor was significantly higher than in the colitis groups at the same time point and in the control group after 21 days.

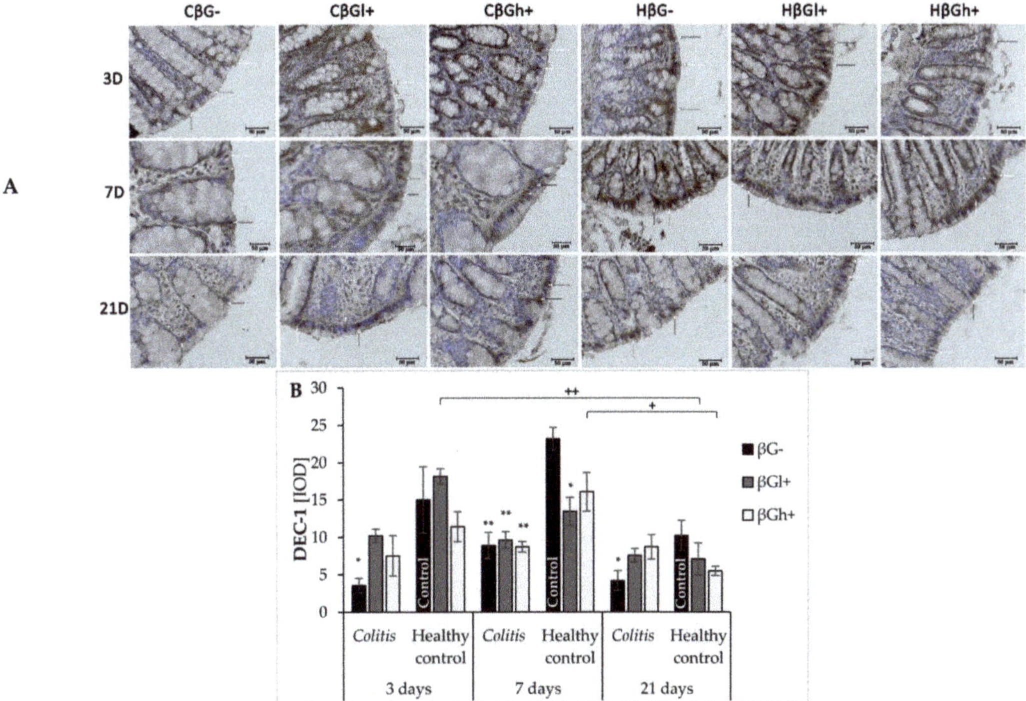

Figure 3. Dectin-1 (DEC-1) expression in colonocytes: results of the immunohistochemical analysis. White arrows indicate colonocytes with high expression of the Dectin-1 (brown precipitate). (**A**)—Light micrographs imaged under the NIKON Eclipse Ti2 microscope ($\times$40 magnification). Dectin-1 antigen is represented by a brown precipitate in the colonocytes. (**B**)— Changes in the expression of DEC-1 in colonocytes presented (mean $\pm$ SE) as integrated optical density (IOD). * Significantly different from the control group (healthy control βG−) at the same time point according to the Dunnett post hoc test (* $p < 0.05$, ** $p < 0.01$). + Significantly different from the same subgroups at another time point according to the Tukey post hoc test (+ $p < 0.05$, ++ $p < 0.01$).

Changes in the expression of TLRs are presented in Figure 4. ANOVA showed a significant interaction between two experimental factors: induced inflammation of the colon and feed supplementation with oat beta-glucans, which influenced the expression level of all the examined TLRs (ANOVA, $p < 0.001$; Figures S2A, S3C and S4C)—TLR 4, TLR 5, and TLR 6. Post hoc analysis showed a significantly lower expression of TLR 4 and TLR 6 in rats from the *colitis* group that consumed feed without beta-glucans (CβG−) compared to the control rats receiving the same feed (HβG−) (Tukey post hoc, $p < 0.01$). In the case of TLR 5, a significantly lower expression was found only in the CβG− group compared to the CβGl+ group (Tukey post hoc, $p < 0.001$).

Statistical analysis also revealed that TLR 4 expression significantly differed between the subgroups only 3 days after *colitis* induction (Figure 4A,D), while a significantly lower expression was found in all *colitis* subgroups compared to the control group (HβG−) (Dunnett post hoc, $p < 0.001$), with the greatest differences observed between the CβG− and HβG− groups. It should be noted that the expression of this protein was also significantly lower in the control group that consumed the feed with low-molar-mass beta-glucans (HβGl+) compared to the rats from the HβG− group. The effect of the dietary intervention length and inflammation on the TLR 4 expression was confirmed by ANOVA ($p < 0.05$; Figure S2B), which showed an increase in the expression of this receptor in the *colitis* group and a decrease in differences observed between the *colitis* groups and the control groups at all time points after the induction of large-intestine inflammation.

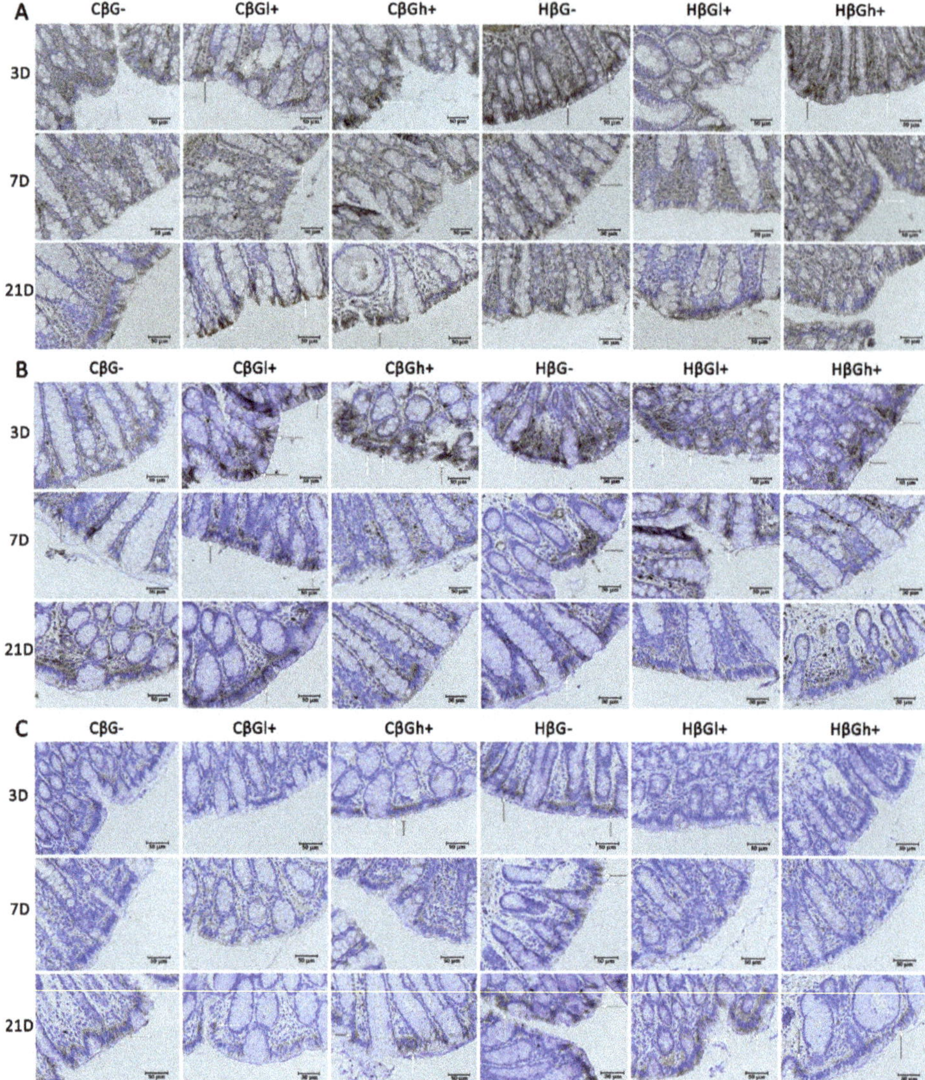

Figure 4. *Cont.*

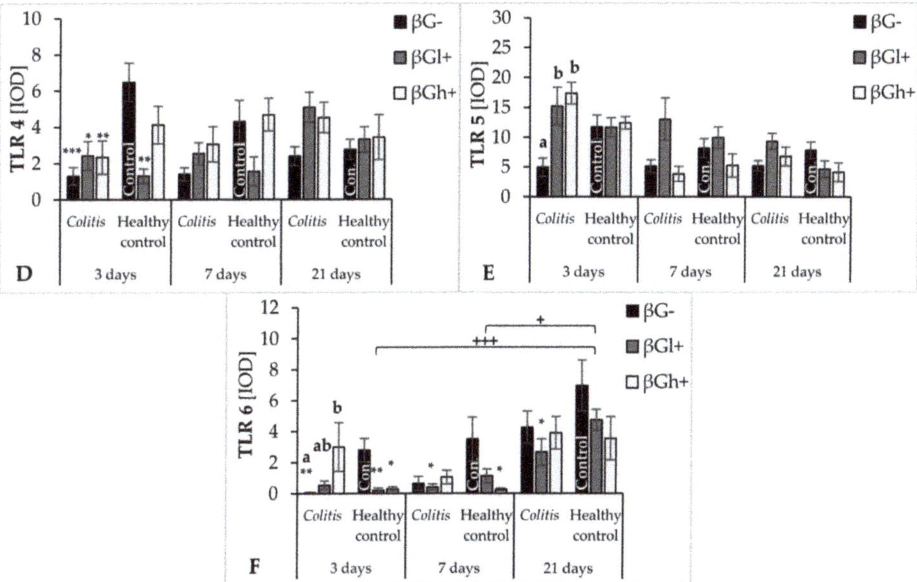

Figure 4. Expression of Toll-like receptors (TLRs) in the colonocytes: results of the immunohistochemical analysis. White arrows indicate colonocytes with high expression of the TLRs (brown precipitate). (**A–C**)—Light micrographs imaged under the NIKON Eclipse Ti2 microscope ($\times$40 magnification). (**D–F**)—Changes in the expression of TLRs (TLR 4, TLR 5, and TLR 6, respectively) presented (mean $\pm$ SE) as integrated optical density (IOD). TLR 4 antigen is represented by a brown precipitate in the colonocytes in (**A**). TLR 5 antigen is represented by a brown precipitate in the colonocytes in (**B**). TLR 6 antigen is represented by a brown precipitate in the colonocytes in (**C**). * Significantly different from the control group (healthy control βG−) at the same time point according to the Dunnett post hoc test (* $p < 0.05$, ** $p < 0.01$, *** $p < 0.001$). + Significantly different from the same subgroups at another time point according to the Tukey post hoc test (+ $p < 0.05$, +++ $p < 0.001$). a,b Different letters denote significant differences in the *colitis*/healthy control group at the same time point according to the Tukey post hoc test ($p < 0.05$).

The expression of TLR 5 was influenced by the time after TNBS administration and feed supplementation with oat beta-glucans (Figure 4B,E). ANOVA showed a significant decrease in the expression of this receptor parallel to the extension of the post-TNBS period (ANOVA, $p < 0.0001$; Figure S3A). The highest expression of TLR 5 was found in rats fed with feed supplemented with low-molar-mass oat beta-glucans, while the lowest expression was found in the group of animals fed with feed without beta-glucans (ANOVA, $p < 0.01$; Figure S3B). Post hoc analysis showed significant differences between the experimental groups only after 3 days of TNBS administration with a lower expression observed in the *colitis* group fed with feed without beta-glucans (CβG−) compared to the *colitis* groups treated with low- and high-molar-mass beta-glucans (CβGl+ and CβGh+) (Tukey post hoc, $p < 0.05$).

TLR 6 expression differed significantly between the dietary groups (Figure 4C,F). One-way ANOVA showed a significant effect of TNBS administration (ANOVA, $p < 0.0001$; Figure S4A) and consumption of feed with or without oat beta-glucans (ANOVA, $p < 0.05$; Figure S4B). After 21 days of induction of colon inflammation, the expression of TLR 6 was approximately four times higher compared to that after 3 or 7 days. The highest expression of this receptor was found in the group of rats fed with feed without oat beta-glucans. ANOVA revealed a statistically significant interaction between three experimental factors: inflammation, length of period after TNBS administration, and the type of feed (ANOVA, $p < 0.001$). Three days after TNBS administration, a significantly lower expression of TLR 6 was noted in CβG−, HβGl+, and HβGh+ groups compared to the control group (HβG−)

(Dunnett post hoc, groups CβG− and HβGl+ vs. group HβG−: $p < 0.01$; group HβGh+ vs. group HβG−: $p < 0.05$). Seven days after the induction of colitis, a lower expression of this receptor was found only in the CβGl+ and HβGh+ groups compared to the HβG− group (Dunnett post hoc, $p < 0.05$). Twenty-one days after the induction of colitis, TLR 6 expression differed significantly only between CβGl+ and HβG− groups (Dunnett post hoc, $p < 0.05$). It should be added that TLR 6 expression in the *colitis* group fed with feed supplemented with high-molar-mass beta-glucans (CβGh+) was at a similar level as in the control group (HβG−) group after 3, 7, and 21 days of TNBS administration.

3.3. Autophagy and Apoptosis Markers Expression in the Large Intestine

Immunohistochemical results showing the expression of selected markers of autophagy (LC3B) and apoptosis (Caspase-3) in IECs are presented in Figure 5. ANOVA showed a significant influence on LC3B expression by all three experimental factors: time since TNBS administration, consumption of oat beta-glucans with feed, and the occurrence of inflammation. Analyzing each factor separately, it was found that LC3B expression decreased with time since TNBS administration (ANOVA, $p < 0.001$; Figure S5A), as well as a significant reduction in its expression, was caused by the induction of colitis (ANOVA, $p < 0.01$; Figure S5C). The consumption of feed with high-molar-mass oat beta-glucans resulted in a significantly higher expression of LC3B compared to other dietary groups (ANOVA, $p < 0.05$; Figure S5B). ANOVA also showed a significant interaction between the three experimental factors ($p < 0.01$), which was reflected by a significantly higher expression of LC3B after 7 days in the HβGh+ group compared with the CβGh+ group (Tukey post hoc, $p < 0.01$), while after 21 days, the expression of this protein was significantly lower in the CβG−, CβGl+, and HβGl+ groups compared to the control group (HβG−) (Dunnett post hoc, $p < 0.001$, $p < 0.01$, and $p < 0.01$, respectively).

All three investigated factors significantly influenced the expression of Caspase-3 (ANOVA, $p < 0.0001$; Figure S6A–C). The expression of this apoptotic protein 3 days after TNBS administration was at a very low level, but it was significantly lower in the *colitis* groups and HβGl+ group compared to the control group (HβG−) (Dunnett post hoc, CβGl+ vs. HβG−: $p < 0.05$; CβGh+ and CβG− vs. HβG−: $p < 0.01$; HβGl+ vs. HβG−: $p < 0.001$), while in the HβGh+ group it was at the same level as in the HβG− group. After 7 days of the experiment, the expression of Caspase-3 in the *colitis* group fed with feed without beta-glucans (CβG−) in relation to other experimental groups was very high, while a significantly lower expression was found in the *colitis* subgroups fed with feed supplemented with both forms of beta-glucans (CβGl+ and CβGh+) compared to the CβG− group (Tukey post hoc, $p < 0.05$). However, a very low expression of this enzyme was observed in all control groups. Caspase-3 expression in the CβGh+ group after 21 days was at a similar level as the expression found after 7 days, with a significantly higher level compared to other *colitis* groups (Tukey post hoc, $p < 0.05$) and control group (HβG−) (Dunnett post hoc, $p < 0.01$).

Figure 6 shows the changes in the expression of autophagy- (LC3-II and Beclin-1) and apoptosis-related proteins (Caspase-3) in the colon wall determined by Western blot. As the results of ANOVA indicate, only time elapsed after TNBS administration had a significant effect on the reduction of LC3-II expression, while the induction of inflammation had a significant effect on Beclin-1 expression with a higher expression observed in the *colitis* group (ANOVA, $p < 0.05$; Figure S7). However, post-hoc analysis showed no significant differences between the experimental subgroups.

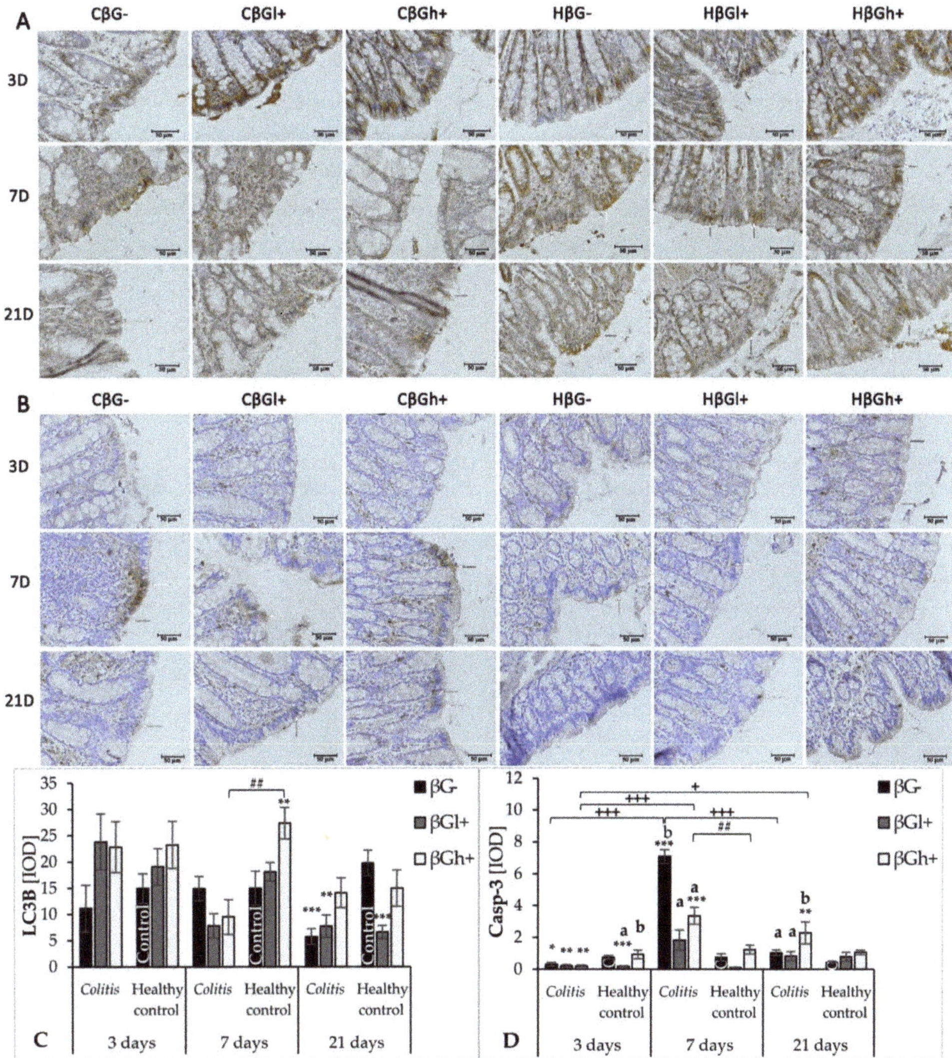

Figure 5. Expression of autophagy and apoptosis markers in the colonocytes: results of the immunohistochemical analysis. White arrows indicate colonocytes with high expression of the LC3B protein (**A**) and Caspase-3 protein (**B**) (brown precipitate). (**A,B**)—Light micrographs imaged under the NIKON Eclipse Ti2 microscope (×40 magnification). (**C,D**)—Changes in the expression of autophagy and apoptosis markers (LC3B and Caspase-3 (Casp-3), respectively) presented (mean ± SE) as integrated optical density (IOD). Autophagy-related protein LC3B antigen is represented by a brown precipitate in the colonocytes in (**A**). Apoptosis-related protein Casp-3 antigen is represented by a brown precipitate in the colonocytes in (**B**). * Significantly different from the control group (control βG−) at the same time point according to the Dunnett post hoc test (* $p < 0.05$, ** $p < 0.01$, *** $p < 0.001$). # Significantly different between the *colitis* and control groups at the same time point and the same feed according to the Tukey post hoc test (## $p < 0.01$). +Significantly different from the same subgroups at another time point according to the Tukey post hoc test (+ $p < 0.05$, +++ $p<0.001$). [a,b] Different letters denote significant differences in the *colitis*/control group at the same time point according to the Tukey post hoc test ($p < 0.05$).

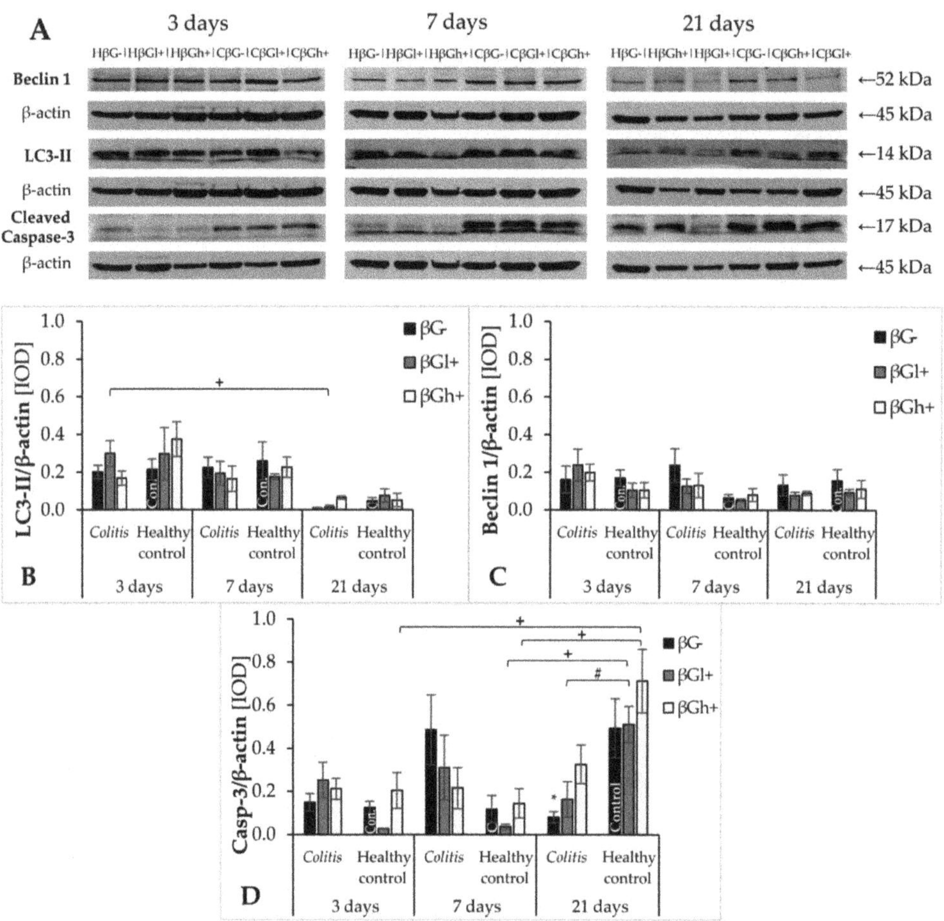

Figure 6. Results of the densitometric analysis for the expression of autophagy and apoptosis markers in the large intestinal wall. (**A**)—Representative immunoblot images. (**B**)—Autophagy-related protein LC3-II. (**C**)—Autophagy-related protein Beclin 1. (**D**)—Apoptosis-related protein cleaved Caspase-3 (Casp-3). * Significantly different from the control group (healthy control βG−) at the same time point according to the Dunnett post hoc test (* $p < 0.05$). # Significantly different between the *colitis* and control groups at the same time point and the same feed according to the Tukey post hoc test (# $p < 0.05$). + Significantly different from the same subgroups at another time point according to the Tukey post hoc test (+ $p < 0.05$).

After 21 days of TNBS administration, Caspase-3 expression was found to be significantly higher (ANOVA, $p < 0.001$; Figure S8A). Two-way ANOVA showed significant interactions between the time since TNBS administration and induced inflammation (ANOVA, $p < 0.001$; Figure S8B). After 7 days of colitis induction, a significantly higher expression of Caspase-3 in the colon wall was found in the *colitis* group (Tukey post hoc, $p < 0.05$), and after 21 days after colitis induction, a higher expression was observed in the noninflammatory group (Tukey post hoc, $p < 0.001$). Post hoc analysis confirmed a significantly higher expression of this enzyme 21 days after TNBS administration in HβGl+ and HβGh+ groups compared to the expression observed in the corresponding *colitis* subgroups after 7 days of TNBS administration (Tukey post hoc, $p < 0.05$), while it confirmed a lower expression in the *colitis* group that consumed feed without beta-glucans (CβG−) compared to the control group that consumed the same feed (HβG−) (Dunnett post hoc, $p < 0.05$).

3.4. Fisher's Linear Discriminant Analysis (FLD)

The results of the FLD analysis obtained for the expression of experimental factors are presented in Figure 7. This analysis was used to find the linear combinations of expression of the analyzed receptors and autophagy and apoptosis markers which allow for the best separation of the groups of animals at three time points selected in the experimental model. The experimental data were divided into three groups depending on the experimental time point, as well as the stage of inflammation corresponding to different periods of the disease (exacerbation and remission period). The FLD method provides the most optimal linear combination of parameters used in the analysis such that the highest possible separation between the data groups is achieved. The results of the linear discriminant analysis at three time points are shown in Figure 7A,C,E. In each figure, six experimental groups are isolated. The data are presented in the space between the linear combinations of parameters (FLDs), marked as FLD_1 and FLD_2, which separate the best-predefined groups.

Figure 7B,D,F shows the correlation vectors of the analyzed parameters in the experiment with FLD_1 and FLD_2 predictors, which determine the direction and strength of the separation of the experimental groups at three time points—3 days (A and B), 7 days (C and D), and 21 days (E and F). The graphs show the parameters that had the greatest impact on the separation of data at particular time points after TNBS administration. The parameter corresponding to a particular vector caused the data to shift in the direction determined by the vector. The performed FLD analysis complemented the ANOVA models by indicating the common set of features that separates the best experimental data (as opposed to the analysis of individual features in the ANOVA model) and allows for an augmented analysis of the obtained results.

FLD analysis performed for the data 3 days after TNBS/saline administration (Figure 7A,B) showed that the factors that most differentiated these experimental groups were Caspase-3 (determined in colonocytes as well as in the whole colon wall), immunohistochemical expression of Dectin-1, TLR 4, and TLR 6 in colonocytes, and expression of Beclin-1 autophagy protein in the colon wall. The FLD analysis allowed determining the combination of the above parameters, which in turn helped in distinguishing the control feed groups supplemented with beta-glucans (HβGl+ and HβGh+) from the other groups in the horizontal plane (FLD_1). The vertical plane (FLD_2) enabled the separation of the control group fed with feed supplemented with low-molar-mass oat beta-glucans (HβGl+) from the other groups (Figure 7A). The expression of Beclin-1 protein was the most correlated with FLD_1, while the expression of TLR 4 and Caspase-3 in colonocytes had the most influence on FLD_2, and that of Dectin-1, TLR 6, and Caspase-3 in the colon wall was important for both FLDs (Figure 7B). Based on the linear discriminant analysis, it can be concluded that the control group supplemented with βGl+ was characterized by a significantly higher expression of Dectin-1 receptor and a lower expression of TLR 6 receptor and Caspase-3 protein in the colon wall compared to the *colitis* groups. The highest Caspase-3 expression was observed in the control group fed with feed supplemented with high-molar-mass oat beta-glucans (HβGh+).

After 7 days of *colitis* induction (Figure 7C,D), the separation of the *colitis* group from control groups was much more visible than after 3 days. The horizontal plane (FLD_1) allowed clear separation of control groups from the *colitis* groups, as well as the separation of *colitis* group (CβG−) and *colitis* groups that consumed the feed supplemented with beta-glucans (CβGl+ and CβGh+). Animals from the groups fed feed with a low-molar-mass oat beta-glucans (HβGl+ and CβGl+) were separated from the other groups, in particular from those receiving feed without beta-glucans supplementation (HβG− and CβG−), also in the vertical FLD direction (FLD_2), however to a lesser extent. Expression of Caspase-3 (in colonocytes and colon wall) and Beclin-1 in the colon wall had the greatest influence on FLD_1, while TLR 5 expression had the most impact on FLD_2. Both FLDs were influenced by the expression of Dectin-1 receptor and LC3B protein in colonocytes. Therefore, it can be concluded that the *colitis* group fed with feed without beta-glucans had the highest expression of Caspase-3 and Beclin-1, while the lowest expression of both these proteins was observed in the control groups. In addition, HβG− and HβGh+ groups showed the highest expression of Dectin-1 receptor and LC3B protein. The highest TLR 5 expression was found in the βGl+-fed groups, both with and without *colitis*.

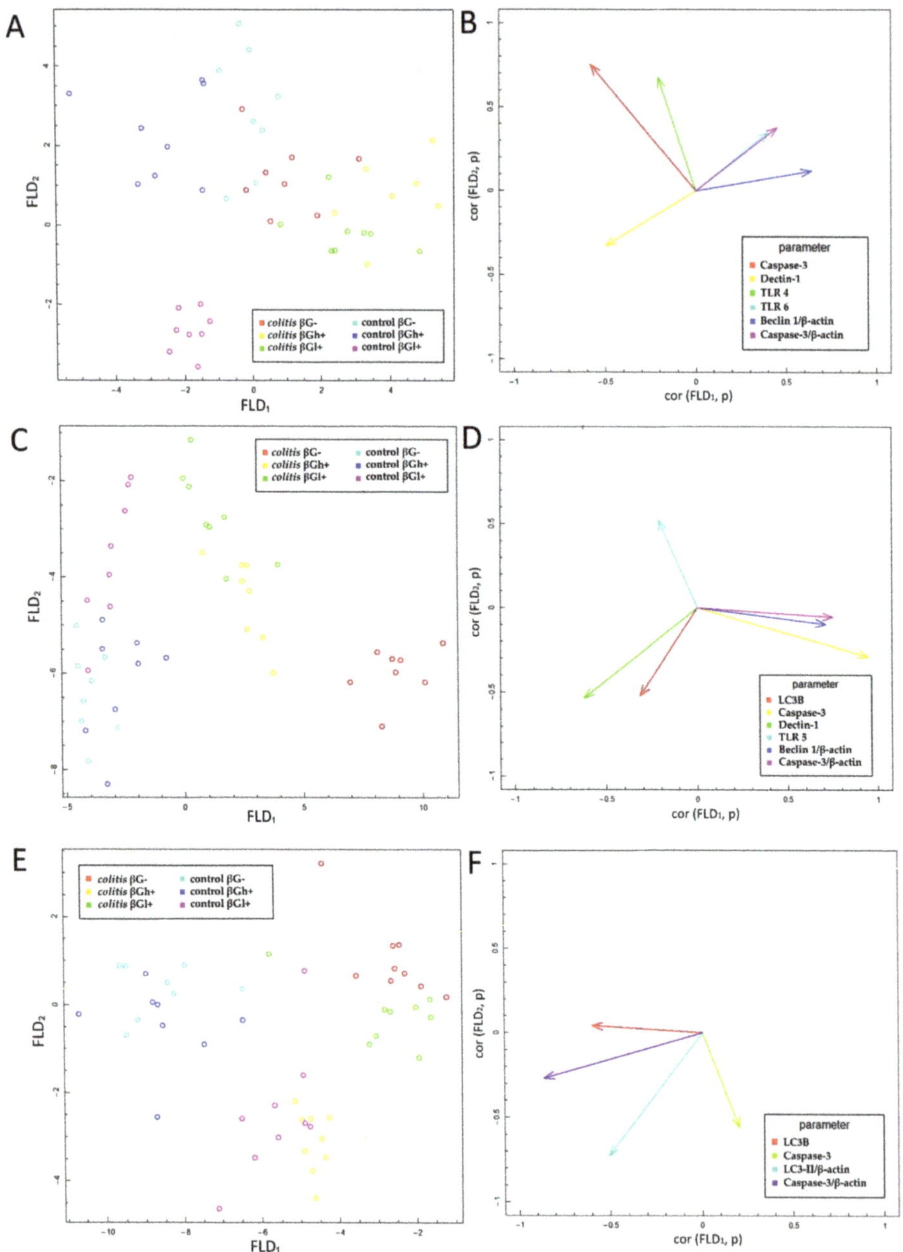

Figure 7. Fisher's linear discriminant (FLD) analysis. (**A,C,E**)—Experimental data on the plane spanned by two of the most data-separating FLDs. (**B,D,F**)—Parameters contributing the most to the FLDs. Scheme (**A,B**): 3. days; (**C,D**): 7 days; (**E,F**): 21 days.

After 21 days of TNBS administration (Figure 7E,F), the analyzed parameters differentiated the experimental groups to a much lesser extent than at earlier time points (3 and 7 days). After 21 days of *colitis* induction, only the expression of LC3B/LC3-II and Caspase-3 in the colonocytes and colon wall allowed separating the experimental groups. $C\beta G-$ group together with $C\beta Gl+$ group was separated by FLD_1 from other subgroups, while $H\beta Gl+$ and $C\beta Gh+$ groups were separated from other subgroups along the vertical axis (FLD_2). LC3B expression in colonocytes and Caspase-3 expression in the colon wall had the greatest impact on FLD_1, whereas Caspase-3 expression in colonocytes influenced FLD_2. Expression of LC3-II in the colon wall had an effect on both FLDs (Figure 7F). FLD analysis showed that LC3B expression in the colonocytes and that of Caspase-3 in the colon wall were slightly higher in the control groups fed with $\beta Gh+$ and $\beta G-$ feed compared to the $C\beta Gl+$ and $C\beta G-$ groups. Expression of Caspase-3 in colonocytes and that of LC3-II in the colon wall were slightly higher in the $H\beta Gl+$ and $C\beta Gh+$ groups compared to the $C\beta G-$ and $C\beta Gl+$ groups.

4. Discussion

Crohn's disease belongs to the group of inflammatory bowel diseases and is characterized by the presence of chronic inflammation within the gut. The course of this disease is associated with alternating periods of exacerbation and remission. Food is one of the significant factors influencing the course of CD through many mechanisms. Another influencing factor is the modulation of cellular processes, including cell death, which maintains homeostasis and eliminates abnormal and damaged cells [2]. In this study, we determined the effects of the consumption of low- and high-molar-mass oat beta-glucans on the expression of selected markers of apoptosis and autophagy in colonocytes in TNBS colitis-induced rats. In addition, we analyzed the expression of colon wall receptors, including TLRs and Dectin-1, which are involved in the recognition of molecular patterns of pathogens in colon epithelial cells. Rectal administration of a TNBS ethanol solution in animals caused transmural colitis, which is a well-described intestinal inflammation animal model with predominantly Crohn's disease symptoms [31,33]. According to Antoniou and coworkers, TNBS-induced colitis is a good model for studying immunopathogenesis and potential treatments for Crohn's disease [30]. In this model intestinal inflammation is achieved by a local administration of TNBS in 50% ethanol which involves both chemical damage and T cell immune reactivity. Additionally, TNBS administration results in acute necrosis of the colon wall due to oxidative damage, along with transmural inflammation that closely resembles the histopathological lesions developed in human CD [30,34,35]. The ethanol breaks the mucosal barrier allowing the penetration of the reagent. The main symptoms in animals with TNBS-induced colitis are bloody diarrhoea, weight loss and intestinal wall thickening. The presence of the features typical of CD was confirmed by histological evaluation, which revealed widespread inflammation of the colon, including not only the mucosa but also the deeper layers of the intestinal wall. The induction of acute inflammation in the colon wall was also confirmed by the results obtained from the analysis of the concentration of proinflammatory cytokines and the lymphocyte profile performed on the same biological material [28] as well as by the results of pro-inflammatory cytokines gene expression. Here, the effects of beta-glucans were assessed at three time points: after 3, 7, and 21 days of TNBS administration. This allowed evaluating the analyzed parameters in animals with different intensities of inflammatory changes in the intestine. It also reflects the periods of exacerbation and remission of colitis in people with CD and enabled determining the effectiveness of beta-glucan supplementation in these periods of the disease. The examined rats developed local acute inflammation immediately after TNBS administration. The inflammation was significantly less severe 1 and 3 weeks after TNBS administration, which indicated remission of the induced inflammation. This was confirmed by macro- and microscopic examinations of changes in the colon, including swelling of the mucosa, microbleeding, and necrosis [28]. It should be noted that the changes were transient and minimally invasive, indicating an early stage of CD development. This was evidenced by the good health of rats, decreased feed consumption only five days after the administration of TNBS, lower weight gain in rats with induced colitis, and no increase in the activity of plasma liver transaminases [29].

In this study, we also examined changes in the intensity of the autophagy process. It contributes to the adaptation of cells and the maintenance of intracellular homeostasis enabling cells to survive under stressful conditions. The autophagy marker in the colon wall and IECs investigated in the study was the expression of the LC3B protein, which participates in the formation and maturation of autophagosomes [8]. A decrease in this protein was found in the colon wall after TNBS administration, which indicates intense repair processes of the intestinal epithelium accompanying/preceding the period of CD remission. In the initial period of the experiment, autophagy was slightly increased in the colon wall in all experimental groups, which could be due to the mechanical irritation caused by rectal administration of TNBS or saline solution. A significant effect of colitis on the reduction of LC3B expression in IECs was found after 21 days. At this time point, the expression of LC3B in colonocytes was approximately four times lower in the colitis rats fed with feed without beta-glucans (CβG−) compared to the βG− control group. A similar effect of colitis was observed in induced rats by Xiong et al. (2019), who showed a decrease in the expression of LC3-II in the epithelium of the large intestine in colitis-induced animals compared to the control group [36]. In humans, CD is also characterized by impaired autophagy [37]. A study recognized polymorphism in the ULK1 kinase gene involved in the autophagy process and mutation of the protein containing the nucleotide-binding oligomerization domain 2 (NOD2) as a direct mechanism responsible for the autophagy impairment in CD [8].

Our results showed that supplementation with high-molar-mass oat beta-glucans reduced the negative effects of inflammation on the expression of LC3B protein. After 21 days of the experiment, it was observed that the expression of LC3B in IECs in the colitis group fed with feed supplemented with high-molar-mass oat beta-glucans was similar to that in the βG− control group. The increase in the expression of this protein under the influence of high-molar-mass beta-glucans in animals without colitis after 7 days of the experiment was also significant compared to other experimental groups. This may indicate the autophagy-enhancing effect of these polysaccharides in IECs in animals with and without colitis. The results of a previous study conducted by our team also indicated the effect of supplementing animal feed with oat beta-glucans on the expression of autophagy-related proteins. In an experiment carried out on rats in which enteritis was induced by intravenous administration of LPS, the transcriptomic analysis showed that the consumption of oat beta-glucans with feed increased the expression of the *Atg10* gene belonging to the *ATG* (autophagy-related genes) family. In addition, high-molar-mass oat beta-glucans caused a significant upregulation of the expression of the *Atg13* gene, another gene of the *ATG* family [38]. Therefore, the above data indicate a significant effect of oat beta-glucans in restoring the activity of the autophagy process in inflamed IECs, and a stronger effect of oat beta-glucans with a high molar mass, which also increased the activity of autophagy in colon tissue of the control animals without colitis.

Disturbances in the apoptotic process are believed to play an important role in the pathogenesis of CD. On one hand, excessively intensified apoptosis of epithelial cells leads to their increased elimination and damage to the intestinal barrier [39]. On the other hand, the low intensity of apoptosis of inflammatory cells promotes their accumulation in the gastrointestinal wall and maintenance of inflammation and promote tumor progression [24,40]. The present study showed that in the early stages of the development of acute colitis in animals with colitis, the expression of Caspase-3, the executive enzyme of apoptosis, was very low. It indicates a strong inflammation caused by the rectal administration of the TNBS ethanol solution. The highest expression of Caspase-3 protein was observed in the control group fed with feed supplemented with high-molar-mass oat beta-glucans (HβGh+), which indicates the beneficial effect of this beta-glucan fraction in this respect. The physical properties of high-molar-mass beta-glucans favor the formation of a protective layer on the inner wall of the intestine, and hence this beta-glucan fraction effectively supports the development of beneficial microbiota producing short-chain fatty acids which favor the faster regeneration of epithelium damaged by saline administration [23,41,42].

After 7 days of TNBS administration, Caspase-3 expression in colitis-induced animals was approximately eight times higher than in the control group, which indicates an increase in the apoptosis process in response to intestinal inflammation. This was also confirmed by the FLD analysis, in which the control groups were clearly separated from the colitis groups. The clear separation of the control groups was mainly due to the greater Caspase-3 expression in the colon epithelium and colon wall in inflammatory animals. An intensified process of apoptosis is also observed in people with CD and UD. This is associated with the disturbance of the intestinal barrier integrity, which causes the transfer of the commensurate microbiota to the intestinal lamina propria. This in turn leads to an increase in inflammation and the level of proinflammatory cytokines such as TNF-α [9,43]. Oat beta-glucans may reduce the extent of apoptosis in the colon tissue and, consequently, improve the integrity of the intestinal barrier. In our study, consumption of feed supplemented with high-molar-mass oat beta-glucans by colitis animals resulted in approximately two times lower Caspase-3 expression after 7 days of inflammation induction. Consumption of feed with low-molar-mass oat beta-glucans resulted in the expression of this enzyme in colitis animals at a similar level as in the control group, which was also confirmed by the results of the FLD analysis. This proves that both forms of beta-glucans are effective in inhibiting apoptotic cell death, reducing inflammation markers, and inducing the remission period. In colitis rats fed with feed supplemented with oat beta-glucans of high molar mass after 7 and 21 days, the expression level of Caspase-3 was similar, but after 21 days it was higher compared to the other feeding groups. This confirms the significant effect of oat beta-glucans with a low and high molar mass in reducing the expression of this apoptotic enzyme during ongoing inflammation, with low-molar-mass beta-glucans having a stronger effect.

Oat beta-glucans influence the activity of the autophagy process and inhibition of apoptosis in inflamed IECs. It is probably related to the interaction of oat beta-glucans on the Dectin-1 receptor and the pattern recognition TLRs: TLR 4, TLR5, and TLR6. These receptors are responsible, among others, for the stimulation of the immune response to pathogenic factors, as well as the regulation of proliferation, autophagy, and apoptosis of intestinal cells [15,17]. In our experiment, the expression of Dectin-1 across all experimental time points was found to be reduced compared to the control group, due to the induced inflammation. It should be noted, that de Vries et al. showed increased Dectin-1 expression in an animal model of DSS-induced colon and mesenteric lymph nodes inflammation [44], while Van Hung and Suzuki showed no significant changes in the expression of this receptor in the DSS-induced small intestine inflammation [45]. Similarly, in patients with Crohn's disease, increased expression of the Dectin-1 receptor in macrophages, neutrophils, and other immune cells has been reported [46]. The observed inconsistency is likely caused by the differences in colitis induction and the type of cells examined for Dectin-1 expression assessment. According to our knowledge, there are no reports describing the effect of TNBS-induced colitis on Dectin-1 expression in intestinal epithelial cells. The decrease in Dectin-1 expression observed in our study may result from a different method of inducing colitis. In addition, we analyzed the Dectin-1 expression in the intestinal epithelial cells, not in the colon wall or the immune cells. In this case decrease in Dectin 1 expression in colonocytes noticed in our study may be related to disrupted intestinal barrier integrity by the ethanolic TNBS solution, that as a consequence, causes infiltration of pathogens/antigens into the deeper layers of the colon wall and allows their direct contact with cells of the immune system. The subsequent activation of immune cells results in an increased expression of Dectin-1 in these cells [30,47]. Moreover, in our study, it is important to point that the increase in the expression of this receptor in IECs was observed in the control groups after 3 and 7 days of saline solution administration, which could have been caused by mechanical damage to these cells evidenced by the lack of similar changes after 21 days. This confirmed that the administration of the saline solution did not cause the injury of the intestinal barrier, that was observed after TNBS administration. We also observed simultaneous reduction of colonocyte Dectin-1 in the colitis group and its increase in control groups caused by the mechanical stimulus that resulted in a significant difference in the expression of this receptor between these groups. Consumption of the feed supplemented with oat beta-glucans reduced the difference between colitis and control groups in the experimental animals on days 3 and 21 after TNBS administration. The results showed an increase in the expression of Dectin-1 in inflamed IECs under the influence of oat beta-glucans. It should be noted that these polysaccharides have a similar effect on the expression of LC3B protein, with oat beta-glucans with a high molar mass having a stronger effect. FLD analysis showed that after 7 days of TNBS administration, the increased expression of Dectin-1 receptor was accompanied by an increased expression of the LC3B protein, the increase being characteristic of the control group fed with feed without beta-glucans and the control group fed with feed with beta-glucans with a high molar mass, which indicates the intensification of autophagy processes stimulating epithelial repair. Sakaguchi et al. (2018) described autophagy activation caused by the attachment of polysaccharide ligands to the Dectin-1 receptor as the mechanism of action of fungal beta-glucans administered to mice with induced colitis. These authors demonstrated that administration of beta-glucans decreased the expression of inflammatory cytokine mRNA in the colon (*Tnf*, *Il1b*, *Il6*) and the expression of TNFR1 receptor in IECs through the interaction of these polysaccharides with the Dectin-1 receptor. In addition, activation of Dectin-1 by beta-glucans reduced the translocation of NF-κB responsible for the induction and maintenance of inflammation [21].

Our other results showed that the concentration of TNF-α and other proinflammatory cytokines in the colon wall of animals with TNBS-induced inflammation was significantly increased at all time points. Consumption of oat beta-glucans reduced the concentration of these inflammatory factors (data not published [28]). TNF-α is one of the inflammatory mediators that directly stimulate the apoptotic process. The extrinsic pathway of apoptosis is induced by the binding of this factor to the TNFR1a receptor [48]. Pott et al. (2018) showed that increased autophagy in the inflamed intestinal epithelium protected cells against TNF-α-induced apoptosis, which in turn helped to maintain the integrity of the intestinal barrier and reduce inflammation. Thus, autophagy disorders cause exacerbation of inflammation and intensification of apoptosis in IECs induced by proinflammatory cytokines, mainly TNF-α [39]. This is in line with the results of another study in which the authors used a model of enteritis induced by murine norovirus infection and administration of DSS (dextran sulfate sodium). Matsuzawa-Ishimoto et al. (2017) showed that autophagy deficiency caused by deletion of the *ATG16L1* gene (encoding a component of the large protein complex essential for autophagy) made IECs in the small intestine more susceptible to TNF-α-induced apoptosis [49]. Currently, the treatment of CD in humans is based on drugs that stimulate autophagy and reduce TNF-α concentration in the inflamed parts of the gastrointestinal tract [50]. In summary, the activation of the inflamed Dectin-1 receptor in IECs by oat beta-glucans reduces the proapoptotic effect of the TNF-α complex with TNFR1 receptor. As a consequence of these changes, the autophagy process is intensified, which entails the protection of cells against excessive apoptosis as well as helps to maintain the integrity of the intestinal barrier and alleviate inflammation.

The antiapoptotic effect of oat beta-glucans in colitis is probably also related to their influence on the TLR expression. The results of this study showed that, after 3 days of TNBS administration, the expression of TLR 4 and TLR 6 receptors in colonocytes was significantly lower in the colitis group receiving feed without beta-glucans as compared to the control group fed with the same feed. In addition, in the same group of rats (CβG−), TLR 5 expression was lower compared to the colitis groups fed with feed supplemented with low- and high-molar-mass oat beta-glucans. This difference disappeared with the passage of time after TNBS administration, which proves that the decrease in the expression of these receptors is mostly influenced by acute intestinal inflammation. Nevertheless, oat beta-glucans caused a significant increase in the expression of these receptors, especially TLR 5 and TLR 6, in inflamed IECs in the initial period of the experiment. This period was characterized by the presence of active inflammation. Oat beta-glucans with a low molar mass had a stronger effect in increasing TLR 5 expression, while the expression of TLR 6 was influenced only by oat beta-glucans with a high molar mass.

Deficiency of TLRs, including TLR 4 and TLR 5, increases the susceptibility to induced inflammation resulting in increased permeability of the intestinal epithelium, inhibition of proliferation, increased apoptosis, and delayed IEC differentiation, which in turn leads to increased damage and ineffective reconstruction of the intestinal epithelium [51,52]. On the other hand, activation of TLRs alleviates the symptoms of colitis by stimulating the synthesis of cytoprotective and function-modulating factors in mesenchymal stem cells and immune cells that migrate to the sites of active inflammation. Furthermore, TLRs take part in the repair of the intestinal epithelium by inducing the synthesis of TFF3 (trefoil factor 3), amphiregulin, and prostaglandin E2, which increase the migration, survival, and proliferation of epithelial cells [22]. TLR 4 overexpression in patients with chronic IBD increases the risk of colorectal cancer due to increased proliferation and decreased apoptosis of IECs that are continuously exposed to proinflammatory cytokines and ROS [24,25].

Discriminant analysis by Fisher showed a positive correlation between the expression of TLR 4 and Caspase-3 in colon epithelial cells after 3 days of the experiment. A higher expression of TLR 4 and Caspase-3 was observed in the control groups fed with feed supplemented with high-molar-mass beta-glucans and beta-glucan-free feed compared to all colitis-induced groups, as well as in the control group consuming feed supplemented with low-molar-mass beta-glucans. This may indicate the activation of the apoptotic pathway with TLR 4 participation, which could have been caused by mechanical damage to the intestinal epithelium after rectal administration of physiological saline. Such damage increases the possibility of pathogen contact with the receptors present on the IEC membranes and consequently induces apoptosis. Some results indicate that bacterial LPS, a component of the cell membrane of some pathogenic bacteria, is one of the factors that reduce the proliferation and increase the apoptosis of cells of the small intestine and colon by activating TLR 4 [53,54]. The significant effect of the consumption of low-molar-mass oat beta-glucans by rats from the control group, which showed lower TLR 4 and Caspase-3 expression compared to other control groups, may indicate the antiapoptotic effect of this polysaccharide. As shown by the results of our research, the mechanism of the antiapoptotic action of beta-glucans is associated with decreased expression of TLR 4. It should be noted, however, that the intensity of the apoptosis process was relatively lower in the control groups, especially compared to the colitis groups after 7 days of TNBS administration.

Flagelin is a TLR5-specific ligand and the major protein of the surface structures of bacterial cells. Expression of TLR 5 in IECs regulates the composition and localization of the intestinal microbiota, preventing the development of intestinal inflammation [55]. The intestinal epithelial layer of TLR 5-deficient mice, as indicated in a review by Burgueño and Abreu (2020), was much more colonized by commensal microorganisms. TLR 5 deficiency, accompanied by an abnormal immune response to commensal antigens, led to the spontaneous development of colitis or exacerbation of the existing inflammation in mice [22]. In our study, we observed a significant effect of the consumption of oat beta-glucans on the increase in TLR 5 expression in acutely inflamed IECs, which consequently resulted in equating the expression of this receptor in colitis animals with the expression found in the βG− control group. There are no reports directly linking the effect of beta-glucans with TLR 5 expression. These polysaccharides are not typical ligands for TLRs, so it should be assumed that their likely mechanism of interaction with these receptors is indirect, possibly resulting from the influence on the expression of MyD88 (myeloid differentiation primary response 88) and TRIF (TIR-domain-containing adapter-inducing interferon-β). Studies by other authors have indicated that TLR 5 is an important factor regulating the composition of the intestinal microbiota. However, microbiota does not have a direct impact on the TLR 5 transcript, but on the MyD88 and TRIF adapters, which are among the main signaling molecules modulating the activity of TLRs by promoting epithelial reconstruction and repair as well as the production of cytoprotective factors [51,56].

Interactions between activated Dectin-1 and TLRs also play an important role. As indicated by the results from other authors, the receptors activated by soluble and insoluble beta-glucans, such as Dectin-1, TLR 2, TLR 4, and TLR 6, in macrophage membranes are important modulators of cytokine synthesis and participate in the activation of intracellular metabolic pathways. The main mediator of the activation of Dectin-1-dependent pathways in immune cells is the Syk tyrosine kinase, which, along with the Dectin-1 receptor, activates NF-κB through the CARD9-Bcl10-MALT1 complex [13]. Williams et al. demonstrated that NF-κB1-deficient mice exhibited increased IEC apoptosis in response to TNF-α. This is due to the function of the TNFR1 receptor. In addition to initiating apoptosis, this receptor also induces the expression of antiapoptotic genes by NF-κB [57]. Through their ability to simultaneously interact with the Dectin-1 receptor, TLR 2, and TLR 4, beta-glucans can induce NF-κB in the MyD88 protein-dependent signaling cascade. However, the binding of Dectin-1 by these polysaccharides with a parallel blocked TLR 5 did not have such an effect [13]. These data suggest that TLR 5 binding by beta-glucans does not further modulate the cytokines produced. It should be noted, however, that water-insoluble beta-glucans have a stronger Dectin-1 receptor-activating effect [13]. Such signaling synergy was also demonstrated by Patidar et al., who described the co-localization and clustering of Dectin-1 and TLR 2 in peritoneal macrophages stimulated by barley or yeast beta-glucans (zymosan), with zymosan showing a stronger effect. The phosphorylation pattern in the present study indicates that Dectin-1 and TLR 2 stimulation can activate downstream signaling pathways in a variety of ways. Dectin-1 follows the Syk kinase signaling pathway, while TLR 2 follows the signaling pathway of IκB kinase (IKK-IκB) [58]. Another in vitro study, in which the authors showed increased activation of NF-κB in human dendritic cells by oat beta-glucans via the Dectin-1 receptor, confirms the activation of the Dectin-1/NF-κB signaling pathway by water-soluble beta-glucans. A significant finding in their study was that they noted a stronger stimulatory effect of beta-glucans subjected to enzymatic digestion, which was probably associated with a greater number of β-(1,3) bonds available for Dectin-1 receptors [17].

The mechanism by which beta-glucans influence apoptosis and autophagy in inflamed IECs is related to their effect on the expression of receptors not only in these but also in the cells of the immune system, such as macrophages or dendritic cells. The research results cited above indicate that beta-glucans modulate the synthesis of inflammatory cytokines by the immune cells, which translates into the modification of autophagy and apoptosis signaling pathways in IECs.

In our study, we described for the first time the effects of beta-glucans at three time points—3, 7, and 21 days after TNBS administration, reflecting the periods of exacerbation and remission that occur in people with CD. The results showed significant correlations between the expression of TLRs and Dectin-1 receptors and that of LC3B, Caspase-3, and Beclin-1 protein. Fisher's discriminant analysis showed such correlations only on days 3 and 7 after induction of colitis, while after 21 days, the analysis showed only the effect of expression of autophagy and apoptosis proteins in the colon wall and epithelium on the differentiation of the experimental groups. The ANOVA and FLD results as well as the results of our previous study indicated a significant effect of time on the effectiveness of beta-glucans in the model of colitis characteristic of CD. In the initial period, the direction of changes in the studied parameters indicates acute local inflammation, while over time, remission occurs [29]. It should be noted that the effect of oat beta-glucans on apoptosis and autophagy through TLRs and Dectin-1 receptors was dependent on the severity of inflammation, which was mainly confirmed by the FLD analysis. With the passage of time after TNBS administration and the progression of inflammatory remission, the immune and metabolic responses to these polysaccharides decreased at the cellular level.

5. Conclusions

In summary, oat beta-glucans were found to have the ability to alleviate the course of induced inflammation. Their influence on the course of apoptosis and autophagy seems to be particularly significant. The observed reduction in the activity of apoptosis and increased activity of autophagy, in combination with the immunomodulatory activity of beta-glucans, suggest their beneficial therapeutic effect. Depending on the molar mass, these polysaccharides may act via different signaling pathways, but both consequently reduce inflammation and accelerate remission by, among others, protecting the integrity of the intestinal barrier. The presented results do not clearly indicate which beta-glucan fraction has a more effective protective effect on IECs; however, it seems that oat beta-glucans with a low molar mass have a slightly stronger effect in alleviating the induced colitis by greatly influencing the apoptosis process. In addition, the results published earlier by our team indicated that oat beta-glucans with low molar mass are more effective in removing the systemic effects of colitis, including anti-inflammatory and indirect antioxidant effects [28,29]. These properties of oat beta-glucans, in particular those of low molar mass, indicate their utility as preparations added during the production of food for special medical purposes for people suffering from inflammatory bowel disease, especially Crohn's disease.

Supplementary Materials: The following are available online at https://www.mdpi.com/2072-664 3/13/2/321/s1, Figure S1: DEC-1 Changes expression in colonocytes expressed (mean ± SE) as integrated optical density.; Figure S2: Changes expression TLR 4 in colonocytes expressed (mean ± SE) as integrated optical density.; Figure S3: Changes expression TLR 5 in colonocytes expressed (mean ± SE) as integrated optical density.; Figure S4: Changes expression TLR 6 in colonocytes expressed (mean ± SE) as integrated optical density.; Figure S5: Changes expression LC3B in colonocytes expressed (mean ± SE) as integrated optical density.; Figure S6: Changes expression Caspase-3 in colonocytes expressed (mean ± SE) as integrated optical density.; Figure S7: Results of the densitometric analysis for Beclin 1 expression in the large intestinal wall expressed (mean ± SE) as integrated optical density.; Figure S8: Results of the densitometric analysis for Caspase-3 expression in the large intestinal wall expressed (mean ± SE) as integrated optical density.; Figure S9: Relative expression of cytokine genes vs. Ldha mRNA.

Author Contributions: Conducted research and performed experiments, Ł.K., K.D., M.G., M.O., K.M.-K., T.K.; writing—prepared the original draft, Ł.K., K.D., M.G.; statistical analysis, results description, visualization, Ł.K.; project administration, Fisher's Linear Discriminant Analysis, K.D.; writing—results discussion, Ł.K., K.D., J.G.-O.; writing—review and editing, K.D., M.G., J.G.-O.; conceptualization, funding acquisition and supervision, J.G.-O. All authors have read and agreed to the published version of the manuscript

Funding: The authors acknowledge the financial support of the National Science Centre, Poland through grant number 2015/17/B/NZ9/01740.

Institutional Review Board Statement: The animal experiment was conducted after the approval of the II Local Animal Care and Use Committee in Warsaw (Resolution # 60/2015; 29.06.2015).

Informed Consent Statement: Not applicable.

Data Availability Statement: The data that support the findings of this study are available on request from the corresponding author [K.D.].

Acknowledgments: The authors wish to thank MS Ewa Żyła and Wilczak from WULS for assistance and supervision over in vivo study and thank Rafał Sapierzyński from WULS for assistance in histology analysis. Furthermore, thank Joanna Harasym from Wroclaw University of Economics and Business for prepare pure extract of oat beta-glucans.

Conflicts of Interest: The authors declare no conflict of interest.

References

1. Actis, G.C.; Pellicano, R.; Fagoonee, S.; Ribaldone, D.G. History of inflammatory bowel disease. *J. Clin. Med.* **2019**, *8*, 1970. [CrossRef]
2. Ramos, G.P.; Papadakis, K.A. Mechanisms of Disease: Inflammatory Bowel Diseases. *Mayo Clin. Proc.* **2019**, *94*, 155–165. [CrossRef]
3. Van Der Sloot, K.W.J.; Amini, M.; Peters, V.; Dijkstra, G.; Alizadeh, B.Z. Inflammatory bowel diseases: Review of known environmental protective and risk factors involved. *Inflamm. Bowel Dis.* **2017**, *23*, 1499–1509. [CrossRef]
4. Gajendran, M.; Loganathan, P.; Catinella, A.P.; Hashash, J.G. A comprehensive review and update on Crohn's disease. *Disease-a-Month* **2018**, *64*, 20–57. [CrossRef]
5. Gajendran, M.; Loganathan, P.; Jimenez, G.; Catinella, A.P.; Ng, N.; Umapathy, C.; Ziade, N.; Hashash, J.G. A comprehensive review and update on ulcerative colitis. *Disease-a-Month* **2019**, *65*, 100851. [CrossRef] [PubMed]
6. Nunes, T.; Bernardazzi, C.; De Souza, H.S. Cell death and inflammatory bowel diseases: Apoptosis, necrosis, and autophagy in the intestinal epithelium. *Biomed. Res. Int.* **2014**, *2014*, 218493. [CrossRef]
7. Patel, K.K.; Stappenbeck, T.S. Autophagy and intestinal homeostasis. *Annu. Rev. Physiol.* **2013**, *75*, 241–262. [CrossRef] [PubMed]
8. Saha, S.; Panigrahi, D.P.; Patil, S.; Bhutia, S.K. Autophagy in health and disease: A comprehensive review. *Biomed. Pharmacother.* **2018**, *104*, 485–495. [CrossRef]
9. Blander, J.M. Death in the intestinal epithelium—basic biology and implications for inflammatory bowel disease. *FEBS J.* **2016**, *283*, 2720–2730. [CrossRef]
10. Green, D.R.; Llambi, F. Cell death signaling. *Cold Spring Harb. Perspect. Biol.* **2015**, *7*, a006080. [CrossRef]
11. Koukourakis, M.I.; Kalamida, D.; Giatromanolaki, A.; Zois, C.E.; Sivridis, E.; Pouliliou, S.; Mitrakas, A.; Gatter, K.C.; Harris, A.L. Autophagosome proteins LC3A, LC3B and LC3C have distinct subcellular distribution kinetics and expression in cancer cell lines. *PLoS ONE* **2015**, *10*, e0137675. [CrossRef] [PubMed]
12. Schläfli, A.M.; Berezowska, S.; Adams, O.; Langer, R.; Tschan, M.P. Reliable LC3 and p62 autophagy marker detection in formalin fixed paraffin embedded human tissue by immunohistochemistry. *Eur. J. Histochem.* **2015**, *59*, 2481. [CrossRef] [PubMed]
13. Kanjan, P.; Sahasrabudhe, N.M.; de Haan, B.J.; de Vos, P. Immune effects of β-glucan are determined by combined effects on Dectin-1, TLR2, 4 and 5. *J. Funct. Foods* **2017**, *37*, 433–440. [CrossRef]
14. Nakashima, A.; Yamada, K.; Iwata, O.; Sugimoto, R.; Atsuji, K.; Ogawa, T.; Ishibashi-Ohgo, N.; Suzuki, K. β-Glucan in foods and its physiological functions. *J. Nutr. Sci. Vitaminol.* **2018**, *64*, 8–17. [CrossRef] [PubMed]
15. Hug, H.; Mohajeri, M.H.; La Fata, G. Toll-like receptors: Regulators of the immune response in the human gut. *Nutrients* **2018**, *10*, 203. [CrossRef] [PubMed]
16. Zhang, X.; Qi, C.; Guo, Y.; Zhou, W.; Zhang, Y. Toll-like receptor 4-related immunostimulatory polysaccharides: Primary structure, activity relationships, and possible interaction models. *Carbohydr. Polym.* **2016**, *149*, 186–206. [CrossRef]
17. Sahasrabudhe, N.M.; Tian, L.; van den Berg, M.; Bruggeman, G.; Bruininx, E.; Schols, H.A.; Faas, M.M.; de Vos, P. Endo-glucanase digestion of oat β-Glucan enhances Dectin-1 activation in human dendritic cells. *J. Funct. Foods* **2016**, *21*, 104–112. [CrossRef]
18. Rahabi, M.; Jacquemin, G.; Prat, M.; Meunier, E.; AlaEddine, M.; Bertrand, B.; Lefèvre, L.; Benmoussa, K.; Batigne, P.; Aubouy, A.; et al. Divergent roles for macrophage C-type lectin receptors, Dectin-1 and mannose receptors, in the intestinal inflammatory response. *Cell Rep.* **2020**, *30*, 4386–4398. [CrossRef]
19. Tam, J.M.; Mansour, M.K.; Khan, N.S.; Seward, M.; Puranam, S.; Tanne, A.; Sokolovska, A.; Becker, C.E.; Acharya, M.; Baird, M.A.; et al. Dectin-1-dependent LC3 recruitment to phagosomes enhances fungicidal activity in macrophages. *J. Infect. Dis.* **2014**, *210*, 1844–1854. [CrossRef]
20. Cohen-Kedar, S.; Baram, L.; Elad, H.; Brazowski, E.; Guzner-Gur, H.; Dotan, I. Human intestinal epithelial cells respond to β-glucans via Dectin-1 and Syk. *Eur. J. Immunol.* **2014**, *44*, 3729–3740. [CrossRef]
21. Sakaguchi, K.; Shirai, Y.; Itoh, T.; Mizuno, M. Lentinan exerts its anti-inflammatory activity by suppressing TNFR1 transfer to the surface of intestinal epithelial cells through Dectin-1 in an in vitro and mice model. *Immunome Res.* **2018**, *14*, 1000165. [CrossRef]
22. Burgueño, J.F.; Abreu, M.T. Epithelial Toll-like receptors and their role in gut homeostasis and disease. *Nat. Rev. Gastroenterol. Hepatol.* **2020**, *17*, 263–278. [CrossRef] [PubMed]
23. Akkerman, R.; Logtenberg, M.J.; An, R.; Van Den Berg, M.A.; de Haan, B.J.; Faas, M.M.; Zoetendal, E.; de Vos, P.; Schols, H.A. Endo-1,3(4)-β-Glucanase-Treatment of Oat β-Glucan Enhances Fermentability by Infant Fecal Microbiota, Stimulates Dectin-1 Activation and Attenuates Inflammatory Responses in Immature Dendritic Cells Renate. *Nutrients* **2020**, *12*, 1660. [CrossRef] [PubMed]
24. Ullman, T.A.; Itzkowitz, S.H. Intestinal inflammation and cancer. *Gastroenterology* **2011**, *140*, 1807–1816. [CrossRef]
25. Sussman, D.A.; Santaolalla, R.; Bejarano, P.A.; Garcia-Buitrago, M.T.; Perez, M.T.; Abreu, M.T.; Clarke, J. In silico and Ex vivo approaches identify a role for toll-like receptor 4 in colorectal cancer. *J. Exp. Clin. Cancer Res.* **2014**, *33*, 45. [CrossRef]
26. Wilczak, J.; Błaszczyk, K.; Kamola, D.; Gajewska, M.; Harasym, J.P.; Jałosińska, M.; Gudej, S.; Suchecka, D.; Oczkowski, M.; Gromadzka-Ostrowska, J. The effect of low or high molecular weight oat beta-glucans on the inflammatory and oxidative stress status in the colon of rats with LPS-induced enteritis. *Food Funct.* **2015**, *6*, 590–603. [CrossRef]
27. Suchecka, D.; Harasym, J.P.; Wilczak, J.; Gajewska, M.; Oczkowski, M.; Gudej, S.; Błaszczyk, K.; Kamola, D.; Filip, R.; Gromadzka-Ostrowska, J. Antioxidative and anti-inflammatory effects of high beta-glucan concentration purified aqueous extract from oat in experimental model of LPS-induced chronic enteritis. *J. Funct. Foods* **2015**, *14*, 244–254. [CrossRef]

28. Żyła, E.; Dziendzikowska, K.; Wilczak, J.; Harasym, J.; Gromadzka-Ostrowska, J. Beneficial effects of oat beta-glucan dietary supplementation in colitis depend on its molecular weight. *Molecules* **2019**, *24*, 3591. [CrossRef]

29. Kopiasz, Ł.; Dziendzikowska, K.; Gajewska, M.; Wilczak, J.; Harasym, J.; Żyła, E.; Kamola, D.; Oczkowski, M.; Królikowski, T.; Gromadzka-Ostrowska, J. Time-dependent indirect antioxidative effects of oat beta-glucans on peripheral blood parameters in the animal model of colon inflammation. *Antioxidants* **2020**, *9*, 375. [CrossRef]

30. Antoniou, E.; Margonis, G.A.; Angelou, A.; Pikouli, A.; Argiri, P.; Karavokyros, I.; Papalois, A.; Pikoulis, E. The TNBS-induced colitis animal model: An overview. *Ann. Med. Surg.* **2016**, *11*, 9–15. [CrossRef]

31. Catana, C.S.; Magdas, C.; Tabaran, F.A.; Craciun, E.C.; Deak, G.; Magdas, V.A.; Cozma, V.; Gherman, C.M.; Berindan-Neagoe, I.; Dumitrascu, D.L. Comparison of two models of inflammatory bowel disease in rats. *Adv. Clin. Exp. Med.* **2018**, *27*, 599–607. [CrossRef] [PubMed]

32. Parra, R.S.; Lopes, A.H.; Carreira, E.U.; Feitosa, M.R.; Cunha, F.Q.; Garcia, S.B.; Cunha, T.M.; Da Rocha, J.J.R.; Féres, O. Hyperbaric oxygen therapy ameliorates TNBS-induced acute distal colitis in rats. *Med. Gas Res.* **2015**, *5*, 6. [CrossRef] [PubMed]

33. Tian, T.; Wang, Z.; Zhang, J. Pathomechanisms of oxidative stress in inflammatory bowel disease and potential antioxidant therapies. *Oxid. Med. Cell. Longev.* **2017**, *2017*, e4535194. [CrossRef] [PubMed]

34. El-Salhy, M.; Hatlebakk, J.G. Changes in enteroendocrine and immune cells following colitis induction by TNBS in rats. *Mol. Med. Rep.* **2016**, *14*, 4967–4974. [CrossRef]

35. Motavallian-Naeini, A.; Andalib, S.; Rabbani, M.; Mahzouni, P.; Afsharipour, M.; Minaiyan, M. Validation and optimization of experimental colitis induction in rats using 2, 4, 6-trinitrobenzene sulfonic acid. *Res. Pharm. Sci.* **2012**, *7*, 159–169.

36. Xiong, Y.J.; Deng, Z.B.; Liu, J.N.; Qiu, J.J.; Guo, L.; Feng, P.P.; Sui, J.R.; Chen, D.P.; Guo, H.S. Enhancement of epithelial cell autophagy induced by sinensetin alleviates epithelial barrier dysfunction in colitis. *Pharmacol. Res.* **2019**, *148*, 104461. [CrossRef]

37. Hooper, K.M.; Barlow, P.G.; Stevens, C.; Henderson, P. Inflammatory bowel disease drugs: A focus on autophagy. *J. Crohn's Colitis* **2017**, *11*, 118–127. [CrossRef]

38. Błaszczyk, K.; Gajewska, M.; Wilczak, J.; Kamola, D.; Majewska, A.; Harasym, J.; Gromadzka-Ostrowska, J. Oral administration of oat beta-glucan preparations of different molecular weight results in regulation of genes connected with immune response in peripheral blood of rats with LPS-induced enteritis. *Eur. J. Nutr.* **2018**, *58*, 2859–2873. [CrossRef]

39. Pott, J.; Kabat, A.M.; Maloy, K.J. Intestinal epithelial cell autophagy is required to protect against TNF-induced apoptosis during chronic colitis in mice. *Cell Host Microbe* **2018**, *23*, 191–202. [CrossRef]

40. Dias, C.B.; Milanski, M.; Portovedo, M.; Horita, V.; De Ayrizono, M.L.S.; Planell, N.; Coy, C.S.R.; Velloso, L.A.; Meirelles, L.R.; Leal, R.F. Defective apoptosis in intestinal and mesenteric adipose tissue of Crohn's disease patients. *PLoS ONE* **2014**, *9*, e98547. [CrossRef]

41. Bai, J.; Ren, Y.; Li, Y.; Fan, M.; Qian, H.; Wang, L.; Wu, G.; Zhang, H.; Qi, X.; Xu, M.; et al. Physiological functionalities and mechanisms of β-glucans. *Trends Food Sci. Technol.* **2019**, *88*, 57–66. [CrossRef]

42. Grundy, M.M.L.; Quint, J.; Rieder, A.; Ballance, S.; Dreiss, C.A.; Cross, K.L.; Gray, R.; Bajka, B.H.; Butterworth, P.J.; Ellis, P.R.; et al. The impact of oat structure and β-glucan on in vitro lipid digestion. *J. Funct. Foods* **2017**, *38*, 378–388. [CrossRef] [PubMed]

43. Neurath, M.F. Cytokines in inflammatory bowel disease. *Nat. Rev. Immunol.* **2014**, *14*, 329–342. [CrossRef] [PubMed]

44. De Vries, H.S.; Plantinga, T.S.; van Krieken, J.H.; Stienstra, R.; van Bodegraven, A.A.; Festen, E.A.M.; Weersma, R.K.; Crusius, J.B.A.; Linskens, R.K.; Joosten, L.A.B.; et al. Genetic association analysis of the functional c.714T>G polymorphism and mucosal expression of dectin-1 in inflammatory bowel disease. *PLoS ONE* **2009**, *4*, e7818. [CrossRef] [PubMed]

45. Van Hung, T.; Suzuki, T. Guar gum fiber increases suppressor of cytokine signaling-1 expression via toll-like receptor 2 and dectin-1 pathways, regulating inflammatory response in small intestinal epithelial cells. *Mol. Nutr. Food Res.* **2017**, *61*, 1700048. [CrossRef] [PubMed]

46. Takedatsu, H.; Mitsuyama, K.; Mochizuki, S.; Kobayashi, T.; Sakurai, K.; Takeda, H.; Fujiyama, Y.; Koyama, Y.; Nishihira, J.; Sata, M. A new therapeutic approach using a schizophyllan-based drug delivery system for inflammatory bowel disease. *Mol. Ther.* **2012**, *20*, 1234–1241. [CrossRef]

47. Kiesler, P.; Fuss, I.J.; Strober, W. Experimental models of inflammatory bowel diseases. *Cell. Mol. Gastroenterol. Hepatol.* **2015**, *1*, 154–170. [CrossRef]

48. Günther, C.; Neumann, H.; Neurath, M.F.; Becker, C. Apoptosis, necrosis and necroptosis: Cell death regulation in the intestinal epithelium. *Gut* **2013**, *62*, 1062–1071. [CrossRef]

49. Matsuzawa-Ishimoto, Y.; Shono, Y.; Gomez, L.E.; Hubbard-Lucey, V.M.; Cammer, M.; Neil, J.; Dewan, M.Z.; Lieberman, S.R.; Lazrak, A.; Marinis, J.M.; et al. Autophagy protein ATG16L1 prevents necroptosis in the intestinal epithelium. *J. Exp. Med.* **2017**, *214*, 3687–3705. [CrossRef]

50. Azzman, N. Crohn's disease: Potential drugs for modulation of autophagy. *Medicina* **2019**, *55*, 224. [CrossRef]

51. Price, A.E.; Shamardani, K.; Lugo, K.A.; Deguine, J.; Roberts, A.W.; Lee, B.L.; Barton, G.M. A map of Toll-like receptor expression in the intestinal epithelium reveals distinct spatial, cell type-specific, and temporal patterns. *Immunity* **2018**, *49*, 560–575. [CrossRef] [PubMed]

52. Abreu, M.T. Toll-like receptor signalling in the intestinal epithelium: How bacterial recognition shapes intestinal function. *Nat. Rev. Immunol.* **2010**, *10*, 131–143. [CrossRef]

53. Neal, M.D.; Sodhi, C.P.; Jia, H.; Dyer, M.; Egan, C.E.; Yazji, I.; Good, M.; Afrazi, A.; Marino, R.; Slagle, D.; et al. Toll-like receptor 4 is expressed on intestinal stem cells and regulates their proliferation and apoptosis via the p53 up-regulated modulator of apoptosis. *J. Biol. Chem.* **2012**, *287*, 37296–37308. [CrossRef] [PubMed]
54. Naito, T.; Mulet, C.; De Castro, C.; Molinaro, A.; Saffarian, A.; Nigro, G.; Bérard, M.; Clerc, M.; Pedersen, A.B.; Sansonetti, P.J.; et al. Lipopolysaccharide from Crypt-Specific Core Microbiota Modulates the Colonic Epithelial Proliferation-to-Differentiation Balance. *Am. Soc. Microbiol.* **2017**, *8*, e01680-17. [CrossRef] [PubMed]
55. Chassaing, B.; Ley, R.E.; Gewirtz, A.T. Intestinal epithelial cell toll-like receptor 5 regulates the intestinal microbiota to prevent low-grade inflammation and metabolic syndrome in mice. *Gastroenterology* **2014**, *147*, 1363–1377. [CrossRef] [PubMed]
56. Brandão, I.; Hörmann, N.; Jäckel, S.; Reinhardt, C. TLR5 expression in the small intestine depends on the adaptors MyD88 and TRIF, but is independent of the enteric microbiota. *Gut Microbes* **2015**, *6*, 202–206. [CrossRef]
57. Williams, J.M.; Duckworth, C.A.; Watson, A.J.M.; Frey, M.R.; Miguel, J.C.; Burkitt, M.D.; Sutton, R.; Hughes, K.R.; Hall, L.J.; Caamaño, J.H.; et al. A mouse model of pathological small intestinal epithelial cell apoptosis and shedding induced by systemic administration of lipopolysaccharide. *DMM Dis. Model. Mech.* **2013**, *6*, 1388–1399. [CrossRef]
58. Patidar, A.; Mahanty, T.; Raybarman, C.; Sarode, A.Y.; Basak, S.; Saha, B.; Bhattacharjee, S. Barley beta-Glucan and Zymosan induce Dectin-1 and Toll-like receptor 2 co-localization and anti-leishmanial immune response in Leishmania donovani-infected BALB/c mice. *Scand. J. Immunol.* **2020**, *92*, e12952. [CrossRef]

Article

Effect of Oat β-Glucan on Affective and Physical Feeling States in Healthy Adults: Evidence for Reduced Headache, Fatigue, Anxiety and Limb/Joint Pains

Thomas M. S. Wolever [1,*], Maike Rahn [2,†], El Hadji Dioum [2], Alexandra L. Jenkins [1], Adish Ezatagha [1], Janice E. Campbell [1] and YiFang Chu [2]

[1] Formerly GI Labs, INQUIS Clinical Research, Ltd., Toronto, ON M5C 2N8, Canada; alexandrajenkins@inquis.com (A.L.J.); aezatagha@inquis.com (A.E.); jcampbell@inquis.com (J.E.C.)
[2] Quaker Oats Center of Excellence, PepsiCo R&D Nutrition, Barrington, IL 60010, USA; mrinny96@gmail.com (M.R.); ElHadji.Dioum@pepsico.com (E.H.D.); yifang.chu@pepsico.com (Y.C.)
* Correspondence: twolever@inquis.com; Tel.: +1-416-861-9177
† Current address: DSM Nutritional Products, North America Nutrition Science and Advocacy, Parsippany, NJ 07054, USA.

Citation: Wolever, T.M.S.; Rahn, M.; Dioum, E.H.; Jenkins, A.L.; Ezatagha, A.; Campbell, J.E.; Chu, Y. Effect of Oat β-Glucan on Affective and Physical Feeling States in Healthy Adults: Evidence for Reduced Headache, Fatigue, Anxiety and Limb/Joint Pains. *Nutrients* **2021**, *13*, 1534. https://doi.org/10.3390/nu13051534

Academic Editors: Seiichiro Aoe, Tatsuya Morita and Naohito Ohno

Received: 12 March 2021
Accepted: 29 April 2021
Published: 1 May 2021

Publisher's Note: MDPI stays neutral with regard to jurisdictional claims in published maps and institutional affiliations.

Abstract: The gastrointestinal (GI) side-effects of dietary fibers are recognized, but less is known about their effects on non-GI symptoms. We assessed non-GI symptoms in a trial of the LDL-cholesterol lowering effect of oat β-glucan (OBG). Participants ($n = 207$) with borderline high LDL-cholesterol were randomized to an OBG (1 g OBG, $n = 104$, $n = 96$ analyzed) or Control ($n = 103$, $n = 95$ analyzed) beverage 3-times daily for 4 weeks. At screening, baseline, 2 weeks and 4 weeks participants rated the severity of 16 non-GI symptoms as none, mild, moderate or severe. The occurrence and severity (more or less severe than pre-treatment) were compared using chi-squared and Fisher's exact test, respectively. During OBG treatment, the occurrence of exhaustion and fatigue decreased versus baseline ($p < 0.05$). The severity of headache (2 weeks, $p = 0.032$), anxiety (2 weeks $p = 0.059$) and feeling cold (4 weeks, $p = 0.040$) were less on OBG than Control. The severity of fatigue and hot flashes at 4 weeks, limb/joint pain at 2 weeks and difficulty concentrating at both times decreased on OBG versus baseline. High serum c-reactive-protein and changes in c-reactive-protein, oxidized-LDL, and GI-symptom severity were associated with the occurrence and severity of several non-GI symptoms. These data provide preliminary, hypothesis-generating evidence that OBG may reduce several non-GI symptoms in healthy adults.

Keywords: randomized clinical trial; humans; symptoms; gastrointestinal tract; musculo-skeletal system; oats; oatmeal; dietary fiber; beta-glucan

1. Introduction

To protect against malnutrition and non-communicable diseases global nutrition recommendations include advice to consume more whole grains to help maintain an adequate intake of dietary fiber [1]. Oats are a whole grain containing the soluble dietary fiber β-glucan [2]. Oat β-glucan (OBG) reduces serum LDL-cholesterol [3,4] and postprandial glycemic responses [5], physiological effects associated with reduced risk for cardiovascular disease [6] and type 2 diabetes [7], respectively. Dietary fibers such as OBG are not absorbed in the small intestine and reach the colon where they increase the mass of the colonic contents and may be partly or completely fermented by colonic microbiota, effects which have desirable and undesirable consequences as explained below.

The undesirable side-effects of dietary fibers include abdominal bloating and pain, flatus and diarrhea which, although transitory in nature, may limit the acceptability of high fiber foods [8]. Consistent with this were our findings in healthy subjects who consumed 3 g/day OBG or Control for 4 weeks. After 2 weeks, flatulence and abdominal discomfort had increased from baseline on both the OBG and Control treatments; but by 4 weeks these symptoms had begun to return to baseline [9]. In a cross-sectional study of adults with coeliac disease, higher long-term oat intake correlated with lower gastrointestinal symptom scores [10].

The desirable effects of increased fiber intake include increased stool bulk which may be beneficial for conditions such as constipation, irritable bowel syndrome [11] and diverticulosis [12] although not all types of fiber are equally effective. Oats and oat bran appear to increase stool weight per gram of dietary fiber to a similar extent as other sources of fiber [13]. Oats also decrease fecal pH (a marker of increased short-chain fatty acid (SCFA) production) and increase the growth of beneficial gut microbiota [14]. Plausible mechanisms have been proposed whereby the stimulation of SCFA production and alteration of the colonic microbiome can influence systemic oxidative stress and inflammation and thereby influence both physical and mental functioning [15–17]. In this context, it is of interest that higher self-reported fiber intake was associated with higher positive affect in healthy children [18] and, in a randomized, cross-over study, consumption of a high fiber breakfast cereal for 2 weeks was associated with less fatigue compared to control in healthy adults [19]. We found that participants with type 2 diabetes treated by diet alone who consumed a low glycemic index diet containing 37 g/day fiber for 1 year experienced less severe headaches, less severe pains in joints or limbs and less severe gloomy thoughts than those on the lower fiber (21–23 g/day) diets [20]. OBG has been shown to enhance the endurance capacity of rats and influence serum metabolites in such a way as to suggest anti-fatigue properties [21], but there is a paucity of information about the effect of OBG on non-gastrointestinal symptoms in humans. To address this, we assessed a panel of non-gastrointestinal symptoms as tertiary objectives in a study the primary purpose of which was to determine if a novel oat product containing OBG would reduce serum LDL-cholesterol [9].

2. Materials and Methods

We performed a randomized, double-blind, placebo controlled, parallel arm design clinical trial at INQUIS Clinical Research, a contract research organization. The protocol was approved by the Western Institutional Review Board® (Puyallup, WA, USA) and all participants provided informed consent by signing the IRB approved consent form. The study was registered at www.clinicaltrials.gov (accessed on 11 April, 2019) with identifier: NCT03911427.

Participants were healthy males and non-pregnant, non-lactating females without diabetes, aged 18–65 year, with fasting calculated LDL-cholesterol between 3.00 and 5.00 mmol/L, inclusive, and BMI between 18.5 to <40 kg/m^2. Details of the inclusion/exclusion criteria can be found in Supplementary Information.

After signing the consent form, participants were screened over a period of 2–4 weeks involving 2 visits to the clinic. First they answered questions about their medical and drug history and had their height, weight and blood pressure measured, then they gave a fasting blood sample and were given instructions about how to record a 3-day diet record (3DDR) [22–24]. Participants who were eligible based on the results of the screening blood sample were contacted and asked to start recording a 3DDR and attend at the clinic to review the 3DDR and fill out the symptoms questionnaire (second screening visit, Screen 2). To avoid the potential confounding effects of high intakes of saturated fat or dietary fiber on the results, participants were excluded if their intake of saturated fat was ≥15% of energy or they were consuming >14 g/1000 kcal dietary fiber. Eligible participants were contacted by telephone or e-mail, advised to follow their usual dietary and exercise habits and to refrain from consuming oat, barley and psyllium products for the duration of the study and an appointment was made for the baseline visit.

2.1. Study Visits

Eligible participants visited the clinic after 10–14 h overnight fasts at baseline (week 0) and at 2 and 4 weeks, having been asked to avoid alcohol and unusual levels of physical activity and food intake for 24 h before each visit. At each visit medications and adverse events were reviewed, weight and blood pressure were measured, a fasting blood sample obtained and a symptoms questionnaire filled out. At baseline, participants were randomly assigned to either the Test or Control treatment. After the blood sample at each visit participants consumed the first daily treatment sachet and were given a snack or light breakfast. At baseline and week 2 participants were provided with a sufficient supply of treatment sachets to last until the next visit and a shaker cup and whisk with which to mix their assigned treatment and were given a study diary in which to record sachet consumption. Compliance was checked at week 2 and week 4 by counting unused sachets and, at week 2, participants were provided with forms and instructions to record another 3DDR before the next visit. Symptom questionnaires were filled out during each of the visits at Screen 2, Baseline, 2 weeks and 4 weeks.

2.2. Interventions

Interventions were provided in color-coded sachets: the Test intervention consisted of an oat ingredient delivering 1 g of β-glucan sachet; the Control intervention was a rice milk powder containing 0 g β-glucan. Each sachet contained 0.9 g fat (0 g saturated), 1.9 g protein, ~14 g available carbohydrate and 1.9 g (Test) or 0.3 g (Control) dietary fiber. Detailed nutrition information is given in Supplementary Table S1. Each sachet was mixed with 8 oz (240 mL) of cold water, shaken for 30 sec and consumed immediately. The Test and Control beverages had similar look, taste and smell. Subjects were asked to consume 3 sachets daily on 3 separate occasions separated by at least 3 h and preferably immediately before or within 10 min of each main meal (breakfast, lunch and dinner).

2.3. Symptoms Questionnaire

Participants were asked to rate the severity of each of 27 symptoms (11 gastrointestinal (GI) symptoms and 16 non-GI symptoms) they experienced over the previous 2 weeks on a 4-point scale: none = 0, mild = 1, moderate = 2 or severe = 3. We used this questionnaire in 2 previous studies in patients with diabetes [20,22,23]. The results for all 16 GI symptoms have been reported elsewhere [9] but the sum of the scores for the 5 most prevalent (major) GI symptoms (flatulence, diarrhea, constipation, abdominal distention and abdominal pain) are included below. Of the 16 non-GI symptoms the following 4 were experienced by a maximum of less than 10% of participants at any visit (baseline, 2 weeks and 4 weeks, respectively) and the results are not presented here: Increased appetite (6%, 9% and 5%), Palpitation/Throbbing of heart (4%, 6%, 4%), Balance disturbances (5%, 3%, 5%) and Numbness/burning/itching hands feet (8%, 4%, 5%). The remaining 12 symptoms were as follows: Headache, Fatigue, Lack of appetite, Tend to become exhausted (Exhaustion), Feelings of anxiety, Lack of energy, Pains in joints or limbs (Limb/joint pain), Diminished ability to concentrate (Reduced ability to concentrate), Feeling cold, Hot flashes/sensation of rising heat (Hot flashes), Gloomy thoughts and Inner tension.

2.4. Data Analysis, Management and Calculations

Blood samples for blood lipids and lipoproteins, glucose and insulin were collected at Baseline, 2 weeks and 4 weeks, and for glycated albumin, high sensitivity c-reactive protein (CRP) and oxidized LDL (oxLDL) at Baseline and 4 weeks and the results presented elsewhere [9]. However, results for CRP (analyzed by LifeLabs, Inc., Mississauga, ON, USA) and oxLDL (analyzed by ELISA kit, catalog #30-7810, Alpco Diagnostics, Salem, NH, USA) are included here because of their potential association with affective and physical symptoms. Each 3DDR was reviewed with the participant by a dietitian or nutritionist and subsequently analyzed for nutrient content (ESHA Food Processor, ESHA Research,

Salem, OR, USA). The nutrients contained in the treatment sachets were added to those from foods recorded on the 3DDR based on the sachet consumption counts

2.5. Power Analysis

The sample size calculation assumed that the SD of the change in LDL-cholesterol between baseline and 4 weeks would be 0.455 mmol/L [24]. Using this SD, n = 90 subjects per arm provides 80% power to detect an 0.19 mmol/L difference (5%) in LDL-cholesterol change between the test and control arms. To improve the likelihood that at least n = 180 would complete the study we allowed for to 15% dropouts and enrolled 207 subjects.

2.6. Statistical Analysis

Included in this analysis are all 191 subjects who were randomized and completed the study with no protocol violations and with 90% compliance (76/84 sachets over 4 weeks) based on the sachet count. The highly skewed distributions of CRP and oxLDL values were normalized by log transformation. Categorical measures were summarized as frequencies. To determine whether the prevalence of symptoms during treatment differed from that at the Baseline visit, and differed between treatments, the proportions of subjects who experienced any symptom (mild, moderate or severe) were compared using the chi-square test. To determine if the severity of symptoms during treatment differed from those before treatment, the ratio of the number of subjects in each treatment group whose scores at 2 weeks and 4 weeks were greater (more severe, M) or less (less severe, L) than the median score before treatment (defined as the median of the scores at the second screening (Scr2) and Baseline visits) were compared to an expected ratio of 1:1 by chi-square test. To determine if the change in severity of symptoms differed between treatments, the ratios of L:M on Test and Control were compared by Fisher's exact test [20]. Similarly, the severity of symptoms in participants whose serum CRP, serum oxLDL and the sum of the scores for 5 major GI symptoms (flatulence, diarrhea, constipation, abdominal bloating, abdominal discomfort) increased versus those in whom they did not increase during treatment.

3. Results

Recruitment began in April 2019 and the last subject visit occurred in February, 2020. The recruitment questionnaire was completed by 2607 individuals of whom 1690 were invited for screening; 538 attended Screen 1 (243 ineligible, 15 lost and 16 withdrawn), 264 attended for Screen 2 (7 excluded for SFA intake > 15% en, 10 excluded for dietary fiber intake >14 g/1000 kcal, 8 lost and 15 withdrawn) and 224 were randomized (17 withdrawn, 103 received Control and 104 received Test). Four (4) Control participants were withdrawn early for non-compliance (n = 3) or a serious adverse event (not related to treatment, n = 1) and 4 participants were withdrawn after completion for non-compliance (n = 2) or antibiotic use (n = 2) leaving n = 95 Control participants who completed the study per protocol. Similarly, 3 Test participants were withdrawn early for antibiotic use (n = 1) or scheduling difficulties (n = 2) and 5 were withdrawn after completion for non-compliance (n = 2), antibiotic use (n = 2) or insulin use (n = 1) leaving n = 96 Test participants who completed the study per protocol. The study flow chart and further details are given elsewhere [9].

The per-protocol population consisted of 72 males and 119 females of whom 102 were Caucasian, 39 South Asian, 14 African, 14 Hispanic, 8 East Asian, 6 South-east Asian, 4 West Asian, 3 mixed and 1 Indigenous; participant sex and ethnicity did not differ significantly on Test vs. Control. Participants were aged (mean $\pm$ SD) 47.6 $\pm$ 11.4 years, BMI 27.9 $\pm$ 4.6 kg/m^2 (BMI 25.0–<30, n = 191; BMI 30.0–<35, n = 132; BMI 35.0–<40, n = 43), total cholesterol 5.83 $\pm$ 0.70 mmol/L, triglycerides 1.46 $\pm$ 0.69 mmol/L, HDL-cholesterol 1.44 $\pm$ 0.41, LDL-cholesterol 3.73 $\pm$ 0.50, fasting glucose 4.95 $\pm$ 0.51 with no significant differences between Test and Control

Mean energy intake increased from baseline on Test and Control by approximately the amount of energy the sachets contained (Figure 1). Sugars intakes, increased to an equivalent extent on both treatments. Starch intake did not change significantly on Test but increased from baseline on Control. Dietary fiber intake increased during treatment on Test by approximately the amount in the Test sachets and the difference was significant compared both to baseline and to Control (Figure 1). Total fat and protein intakes increased from baseline on Control, but not Test; alcohol intake decreased from baseline on Test.

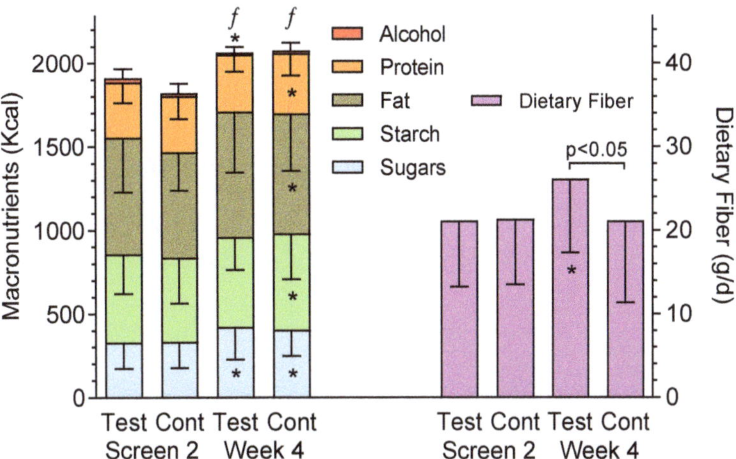

Figure 1. Macronutrient and dietary fiber intakes during the trial. Bars represent mean ± SEM intakes (expressed in Kcal for sugars, starch, fat, protein and alcohol) and grams for dietary fiber before treatment (dietary records obtained at the second screening visit; Screen 2) and during treatment (Week 4) for $n = 96$ on Test and $n = 95$ on Control (Cont). Error bars go down for sugars, starch, fat, protein and dietary fiber and up for alcohol. The only significant difference between Test and Control is for fiber. * Asterisks below the error bars indicate a significant difference from Screen 2 for the respective nutrient, except for alcohol the asterisk is above the error bar, ($p < 0.05$) by paired *t*-test. ƒ Significant difference in energy intake between Screen 2 and Week 4 ($p < 0.05$) by paired *t*-test.

3.1. Gastrointestinal (GI) Symptoms

The occurrence and severity of GI symptoms are presented in detail elsewhere [9]. One or more major GI symptom (flatulence, diarrhea, constipation, abdominal distention, and abdominal pain) was present in 36% of Test and 28% of Control participants at baseline (ns); the prevalence increased to 61% and 52%, respectively, at week 2 and fell to 47% and 42% by week 4 (differences between Test and Control were not significant). Furthermore, the sum of the scores for the 5 major GI symptoms did not differ between Test and Control, respectively, at any visit with mean ± SD as follows: screen 2, 0.95 ± 1.48 vs. 0.81 ± 1.55; baseline, 0.77 ± 1.39 vs. 0.53 ± 1.29; week 2, 1.67 ± 2.12 vs. 1.45 ± 2.09; week 4, 1.20 ± 1.83 vs. 0.91 ± 1.39.

3.2. Non-GI Symptoms

The number of subjects reporting symptom severity of none, mild, moderate or severe for each visit on Test and Control are shown in Supplementary Tables S2 and S3. The prevalence and severity of the 12 most common symptoms are shown in Figure 2.

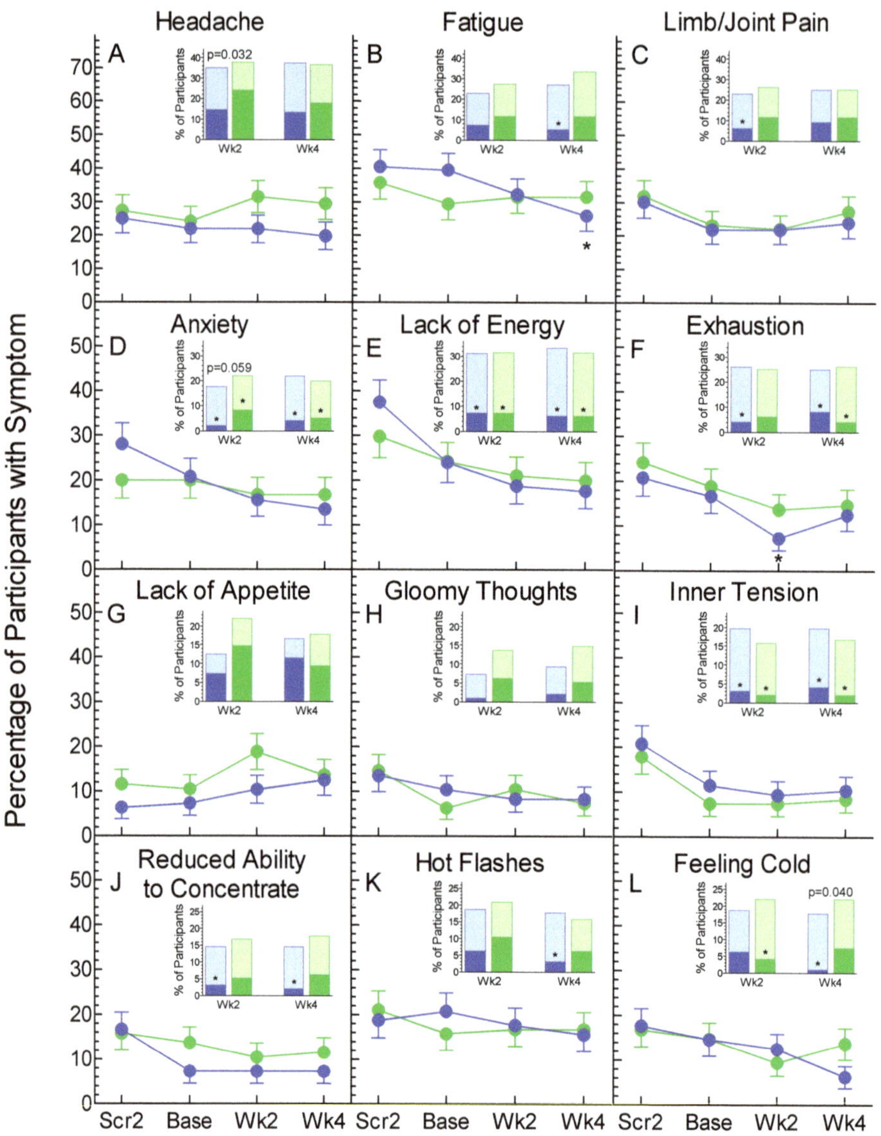

Figure 2. Prevalence and severity of headache (**A**), fatigue (**B**), limb/joint pain (**C**), anxiety (**D**), lack of energy (**E**), exhaustion (**F**), lack of appetite (**G**), gloomy thoughts (**H**), inner tension (**I**), reduced ability to concentrate (**J**), hot flashes (**K**) and feeling cold (**L**). Main panels: Values are means ± SD for $n = 96$ on Test (blue symbols and bars) and $n = 95$ on Control (green symbols and bars). Scr2 = second screening visit; Base = baseline; Week2 = week 2, Wk4 = week 4. * Prevalence significantly different from that at the Baseline visit by chi-squared test ($p < 0.05$). Insets: Percentage of participants in whom the symptom was more severe (dark bars) or less severe (light bars) than the severity before treatment (median severity at Scr2 and Base). The percent of participants with no change in severity is not shown. *p*-values are the significance of the difference between the ratio of M:L on Test vs. Control ($p < 0.05$ by Fisher's exact test). * Ratio of M:L significantly different from 1:1 ($p < 0.05$ by binomial distribution) where M and L are the number of participants in whom the symptom was more severe (M) or less severe (L).

The presence or severity of 7 of the symptoms at Baseline did not differ significantly by age, sex or BMI. The presence of Limb/Joint pain at Baseline differed by sex (females > males) and age (over 48 years > under 48 years) and the severity of Limb/Joint pain was greater in obese compared to lean or overweight subjects. The presence of fatigue and hot flashes were affected by sex (females > males) and age (over 48 years > under 48 years). The severity of lack of energy was greater in females than males, whereas feeling cold was more common in males than females (Supplementary Table S4).

The prevalence of symptoms did not differ significantly between Test and Control for any symptom at any time. However, on Test the prevalence of fatigue was lower at week 4 and exhaustion at week 2 compared to baseline (Figure 2B,F). The change in severity of symptoms did not differ significantly between Test and Control except for headache and anxiety (less severe on Test vs. Control at week 2) and feeling cold at week 4 (insets on Figure 2A,D,L). Compared to before treatment (median of scores at the Screen 2 and Baseline visits), symptoms were less severe as follows: fatigue (Test at week 4), lack of energy (both Test and Control at week 2 and week 4), exhaustion (Test at week 2 and week 4 and Control at week 4), limb/joint pain (Test at week 2), hot flashes (Test at week 4), feeling cold (Control at week 2 and Test and week 4); anxiety and inner tension (Test and Control at week 2 and week 4); and reduced ability to concentrate (Test at week 2 and week 4) (Figure 2). If the severity of symptoms at during treatment is compared to that at the Baseline visit alone (Supplementary Table S5), fewer participants have more severe or less severe symptoms, but the direction of the effects are similar to those shown in Figure 2.

The presence of symptoms at baseline was associated with increased serum CRP (mg/L) (median [25%, 25%]) as follows (absence vs. presence, respectively): lack of appetite, 1.31 [0.66, 3.36], $n = 167$ vs. 2.52 [1.48, 4.15] $n = 24$, $p = 0.013$; exhaustion, 1.30 [0.65, 3.40], $n = 138$ vs. 2.12 [1.07, 3.91], $n = 53$, $p = 0.005$; and limb/joint pain, 1.30 [0.65, 3.03], $n = 127$ vs. 2.11 [1.08, 4.12], $n = 64$, $p = 0.009$. Although the presence of headache at baseline was not associated with higher serum CRP, in participants whose CRP did not increase during the study the severity of headache was greater on Control than Test ($p = 0.049$; Figure 3A). The severity of exhaustion was greater in participants whose CRP increased vs. those in whom it did not, however, changes in CRP did not affect the difference between Test and Control (Figure 3B).

Serum oxLDL was not associated with the presence of any symptom at baseline. The severity of fatigue was similar for participants in whom oxLDL increased during treatment compared to those in whom it did not. However, in participants whose oxLDL did not increase, fatigue was more severe on Control than Test ($p = 0.052$, Figure 3C). The severity of hot flashes was greater in participants in whom oxLDL increased compared to those in whom it did not ($p = 0.049$), but the severity of hot flashes on Test was similar to Control regardless of the change in oxLDL (Figure 4D).

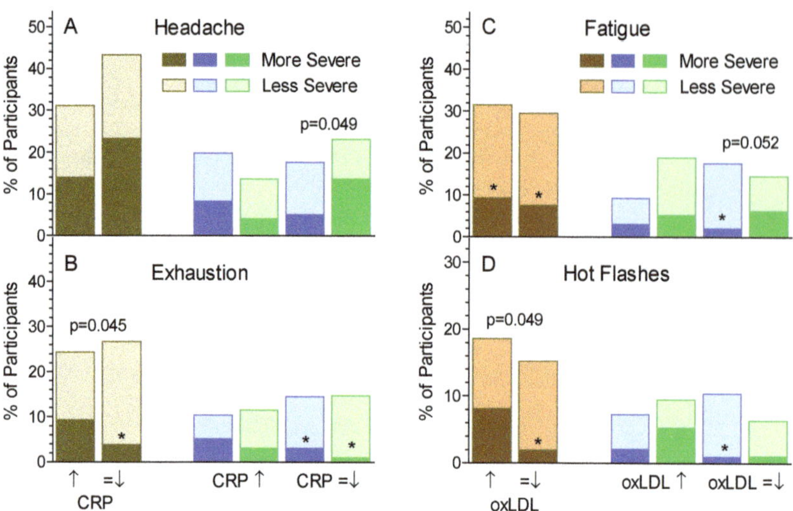

Figure 3. Effect of changes in serum CRP and oxLDL on symptom severity. Percent of participants whose symptoms became more severe (dark bars) or less severe (light bars) during treatment for those whose serum CRP (panels **A** and **B**) or oxLDL (panels **C** and **D**) increased (↑) or did not increase (=↓) after 4 weeks treatment. The percent of participants with no change in severity is not shown. Brown/orange bars are for all participants, blue bars for Test and green bars for Control. *p*-values over 2 bars are the significance of the difference between the ratios of More:Less severe for the 2 bars (e.g., panel **A**, in participants whose CRP did not increase after 4 weeks treatment, Headache was significantly more severe in those on Control vs. Test *p* = 0.049). * Ratio of more:less severe differs significantly from 1:1 by chi-square test (*p* < 0.05).

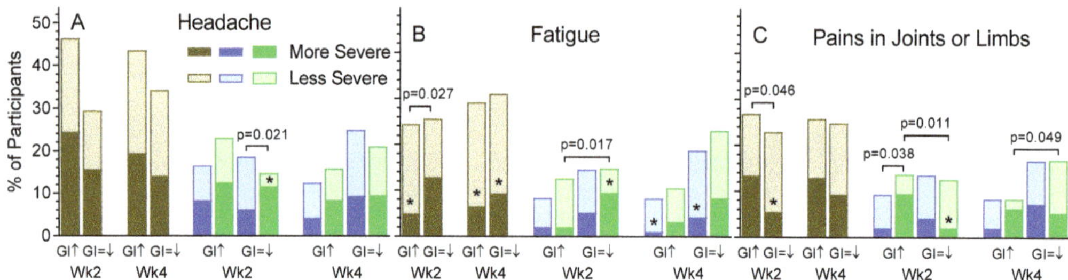

Figure 4. Effect of changes in GI symptoms on the severity of the non-GI symptoms headache (**A**), fatigue (**B**) and pains in joints or limbs (**C**). Bars show the percent of participants whose non-GI symptoms became more severe (dark bars) or less severe (light bars) during treatment for those whose GI symptoms (sum of scores for flatulence, diarrhea, constipation, abdominal distention, and abdominal pain) became more severe (GI↑) or not (GI = ↓) during treatment. The percent of participants with no change in severity is not shown. Brown bars show all participants, blue bars Test and green bars Control. *p*-values are the significance of the difference between the ratios of More:Less severe for the bars indicated (e.g., panel **A**, in participants whose GI symptoms did not become more severe after 2 weeks treatment, Headache was more severe on Control vs. Test *p* = 0.021). * Ratio of more:less severe differs significantly from 1:1 by chi-square test (*p* < 0.05).

At baseline, 11 of the 12 non-GI symptoms were significantly more common at in the *n* = 88 participants with GI symptoms than in the *n* = 103 without GI symptoms (Table 1). Additionally, changes in the severity of GI symptoms during the trial were associated with differences in the severity of several non-GI symptoms. The overall severity of headache was not associated with the severity of GI symptoms at either 2 weeks or 4 weeks. However, at 2 weeks, headache was less severe on Test vs. Control in participants with no increase in the severity of GI symptoms (*p* = 0.021, Figure 4A). The overall severity of fatigue at

2 weeks was lower in participants whose GI symptoms became more severe compared to those in whom they did not ($p = 0.027$). The same was true for participants on the Control treatment ($p = 0.017$). However, the severity of Fatigue on Test treatment was not associated with differences in the severity of GI symptoms (Figure 4B). The overall severity of joint/limb pain was greater at 2 weeks in participants with more severe vs. the same or less severe GI symptoms ($p = 0.046$). The same was true for Control participants at 2 weeks ($p = 0.011$) and 4 weeks ($p = 0.049$) but not for Test participants (ns). At 2 weeks, the severity of joint/limb pain was lower on Test vs. Control in participants with more severe GI symptoms than at baseline ($p = 0.038$, Figure 4C).

Table 1. Association of non-GI symptoms with major GI symptoms.

Non-GI Symptom	Major GI Symptom		p
	Absent ($n = 103$)	Present ($n = 88$)	
Headache	16 (16)	28 (32)	0.008
Fatigue	20 (19)	46 (52)	<0.001
Pains in joints or limbs	8 (8)	35 (40)	<0.001
Feelings of anxiety	12 (12)	27 (31)	0.001
Lack of energy	7 (7)	39 (44)	<0.001
Tend to become exhausted	8 (8)	26 (30)	<0.001
Lack of appetite	6 (6)	11 (13)	0.11
Gloomy thoughts	3 (3)	13 (15)	0.003
Inner tension	4 (4)	14 (16)	0.005
Diminished ability to concentrate	2 (2)	18 (20)	<0.001
Hot flashes	10 (10)	25 (28)	<0.001
Feeling cold	5 (5)	23 (26)	<0.001

Values are number (%) of participants with each non-GI symptom at baseline amongst the participants without any major GI symptom (flatulence, diarrhea, constipation, abdominal distension and abdominal pain) compared to those with one or more major GI symptom at screening or baseline. The right-most column shows the significance of the differences by chi-squared test.

4. Discussion

The present results provide hypothesis-generating evidence that the OBG-enriched oat product tested may influence several affective and physical feeling states in healthy adults with LDL cholesterol between 3 and 5 mmol/L. Furthermore, markers of systemic inflammation and oxidative stress and the severity of GI symptoms modified some of these effects. These findings are consistent with current concepts about how alterations in the gut-brain axis (increased colonic fermentation and alteration of the gut microbiota) influence metabolism, behavior and brain function by a variety of different mechanisms including effects on afferent pathways between the gut and the brain, short-chain fatty acid (SCFA) production and other microbial by-products and metabolites which have local and systemic effects on gut hormones, oxidative stress and inflammation [17,25].

The association between the presence of GI symptoms and the increased prevalence of 11 of the 12 non-GI symptoms is not a new finding and could be ascribed to psychological or physiological factors. For example, it is known that the perception of GI symptoms such as constipation and flatulence are not related to the amount of gas in the colon [26] and are associated with personality and anxiety [27]. Thus, participants who noticed GI symptoms may have been, in general, more aware of other abnormal sensations and feelings they experienced and were more able to remember to report them. However, the association between GI and non-GI symptoms could have a physiological basis via the gut-brain axis. There is abundant evidence from studies in vitro and in animal models that oats and OBG influence SCFA production and the colonic microbiome [14], although the results are inconsistent and there are few studies in humans. OBG-enriched oat bran containing 20 g/d dietary fiber and 10 g/d OBG did not increase fecal SCFA after 4 weeks but did so at 8 and 12 weeks in healthy subjects [28]. Furthermore, fecal SCFA were not increased by either granola containing 6 g/d fiber and 3 g/d OBG daily for 5 weeks [29] or by 60 g oatmeal containing 8.5 g dietary fiber and 4.7 g OBG daily for 1 week [30]. However, the granola increased fecal bifidobacteria and lactobacilli [29] and the oatmeal influenced bacterial metabolism as judged by reduce β-galactosidase and urease concentrations [30]. A recent study showed, in hypercholesterolemic subjects that compared to 80 g white rice, 80 g oatmeal daily for 45 days increased the abundance of fecal Firmicutes [31]. The effect of OBG on the fecal microbiome might also by its molecular weight, since 3 g high molecular weight (1349 kDa) barley β-glucan for 5 weeks significantly increased fecal Bacteriodetes and reduced Firmicutes compared to a β-glucan-free control and to 3 g low molecular weight (288 kDa) barley β-glucan [32]. Thus, there is evidence in humans that the amount of OBG we provided could influence the fecal microbiome, but not fecal SCFA. However, the composition of the fecal microbiome may not reflect that of the mucosa-associated micro-organisms [33]. In addition, there is evidence in humans that increased SCFA production can occur in the absence of a change in fecal SCFA [34] and that a high fecal acetate concentration may reflect reduced acetate absorption rather than increased production [35].

The severity of headache was lower on Test than Control at 2 weeks in all subjects (Figure 2A) and at 4 weeks in those in whom serum CRP did not increase (Figure 3A). The causes of headaches are not well understood, but may include hypoglycemia [36], obesity [37] or inflammation [25]. However, the $n = 67$ participants with headache at baseline had similar (mean $\pm$ SD) BMI, 28.08 $\pm$ 4.70 kg/m^2, and median CRP, 1.52 mg/L, as the $n = 127$ without headache, 27.89 $\pm$ 4.58 kg/m^2, and 1.53 mg/L.

Fatigue became less common and less severe than at baseline after 4 weeks treatment with OBG, as did exhaustion after 2 weeks, but the differences from control were not significant. Lack of energy was less severe on both Test and Control treatments compared to baseline. These results are in line with studies showing that OBG [21] and konjac oligosaccharide [38] increased maximum exercise time in rats or mice and altered metabolites suggestive of reduced fatigue. In humans, consuming cereals high in wheat-fiber for 2 weeks reduced fatigue compared to control in healthy adults [19] and 13 weeks treatment with a prebiotic containing inulin and fructo-oligosaccharides reduced exhaustion in elderly subjects when compared to control [39]. Fatigue is associated with lack of sleep, an association that may be mediated by increased inflammation [40]. At baseline, serum CRP was similar in participants with and without fatigue (not shown) and with or without lack of energy (not shown), but the presence of exhaustion at baseline was associated with a higher CRP. Furthermore, the severity of exhaustion increased from baseline to a greater extent in participants whose CRP increased compared to those in whom it did not (Figure 3B).

Pains in muscles can be due to overuse or to reduced circulation, while pain in joints is most commonly due to intervertebral disc herniation or osteoarthritis, both of which are associated with obesity, chronic inflammation [41–43] and older age. This is consistent with our data in that the $n = 64$ participants with limb/joint pain at baseline were older (mean $\pm$ SD, 52.1 $\pm$ 10.3 vs. 45.3 $\pm$ 11.3 years, $p < 0.001$), and had higher BMI (29.1 $\pm$ 4.5 vs. 27.3 $\pm$ 4.6 kg/m^2, $p = 0.013$), and higher CRP (2.11 [1.08, 4.12] vs. 1.30 [0.65, 3.03] mg/L, $p = 0.009$) than the $n = 127$ without limb/joint pain. After 2 weeks treatment, limb/joint pain was less severe than at baseline in all subjects on OBG (Figure 2C), and, in participants with more severe GI symptoms, less severe on OGB compared to control at 2 weeks (Figure 4C). On Control, but not OBG, having more severe GI symptoms was associated with more severe limb/joint pain at 2 and 4 weeks (Figure 4C). These observations, along with the increased prevalence of limb/joint pain in those with GI symptoms (Table 1) suggest a link between the gut and joint/limb pain. There is some evidence that people with rheumatoid arthritis (RA) have different colonic microbiota and that probiotic treatment can improve the condition in the absence of changes in CRP [44] but this may not apply to our study population, most of whom likely did not have RA.

During the study, the severity of anxiety decreased from baseline on both treatments, with the reduction on OBG being greater than that on Control at 2 weeks ($p = 0.059$, Figure 2D). A link between the colonic microbiota and anxiety is suggested by a study in which a probiotic formulation containing *Lactobacillus helveticus* R0052 and *Bifidobacterium longum* R0175, given to male Wistar rats for 2 weeks, reduced anxiety-like behavior, and given to humans for 30 d reduced psychological distress, the Hospital Anxiety and Depression Scale and urinary free-cortisol excretion [45]. Additionally, compared to control, 5.5 g/d galactooligosaccharides for 3 weeks, reduced the salivary cortisol waking response and increased the processing of positive vs. negative attentional vigilance in healthy subjects, although these effects were not seen with fructooligosaccharides [46].

The prevalence of feeling cold at 4 weeks did not change from baseline on either treatment, but the severity of feeling cold at 4 weeks was less on Test than Control (Figure 2L). Consistent with several studies in which feeling cold was associated with musculoskeletal pain [47–49], we found that feeling cold was present in 41.9% of 43 participants who had joint/limb pain at baseline but only 6.8% of 148 participants who did not ($p < 0.001$). Furthermore, after 4 weeks, the severity of feeling cold had increased from baseline in 20% of the 20 participants who experienced more severe joint/limb pain compared to only 2.3% of the 171 in who did not ($p < 0.001$). However, it is not clear why the prevalence and severity of feeling cold would be less after 4 weeks on OBG vs. Control.

Hot flashes are a common symptom in post-menopausal women. In this study, 93.5% of the 46 participants with hot flashes at baseline were women ($p < 0.001$). Furthermore, although the mean $\pm$ SD age of women with and without hot flashes at baseline were similar, 52.5 ± 9.8 vs. 49.6 ± 11.4 year ($p = 0.16$), a higher percentage of $n = 76$ women aged > 48 year had hot flashes compared to the $n = 43$ aged ≤ 48 year, 46 vs. 19%, $p = 0.003$. Increased oxidative stress is associated with a greater severity of hot flashes [50], but it is not clear if menopausal symptoms are a cause or an effect of oxidative stress [51]. In this study the concentration of oxLDL at baseline (median {95%CI}) did not differ in the 46 participants with hot flashes at baseline, 63 {37, 124} µg/L, compared to the 145 without, 78 {45, 144} µg/L. However, it is of interest that the severity of hot flashes at week 4 compared to baseline increased more in participants in whom oxLDL increased compared to those in whom it did not (Figure 3D).

The major weakness of this assessment of non-GI symptoms is that the study was designed to assess the effect of OBG on serum cholesterol; assessment of symptoms was a tertiary objective not listed in the study registration. Several hundred statistical analyses were performed post hoc with no correction for multiple comparisons; this increases the likelihood of making type 1 errors. Furthermore, validated questionnaires were not used to assess the symptoms. Nevertheless, there was significant amount of internal and external consistency of the results with what might be expected from the literature.

5. Conclusions

These results provide hypothesis-generating evidence that OBG may have a beneficial effect on several affective and physical feeling states in healthy adults. Since there is a paucity of information about the effects of dietary fiber in general, and oats and OBG specifically, on non-GI symptoms in humans, these results provide information which may be useful for designing the studies which would be required to confirm these observations.

Supplementary Materials: The following are available online at https://www.mdpi.com/2072-6643/13/5/1534/s1, Inclusion/exclusion criteria; Table S1: Nutritional composition of intervention products (per sachet); Tables S2 and S3: Occurrence of non-GI symptoms on the symptoms questionnaire; Table S4: Effect of age, sex and BMI on symptoms at Baseline; Table S5: Severity of symptoms when compared to severity at the Baseline visit.

Author Contributions: Conceptualization, T.M.S.W., M.R. and Y.C.; methodology, T.M.S.W.; formal analysis, T.M.S.W.; investigation, A.L.J., A.E. and J.E.C.; resources, M.R., E.H.D. and Y.C.; data curation, A.E. and J.E.C.; writing—original draft preparation, T.M.S.W.; writing—review and editing,

M.R., E.H.D., A.L.J., A.E., J.E.C. and Y.C.; visualization, T.M.S.W.; supervision, T.M.S.W.; project administration, A.L.J., A.E. and J.E.C. All authors have read and agreed to the published version of the manuscript.

Funding: This research was funded by PepsiCo.

Institutional Review Board Statement: The study was conducted according to the guidelines of the Declaration of Helsinki, and approved by the WIRB® (protocol #20190694, 28 March 2019).

Informed Consent Statement: Informed consent was obtained from all subjects involved in the study.

Data Availability Statement: The data presented in this study are available upon request from the corresponding author.

Acknowledgments: The authors wish to thank Kirstin Harris (PepsiCo) and Alison Kamil (PepsiCo) for support in the early stages of the project.

Conflicts of Interest: T.M.S.W. and A.L.J. are employees and part owners of INQUIS; A.E. and J.E.C. are employees of INQUIS; M.R. was an employee at PepsiCo during the clinical and data analysis phase; E.D. and Y.C. are employees of PepsiCo. INQUIS, T.M.S.W., A.E., J.E.C. and A.L.J. do not own any intellectual property which may arise from this study. T.M.S.W., A.E., J.E.C. and A.L.J. have not received any payments directly from PepsiCo and have no financial interest in PepsiCo. The funder assisted in designing the study but had no role in the collection or analysis of the data. The funder decided to publish these results and commented on the manuscript drafts; however, T.M.S.W. had sole responsibility for determining its final content.

Disclaimer: The views expressed in this manuscript are those of the authors and do not necessarily reflect the position or policy of PepsiCo, Inc.

References

1. Healthy Diet; Fact Sheet No. 394, Updated August 2018. World Health Organization. Available online: http://www.who.int/mediacentre/factsheets/fs394?en/ (accessed on 20 November 2020).
2. Wood, P.J. Oat β-glucan: Structure, location and properties. In *Oats: Chemistry and Technology*; Webster, F.H., Ed.; AACC Inc.: St. Paul, MN, USA, 1986; pp. 121–152.
3. Whitehead, A.; Beck, E.J.; Tosh, S.; Wolever, T.M.S. Cholesterol-lowering effects of oat β-glucan: A meta-analysis of randomized controlled trials. *Am. J. Clin. Nutr.* **2014**, *100*, 1413–1421. [CrossRef]
4. Ho, H.V.T.; Sievenpiper, J.L.; Zurbau, A.; Mejia, S.B.; Jovanovski, E.; Au-Yeung, F.; Jenkins, A.L.; Vuksan, V. The effect of oat β-glucan on LDL-cholesterol, non-HDL-cholesterol and apoB for CVD risk reduction: A systematic review and meta-analysis of randomized-controlled trials. *Br. J. Nutr.* **2016**, *116*, 1369–1382. [CrossRef] [PubMed]
5. Zurbau, A.; Noronha, J.; Khan, T.A.; Sievenpiper, J.L.; Wolever, T.M.S. The effect of oat β-glucan on postprandial blood glucose and insulin responses: A systematic review and meta-analysis. *Eur. J. Clin. Nutr.* **2021**. published on line before print. [CrossRef] [PubMed]
6. Delahoy, P.J.; Magliano, D.J.; Webb, K.; Grobler, M.; Liew, D. The relationship between reduction in low-density lipoprotein cholesterol by statins and reduction in risk of cardiovascular outcomes: An updated metaanalysis. *Clin. Ther.* **2009**, *31*, 236–244. [CrossRef] [PubMed]
7. Livesey, G.; Taylor, R.; Livesey, H.F.; Buyken, A.E.; Jenkins, D.J.A.; Augustin, L.S.A.; Sievenpiper, J.L.; Barclay, A.W.; Liu, S.; Wolever, T.M.S.; et al. Dietary glycemic index and load and the risk of type 2 diabetes: A systematic review and updated meta-analysis of prospective cohort studies. *Nutrients* **2019**, *11*, 1280. [CrossRef]
8. Grabistske, H.A.; Slavin, J.L. Gastrointestinal effects of low-digestible carbohydrates. *Crit. Rev. Food Sci. Nutr.* **2009**, *49*, 327–360. [CrossRef]
9. Wolever, T.M.S.; Rahn, M.; Dioum, E.; Spruill, S.E.; Ezatagha, A.; Campbell, J.E.; Jenkins, A.L.; Chu, Y. Impact of an oat β-glucan beverage on LDL-cholesterol and cardiovascular risk in men and women with borderline high cholesterol: A double-blind, randomized, controlled clinical-trial. *J. Nutr.* **2021**, in press.
10. Kaukinen, K.; Collin, P.; Huhtala, H.; Mäki, M. Long-term consumption of oats in adult celiac disease patients. *Nutrients* **2013**, *5*, 4380–4389. [CrossRef]
11. Ford, A.C.; Moayyedi, P.; Lacy, B.E.; Lembo, A.J.; Saito, Y.A.; Schiller, L.R.; Soffer, E.E.; Spiegel, B.M.; Quigley, E.M. American College of Gastroenterology monograph on the management of irritable bowel syndrome and chronic idiopathic constipation. *Am. J. Gastroenterol.* **2014**, *109* (Suppl. 1), S2–S26. [CrossRef] [PubMed]
12. Humes, D.; Smith, J.K.; Spiller, R.C. Colonic diverticular disease. *BMJ Clin. Evid.* **2011**, *2011*, 0405.
13. Cumming, J.H. The effect of dietary fiber on fecal weight and composition. In *CRC Handbook of Dietary Fiber in Human Nutrition*, 3rd ed.; Spiller, G.A., Ed.; CRC Press: Roca Raton, FL, USA, 1985; pp. 183–252.

14. Korczak, R.; Kocher, M.; Swanson, K.S. Effects of oats on gastrointestinal health as assessed by in vitro, animal and human studies. *Nutr. Rev.* **2020**, *78*, 343–363. [CrossRef]

15. Swann, O.G.; Kilpatrick, M.; Breslin, M.; Oddy, W.H. Dietary fiber and its associations with depression and inflammation. *Nutr. Rev.* **2020**, *78*, 394–411. [CrossRef]

16. Luca, M.; Di Mauro, M.; Di Maruo, M.; Luca, A. Gut microbiota in Alzheimer's disease, depression, and type 2 diabetes mellitus: The role of oxidative stress. *Oxid. Med. Cell Longev.* **2019**, *2019*, 4730539. [CrossRef] [PubMed]

17. Morais, L.H.; Schrieber, H.L.; Mazmanian, S.K. The gut microbiota-brain axis in behaviour and brain disorders. *Nat. Rev. Microbiol.* **2021**, *19*, 241–255. [CrossRef]

18. O'Reilly, G.A.; Huh, J.; Schembre, S.M.; Tate, E.B.; Pentz, M.A.; Dunton, G. Association of usual self-reported dietary intake with ecological momentary measures of affective and physical feeling states in children. *Appetite* **2015**, *92*, 314–321. [CrossRef]

19. Smith, A.; Bazzoni, C.; Beale, J.; Elliott-Smith, J.; Tiley, M. High fibre breakfast cereals reduce fatigue. *Appetite* **2001**, *37*, 249–250. [CrossRef]

20. Wolever, T.M.S.; Chiasson, J.-L.; Josse, R.G.; Leiter, L.A.; Maheux, P.; Rabasa-Lhoret, R.; Rodger, N.W.; Ryan, E.A. Effects of changing the amount and source of dietary carbohydrates on symptoms and dietary satisfaction over a 1-year period in subjects with type 2 diabetes: Canadian trial of carbohydrates in diabetes (CCD). *Can. J. Diab.* **2017**, *41*, 164–176. [CrossRef] [PubMed]

21. Xu, C.; Lv, J.; Lo, M.; Cui, S.W.; Hu, X.; Fan, M. Effects of oat β-glucan on endurance exercise and its anti-fatigue properties in trained rats. *Carbohydr. Polym.* **2013**, *92*, 1159–1165. [CrossRef]

22. Chiasson, J.-L.; Josse, R.G.; Hunt, J.A.; Palmason, C.; Rodger, N.W.; Ross, S.A.; Ryan, E.A.; Tan, M.H.; Wolever, T.M.S. The efficacy of acarbose in the treatment of patients with non-insulin-dependent diabetes mellitus. A multicenter controlled clinical trial. *Ann. Int. Med.* **1994**, *121*, 928–935. [CrossRef] [PubMed]

23. Wolever, T.M.S.; Chiasson, J.-L.; Josse, R.G.; Hunt, J.A.; Palmason, C.; Rodger, N.W.; Ross, S.A.; Ryan, E.A.; Tan, M.H. No relationship between carbohydrate intake and effect of acarbose on HbA1c or gastrointestinal symptoms in type 2 diabetic subjects consuming 30–60% of energy from carbohydrate. *Diabetes Care* **1998**, *21*, 1612–1618. [CrossRef]

24. Wolever, T.M.S.; Tosh, S.M.; Gibbs, A.L.; Brand-Miller, J.; Duncan, A.M.; Hart, V.; Lamarche, B.; Thomson, B.A.; Duss, R.; Wood, P.J. Physicochemical properties of oat β-glucan influence its ability to reduce serum LDL cholesterol in humans: A randomized clinical trial. *Am. J. Clin. Nutr.* **2010**, *92*, 723–732. [CrossRef]

25. Hindiyeh, N.; Aurora, S.K. What the gut can teach us about migrane. *Curr. Pain Headache Rep.* **2015**, *19*, 33. [CrossRef]

26. Lasser, R.B.; Bond, J.H.; Levitt, M.D. The role of intestinal gas in functional abdominal pain. *N. Eng. J. Med.* **1975**, *293*, 524–526. [CrossRef] [PubMed]

27. Devroede, G.; Girard, G.; Bouchoucha, M.; Roy, T.; Black, R.; Camerlain, M.; Pinard, G.; Schang, J.-C.; Arhan, P. Idiopathic constipation by colonic dysfunction: Relationship with personality and anxiety. *Dig. Dis. Sci.* **1989**, *34*, 1428–1433. [CrossRef] [PubMed]

28. Nilsson, U.; Johansson, M.; Nilsson, A.; Björk, I.; Nyman, M. Dietary supplementation with β-glucan enriched oat bran increases faecal concentration of carboxylic acids in healthy subjects. *Eur. J. Clin. Nutr.* **2008**, *62*, 978–984. [CrossRef] [PubMed]

29. Connolly, M.L.; Tzounis, X.; Tuohy, K.M.; Lovegrove, J.A. Hypocholesterolemic and probiotic effects of a whole-grain oat-based granola breakfast cereal in a cardio-metabolic "at risk" population. *Front. Microbiol.* **2016**, *7*, 1675. [CrossRef]

30. Valeur, J.; Puaschitz, N.G.; Midtvedt, T.; Berstad, A. Oatmeal porridge: Impact on microflora-associated characteristics in healthy subjects. *Brit. J. Nutr.* **2016**, *115*, 62–67. [CrossRef] [PubMed]

31. Ye, M.; Sun, J.; Chen, Y.; Ren, Q.; Li, Z.; Zhao, Y.; Pan, Y.; Xue, H. Oatmeal induced gut microbiota alteration and its relationship with improved lipid profiles: A secondary analysis of a randomized clinical trial. *Nutr. Metab. (Lond.)* **2020**, *17*, 85. [CrossRef] [PubMed]

32. Wang, Y.; Ames, N.P.; Tun, H.M.; Tosh, S.M.; Jones, P.J.; Khafipour, E. High molecular weight barley β-glucan alters gut microbiota toward reduced cardiovascular disease risk. *Front. Micriobiol.* **2016**, *7*, 129. [CrossRef]

33. Zoetendal, E.G.; von Wright, A.; Vilpponen-Salmela, T.; Ban-Amor, K.; Akkermans, A.D.L.; de Vos, W.M. Mucosa-associated bacteria in the human gastrointestinal tract are uniformly distributed along the colon and different from the community recovered from feces. *Appl. Environ. Microbiol.* **2002**, *68*, 3401–3407. [CrossRef] [PubMed]

34. Vogt, J.A.; Ishii-Schrade, K.B.; Pencharz, P.B.; Wolever, T.M.S. L-Ramnose increases serum propionate after long-term supplementation, but lactulose does not raise serum acetate. *Am. J. Clin. Nutr.* **2004**, *80*, 1254–1261. [CrossRef] [PubMed]

35. Vogt, J.A.; Wolever, T.M.S. Fecal acetate is inversely related to acetate absorption from the human rectum and distal colon. *J. Nutr.* **2003**, *133*, 3145–3148. [CrossRef]

36. Torelli, P.; Manzoni, G.C. Fasting headache. *Curr. Pain Headache Rep.* **2010**, *14*, 284–291. [CrossRef] [PubMed]

37. Bigal, M.E.; Rapoport, A.M. Obesity and chronic daily headache. *Curr. Pain Headache Rep.* **2010**, *16*, 101–109. [CrossRef] [PubMed]

38. Zeng, Y.; Zhang, J.; Zhang, Y.; Men, Y.; Zhang, B.; Sun, Y. Prebiotic, immunomodulating and anti-fatigue effects of konjac oligosaccharide. *J. Food Sci.* **2018**, *83*, 3110–3117. [CrossRef]

39. Buigues, C.; Fernández-Garrido, J.; Pruimboom, L.; Hoogland, A.J.; Navarro-Martínez, R.; Martínez-Martínez, M.; Verdejo, Y.; Mascarós, M.C.; Peris, C.; Cauli, O. Effect of a prebiotic formulation on frailty syndrome: A randomized, double-blind clinical trial. *Int. J. Mol. Sci.* **2016**, *17*, 932. [CrossRef]

40. Thomas, K.S.; Motivala, S.; Olmstead, R.; Irwin, M.R. Sleep depth and fatigue: Role of cellular inflammatory activation. *Brain Behav. Immun.* **2011**, *25*, 53–58. [CrossRef]

41. Lean, M.E.J.; Han, T.S.; Seidell, J.C. Impairment of health and quality of life in people with large waist circumference. *Lancet* **1998**, *351*, 853–856. [CrossRef]

42. Schaible, H.G. Spinal mechanisms contributing to joint pain. *Novartis Found. Symp.* **2004**, *260*, 4–22.

43. Sanghi, D.; Mishra, A.; Sharma, A.C.; Raj, S.; Mishra, R.; Natu, S.M.; Agarwal, S.; Srivastava, R.N. Elucidation of dietary risk factors in osteoarthritis knee: A case-control study. *J. Am. Coll. Nutr.* **2015**, *34*, 15–20. [CrossRef]

44. Bravo-Blas, A.; Wessel, H.; Milling, S. Microbiota and arthritis: Correlations or cause? *Curr. Opin. Rheumatol.* **2016**, *28*, 161–167. [CrossRef] [PubMed]

45. Messaoudi, M.; Lalonde, R.; Violle, N.; Javelot, H.; Desor, D.; Najdi, A.; Bisson, J.; Rougeot, C.; Pichelin, M.; Cazaubiel, M.; et al. Assessment of psychotropic-like properties of a probiotic formulation (Lactobacillus helveticus R0052 and Bifidobacterium longum R0175) in rats and human subjects. *Brit. J. Nutr.* **2011**, *105*, 755–764. [CrossRef] [PubMed]

46. Schmidt, K.; Cowen, P.J.; Harmer, C.H.; Tzortzis, G.; Errington, S.; Burnet, P.W.J. Prebiotic intake reduces the waking cortisol response and alters emotional bias in healthy volunteers. *Psychoparmacology* **2015**, *232*, 1793–1801. [CrossRef] [PubMed]

47. Bang, B.E.; Aasmoe, L.; Aardal, L.; Andorsen, G.S.; Bjørnbakk, A.K.; Egeness, C.; Espejord, I.; Kramvik, E. Feeling cold at work increases the risk of symptoms from muscles, skin and airways in seafood industry workers. *Am. J. Ind. Med.* **2005**, *47*, 65–71. [CrossRef] [PubMed]

48. Sormunen, E.; Remes, J.; Hassi, J.; Pienimäki, T.; Rintamäki, H. Factors associated with self-estimated work ability and musculoskeletal symptoms among male and female workers in cooled food-processing facilities. *Ind. Health* **2009**, *47*, 271–282. [CrossRef]

49. Farbu, E.H.; Skandfer, M.; Nielsen, C.; Brenn, T.; Stubhaug, A.; Höper, A.C. Working in a cold environment, feeling cold at work and chronic pain: A cross-sectional analysis of the Tromsø study. *BMJ Open* **2019**, *9*, e031248. [CrossRef]

50. Sánchez-Rodríguez, M.A.; Zacarías-Flores, M.; Arronte-Rosales, A.; Mendoza-Núñez, V.M. Association between hot flashes severity and oxidative stress among Mexican postmenopausal women: A cross-sectional study. *PLoS ONE* **2019**, *14*, e0214264. [CrossRef] [PubMed]

51. Cervellati, C.; Bergamini, C.M. Oxidative damage and the pathogenesis of menopause related disturbances and diseases. *Clin. Chem. Lab. Med.* **2016**, *54*, 739–753. [CrossRef]

Article

Clinical Outcomes after Oat Beta-Glucans Dietary Treatment in Gastritis Patients

Sylwia Gudej [1,*], Rafał Filip [2], Joanna Harasym [3,4,*], Jacek Wilczak [5], Katarzyna Dziendzikowska [1], Michał Oczkowski [1], Małgorzata Jałosińska [6], Małgorzata Juszczak [7], Ewa Lange [1] and Joanna Gromadzka-Ostrowska [1]

[1] Department of Dietetics, Institute of Human Nutrition Sciences, Warsaw University of Life Sciences (SGGW-WULS), Nowoursynowska 159c, 02-776 Warsaw, Poland; katarzyna_dziendzikowska@sggw.edu.pl (K.D.); michal_oczkowski@sggw.edu.pl (M.O.); ewa_lange@sggw.edu.pl (E.L.); joanna_gromadzka_ostrowska@sggw.edu.pl (J.G.-O.)

[2] Department of Gastroenterology with IBD Unit, Faculty of Medicine, University of Rzeszow, Kopisto 2A Str., 35-315 Rzeszow, Poland; r.s.filip@wp.pl

[3] Adaptive Food Systems Accelerator—Research Centre, Wroclaw University of Economics and Business, Komandorska 118/120, 53-345 Wrocław, Poland

[4] Department of Biotechnology and Food Analysis, Wroclaw University of Economics and Business, Komandorska 118/120, 53-345 Wrocław, Poland

[5] Department of Physiological Sciences, Institute of Veterinary Medicine, Warsaw University of Life Sciences (SGGW-WULS), Nowoursynowska 159, 02-776 Warsaw, Poland; jacek_wilczak@sggw.edu.pl

[6] Department of Food Gastronomy and Food Hygiene, Institute of Human Nutrition Sciences, Warsaw University of Life Sciences (SGGW-WULS), Nowoursynowska 159c, 02-776 Warsaw, Poland; malgorzata_jalosinska@sggw.edu.pl

[7] Department of Medical Biology, Institute of Rural Health, Jaczewskiego 2, 20-090 Lublin, Poland; juszczak.malgorzata@imw.lublin.pl

* Correspondence: sylwia.gudej@imid.med.pl (S.G.); joanna.harasym@ue.wroc.pl (J.H.)

Citation: Gudej, S.; Filip, R.; Harasym, J.; Wilczak, J.; Dziendzikowska, K.; Oczkowski, M.; Jałosińska, M.; Juszczak, M.; Lange, E.; Gromadzka-Ostrowska, J. Clinical Outcomes after Oat Beta-Glucans Dietary Treatment in Gastritis Patients. *Nutrients* **2021**, *13*, 2791. https://doi.org/10.3390/nu13082791

Academic Editors: Seiichiro Aoe, Tatsuya Morita and Naohito Ohno

Received: 7 July 2021
Accepted: 12 August 2021
Published: 14 August 2021

Publisher's Note: MDPI stays neutral with regard to jurisdictional claims in published maps and institutional affiliations.

Abstract: The prevalence of gastritis in humans is constantly growing and a prediction of an increase in this health problem is observed in many countries. For this reason, effective dietary therapies are sought that can alleviate the course of this disease. The objective of this study was to determine the effect of chemically pure oat beta-glucan preparations with different molar masses, low or high, used for 30 days in patients with histologically diagnosed chronic gastritis. The study enrolled 48 people of both genders of different ages recruited from 129 patients with a gastritis diagnosis. Before and after the therapy, hematological, biochemical, immunological and redox balance parameters were determined in the blood and the number of lactic acid bacteria and SCFA concentrations in the feces. Our results demonstrated a beneficial effect of oat beta-glucans with high molar mass in chronic gastritis in humans, resulting in reduced mucosal damage and healthy changes in SCFA fecal concentration and peripheral blood serum glutathione metabolism and antioxidant defense parameters. This fraction of a highly purified oat beta-glucan is safe for humans. Its action is effective after 30 days of use, which sheds new light on the nutritional treatment of chronic gastritis.

Keywords: oat beta-glucan; gastritis; inflammatory process; antioxidant properties; prebiotics; short-chain fatty acids

1. Introduction

Gastritis is usually diagnosed by the histological characteristics of tissue samples after a gastric mucosal biopsy, especially during routine upper esophagogastroduodenoscopy [1]. *Helicobacter pylori* infection is considered the main factor responsible for developing the inflammatory changes in gastric mucosa. Long-lasting gastritis predisposes to the development of gastric atrophy, intestinal metaplasia and dysplasia, which may lead to the development of the intestinal type of gastric cancer [1,2]. Apart from *H. pylori* infection, gastritis risk factors comprise inappropriate diet, non-steroid anti-inflammatory drugs (NSAIDs) and other drugs, excessive alcohol use, smoking, stress and older age [3].

There are no specific recommendations for treating chronic gastritis in the case of *H. pylori*-negative test results. Therefore, treatment regimens vary and depend on clinicians, for example, some empirically prescribe proton pump inhibitors (PPIs), some only mucosa-protecting agents, while others use both [1]. Occasionally, natural supplements or herbal medicines are also recommended to patients. On the other hand, it is still controversial for many clinicians if, for non-ulcer patients with *H. pylori* infection, eradication therapy should be advised [4]. Nevertheless, although many drugs are used to prevent and treat *H. pylori*-negative gastritis, only PPIs have proven their efficacy, especially in high-risk patients [5]. However, since PPIs are no longer considered to be completely safe for long-term use, there is still a great demand for additional treatments or/and prevention.

Beta-glucans are polysaccharides found in the cell walls of cereals, fungi, yeast and algae. They are also considered as the soluble fraction of dietary fiber. Depending on their origin, they show differences in structure. They can create linear, branched and cyclic macroparticles that impact their biological activity. Beta-glucan present in the aleuronic layer of oat grains is a mixture of β-D-glucose unbranched chains linked by β (1-3), β (1-4) glycosidic bonds. Dissolved beta-glucans absorb large amounts of water to form high-viscosity gums [6]. They do not undergo enzymatic degradation in the stomach, creating a mucus layer, protecting against irritation and alleviating inflammation [7]. Many studies have documented their positive effect on lowering postprandial glucose and insulin levels and hypocholesterolemic results [8,9]. Beta-glucans have also been tested as a cancer treatment adjuvant [10]. Many studies have also shown that beta-glucans may be immune stimulators by activating macrophages or the stimulation of the synthesis and activation of cytokines [11,12]. Studies have also confirmed the ability of these bioactive substances to reduce infection and help to lower the death rate of surgical hospital patients due to bacterial infections or postoperative inflammatory conditions of the gastrointestinal tract [13]. However, few studies have evaluated the antioxidative and anti-inflammatory properties of beta-glucans in the case of chronic gastritis. A survey conducted with people suffering from mild chronic gastritis brings new knowledge about the safety and the possibility of using the health-promoting properties of oat beta-glucans in treating this stomach disorder.

2. Materials and Methods

2.1. Study Design

The Research Ethics Board of the Institute of Rural Health in Lublin, Poland, approved this study (Commission Decisions No. 17/2011 and 17/2013). Informed written consent was obtained from all subjects. The study was designed as a randomized, parallel-group, double-blind, 4-week study, conducted at the Gastroenterology Outpatients' Department and Department of GI Endoscopy of the Institute of Rural Health in Lublin (Poland). The study population was either randomly assigned to a placebo (P) group, receiving an oral dose of 100 mL 3% solution of potato starch (placebo), or to one of the treated groups receiving an oral dose of 100 mL of 3% solution of high molar mass beta-glucan (G1; 2,180,000 g/mol, purity—81.9%) or 100 mL of 3% solution of low molar mass beta-glucan (G2; 70,000 g/mol, purity—89.1%) [14]. A total of 4 clinic visits were scheduled, consisting of a screening visit (visit 1), baseline visit (visit 2), end of the treatment visit (visit 3) and follow-up visit 2 weeks after completing the treatment (visit 4).

2.2. Study Subjects

The 129 participants were recruited from patients who reported to the Gastroenterology Outpatients' Department and Department of Gastrointestinal Endoscopy of the Institute of Rural Health in Lublin to undergo an elective esophagogastroduodenoscopy (EDG) due to dyspepsia. All participants underwent a baseline screening assessment which included a medical history and physical examination. All consecutive eligible patients with an initial diagnosis of gastritis within the gastric antrum were invited to participate in the study. Finally, 48 patients of both genders (14 women and 34 men), aged 23–74 years, were recruited, randomized and subsequently treated.

The inclusion criteria were: age between 18 and 75 years and presence of endoscopic characteristics of antral gastritis, which were then confirmed by histological examination of biopsy specimens. Subjects were excluded from the study if they met any of the following exclusion criteria: history of peptic ulcer disease; acute or erosive *gastritis*, autoimmune gastritis; prior esophagal, gastric or duodenal surgery; renal dysfunction; significant liver disease indicated by platelet count below 70,000, ascites or known gastroesophageal varices; active treatment with an H2-receptor antagonist, proton pump inhibitor, antacids or prokinetics, sucralfate; anticoagulants; fiber supplements; ingestion of aspirin or non-steroid anti-inflammatory drugs within the previous 30 days; known alcohol abuse; surreptitious drug-taking; pregnancy or lactation; the presence of alarm features, including weight loss, hematemesis, melena or rectal bleeding; significant coexisting illness or a condition that could limit the ability to participate in the study.

The basic clinical examination was conducted and vital signs were examined at all study visits. All adverse events (AEs) were recorded at all visits. The investigators and medical experts evaluated their severity and relation to treatment and reported them to the respective authority. Participants also assessed their general well-being with a questionnaire.

2.3. Dietary Supplements

To obtain the fractions of 1-3, 1-4-beta-D-glucan from oat with two ranges of molar masses of 30,000–90,000 g/mol and 2,000,000–3,000,000 g/mol, a dedicated method was created. As an effect of the preliminary study, we concluded that a high molar mass fraction of 1-3, 1-4-ß-D-glucan from oat is particularly susceptible to chain length reduction due to the polymer chain's physical degradation when subjected to mechanical stress during milling.

Briefly, to obtain a 1-3, 1-4-beta-D-glucan fraction of high molar mass, dehulled oat grains were subjected to repetitive milling (4–5 passages, with a subsequent mill gap change from 1.5 to 0.5 mm) in a corrugated roller mill. The maximum separation of the grain bran from the endosperm was obtained with minimal mechanical reduction of the bran particles. The output fraction of oat bran obtained using the milling technology was divided into two streams.

One stream was processed according to the alkaline extraction method described in a European patent [15] and led to the recovery of a 1-3, 1-4-beta-D-glucan fraction of molar mass within the range of 2,000,000–3,000,000 g/mol. The second stream was further mechanically processed for particle size reduction in the fraction. Mechanical defragmentation allows for creating controlled and repeatable processing conditions. Due to the characteristics of the oat bran layer taken off during the milling stage on the crushers (small particles of low weight, relatively flexible and not resistant), it was decided to proceed to mill it after prior freezing of the bran fraction. After 24 h at −20 °C, the fraction was subjected to intensive milling in a finger type mill (MN1 Retsch mill). The obtained fraction was frozen again (24 h at −20 °C) and subjected to another reduction by milling. A four-fold freezing and reduction operation made it possible to obtain beta 1-3, 1-4-beta-D-glucan with a fixed molar mass in the range of 30,000–90,000 g/mol by extraction by the method described in a Polish patent and described in our previous paper [16].

Molar mass was determined by an indirect method with a viscometer (Ubbelohde's viscometer) against commercially available mass standards of oat 1-3, 1-4-beta-D-glucan (Megazyme International LTD, Wicklow, Ireland). The obtained fractions were freeze-dried (freeze-dryer LR-1 Delta 1–24 LSC), and their purity was determined by the enzymatic method using lichenase (Megazyme International LTD, Wicklow, Ireland) according to the methodology described previously [14].

The verification of the effect of administered dietary oat beta-glucan was undertaken using a 3% concentration of the pure substance in dietary supplements offered to the patients according to EFSA recommendations towards lowering blood cholesterol effect [17]. Additionally, due to the very high stickiness of beta-glucan powder after absorbing any water, it was decided to administer it in the form of a diluted gel to avoid any clogging of the esophagus during swallowing.

The gel form was prepared by dissolving the appropriate weight in 1 L of cold distilled water; then, the solution was subjected to 5 min, 750 W microwave heating to pre-dissolve beta-glucan. After microwave heating, the solution was stirred intensively with a three-leaf blender, and then hot distilled water was added and it was stirred again intensively. After reaching a homogeneous mixture, the sample was subjected to microwave radiation to adequately hydrate the beta-glucan chains for 10 min/750 W.

Then, the solution was poured into 50 mL standing test tubes (autoclavable Falcon type) and sterilized in an autoclave for 10 min at 121 °C. The two test tubes supplied 100 mL of 3% high or low molar mass beta-glucan or placebo (potato starch).

The packages given to the patients consisted of 50 mL Falcon tubes packed in boxes, closed and secured with a label marked "For testing purposes" and indicating the best-before date. Due to the lack of any preservatives that could cause additional effects in the gastric and intestinal mucosa, the shelf life was established as six weeks in preliminary tests.

For the month-long treatment of one patient, three boxes were prepared, having 20 Falcon tubes of 50 mL each. After obtaining a patient's consent to carry out a month-long treatment, the treatments were delivered to the patients. After the end of the treatment, patients returned the packages which were intended for disposal.

Patients were instructed not to take any drugs 1 h before and 1 h after dosing. Patients were not allowed to take any fiber supplements or parapharmaceuticals during the study period. Patients were instructed to drink the delivered supplement 15–30 min before breakfast and supper. For both study groups, the diet (easily digestible diet, without beta-glucan products) was standardized during the treatment period.

2.4. Esophagogastroduodenoscopy and Histopathological Examination

For each subject, the same endoscopist performed the evaluations. Investigators and other personnel participating in the endoscopic assessment were blinded to the subjects' treatment allocation. The appearance of the mucosa of the stomach and duodenum was evaluated and scored with the use of esophagogastroduodenoscopy scores (EGs) which were partially based on the Sydney grading/scoring system [18]. The initial diagnosis of chronic gastritis was categorized as superficial gastritis, erosive gastritis and atrophic gastritis according to endoscopic appearance. Two biopsies for histology were taken from the gastric antrum within 2 cm from the pylorus, one from the distal lesser curvature and the other from the distal greater curvature. One more biopsy specimen was taken for rapid urease testing (GUT Plus, Gatrex, Warsaw, Poland), confirming or excluding the presence of *H. pylori*.

Endoscopy with biopsies and histological examinations were performed before treatment (inclusion criteria) and the day after discontinuation (safety measure) of the administration. Histopathological evaluation was conducted in the ALAB Laboratory Sp. z o.o., Warsaw (Poland), and a description of each specimen was obtained.

2.5. Blood Sample Collection, Hematological and Biochemical Assays

Samples for biochemical analysis were collected at visits 2 and 3. After 12 h of fasting, blood samples were collected between 08:00 and 11:00 at the Department of Gastrointestinal Endoscopy of the Institute of Rural Health in Lublin. They were centrifuged, and serum (0.5 mL each portion) was kept frozen at −70 °C until the analysis.

Serum creatinine, total bilirubin concentrations, alkaline phosphatase (ALP) and gamma-glutamyltransferase (GGT) activities and basic hematological parameters were analyzed using routine laboratory techniques in the medical laboratory of Synevo Sp. z o.o.

in Lublin, Poland. Plasma TNF-alpha and C-reactive protein (CRP) levels were measured using ELISA kits (Merck Millipore, Darmstadt, Germany). Plasma total antioxidant status (TAS) and the activity of glutathione peroxidase (GPx), glutathione reductase (GR) and superoxide dismutase (SOD) in blood serum were examined using Randox reagent kits (Crumlin, United Kingdom). The levels of whole blood reduced (GSH) and oxidized (GSSG) glutathione were measured according to Rebrin, Forster and Sohal [19] using a 4-channel electrochemical array for simultaneous detection. The GSH:GSSG ratio was also calculated.

2.6. Feces Sample Collection, Determination of the Number of Lactic Acid Bacteria (LAB) and Analysis of Short-Chain Fatty Acids (SCFAs)

Stool samples were collected in the morning after an overnight rest on the same day as the blood samples and stored at $-80\ ^\circ$C until assayed.

Microbiological analysis was performed using a TEMPO® (Biomerieux, Marcy l'Etoile, France) system to enumerate quality indicator organisms in the samples. Feces samples were homogenized in peptone water (Merck, Darmstadt, Germany) in a ratio of 1:9 using stomacher homogenizer. Then, serial dilutions of the samples were prepared, and a selective liquid culture medium for LAB was inoculated with bacteria from the homogenized feces. After incubation, the bacterial suspension was placed in TEMPO® enumeration cards containing 48 wells across three different dilutions to automatically determine the most probable number (MPN) of lactic acid bacteria. Filled cards were incubated aerobically at $37\ ^\circ$C for 48 h.

Analysis of short-chain fatty acids in feces was performed using HPLC with UV detection. The mobile phase for the isocratic elution of SCFAs was 15 mM monobasic sodium phosphate:methanol (80:20) on Hypersil BDS 150 $\times$ 4.6 mm, 5 µm column (Sigma-Aldrich (St. Louis, MO, USA)) at a mobile phase flow rate of 1.2 mL/min and detection at 224 nm. The quantitative and qualitative analyses of propionic, butyric, acetic, formic and lactic acid were performed using pure chemical standard SCFAs (Sigma-Aldrich (St. Louis, MO, USA)).

2.7. Statistical Analysis

The data were analyzed using Statistica 13.3 PL (TIBCO Software Inc., Tulsa, OK, USA). Normal distribution and equality of variances were tested for in all the data. Data that presented non-normal distribution were transformed with a common logarithm. A two-way repeated measures ANOVA was used for comparisons between groups (placebo, beta-glucan with high molar mass, beta-glucan with low molar mass) and the two time points (before and after dietary intervention) of the experiment (group $\times$ time point 3 $\times$ 2; with repeated measures on time points). Fisher's test was used for post hoc comparisons. Statistical significance was accepted for $p < 0.05$. The obtained results are presented as mean $\pm$ standard error of the mean (SEM).

Fisher's linear discriminant analysis (F-LDA) using R statistical software v. 3.3.3. (www.rproject.org/ (accessed on 23 May 2021)) (R: The R Project for Statistical Computing) analyzing the interaction between investigated parameters at the two time points was also performed.

3. Results

3.1. Study Population

A total of 129 patients were screened, of which 48 started in the study. The included patients consisted of women and men aged 23–74 with BMI 17.0–38.8 kg/m^2, who were randomly assigned to three following dietary groups, with 16 patients each: placebo group (P), oat beta-glucan with a high molar mass group (G1) and oat beta-glucan with low molar mass (G2). Dietary intervention with 3% of an aqueous suspension of potato starch (P group) or oat beta-glucan with high molar mass (G1 group) or oat beta-glucan with low molar mass (G2 group) was recommended for 30 days. Compliance with the treatment regimen was calculated from the number of study products dispensed to the subjects, and the number of study products returned to the study site. No difference in compliance was observed.

The age and anthropometric parameters of the study population are shown in Table 1. None of the examined parameters differed between the groups, between men and women or before vs. after dietary intervention.

Table 1. Age and anthropometric parameters of study population (mean ± SEM).

Parameters	Group	Mean ± SEM	Min	Max
Age (years)	Placebo	50.75 ± 3.61	23.00	68.00
	G1	46.00 ± 3.82	23.00	74.00
	G2	50.41 ± 3.24	27.00	68.00
Height [cm]	Placebo	170.88 ± 2.19	156.00	183.00
	G1	169.93 ± 2.10	152.00	178.00
	G2	171.41 ± 2.19	156.00	184.00
Weight [kg]	Placebo	86.63 ± 2.37	73.20	109.60
	G1	79.93 ± 3.95	54.00	116.50
	G2	79.17 ± 3.56	64.90	113.10
BMI [kg/m^2]	Placebo	29.82 ± 1.00	23.43	38.83
	G1	27.72 ± 1.26	17.04	36.80
	G2	26.88 ± 0.93	20.97	36.51

3.2. Well-Being and Adverse Effects during the Study

There was no significant difference in subjective assessment of the well-being of patients during the study. There were no statistically significant differences in well-being between the dietary groups. From 75% (placebo group) to 78% (G1 group) and 79% (G2 group) of patients declared good and very good well-being, none of the patients declared bad or very bad well-being. The overall incidence of adverse events (AEs) was similar across the study groups. None of the AEs led to the discontinuation of the study and no serious adverse event (SAE) was recorded. The most commonly occurring AEs during the treatment period were dyspepsia (seven in G2, six in G1 and five in the placebo group), upper abdominal pain or discomfort (eight in G2, four in G1 and ten in the placebo group), upper pulmonary tract infections (one in the G1 group and one in the placebo group) and an increase in arterial pressure (two in the G1 group and two in the placebo group). Both dyspepsia and upper abdominal pain and discomfort were usually mild, and the symptoms disappeared within a few days. Therefore, dyspepsia did not generally lead to discontinuation of the study. The increase in arterial blood pressure was mild and occurred episodically in subjects with a history of arterial hypertension. No clinically relevant treatment-related events were recorded during the study.

3.3. Histological Examination of Biopsy Specimens

In each case, the following features were microscopically evaluated using the visual analogue scales (according to the updated Sydney System): inflammation (presence of lymphocytes and plasma cells within the lamina propria) and its activity (presence of neutrophils), glandular atrophy, intestinal metaplasia (complete or incomplete) and presence of *Helicobacter pylori* in the superficial mucus [9]. Evaluated featured were scored in the range 0–3 (0, none; 1 mild; 2, moderate; 3, severe).

Before the dietary intervention, all patients participating in the study had histologically diagnosed chronic gastritis. The changes found in the gastric mucosa include lymphocytes infiltration within the lamina propria (score 1–2), the presence of neutrophils (score 1–2) and the presence of *Helicobacter pylori* (80% of patients were negative). After 30 days of beta-glucans, or placebo administration only in the G1 group, a reduction in inflammation manifested by a decrease in the infiltration of lymphocytes (score 0–1) and neutrophils (score 0) was found.

3.4. Peripheral Blood Hematology and Biochemical Parameters

No statistical differences (ANOVA) related to the type of dietary intervention or duration of treatment were found in the hematological parameters, except platelet count with a higher value in group G2 (Table 2), as well as the chosen biochemical blood serum parameters (Table 3).

Table 2. Hematological parameters of study population (mean $\pm$ SEM).

Parameters	Group	Before Treatment	After Treatment
WBC $\times 10^3/\mu L$	Placebo	6.22 ± 0.34	5.81 ± 0.27
	G1	6.36 ± 0.45	6.25 ± 0.47
	G2	5.90 ± 0.20	6.20 ± 0.30
RBC $\times 10^6/\mu L$	Placebo	4.94 ± 0.05	4.91 ± 0.05
	G1	5.03 ± 0.10	4.92 ± 0.07
	G2	4.86 ± 0.08	4.91 ± 0.08
Hemoglobin [g/dL]	Placebo	14.51 ± 0.38	14.48 ± 0.27
	G1	15.25 ± 0.29	14.97 ± 0.29
	G2	14.71 ± 0.27	14.86 ± 0.30
Hematocrit [%]	Placebo	42.84 ± 0.87	42.69 ± 0.52
	G1	44.18 ± 0.68	43.37 ± 0.77
	G2	42.60 ± 0.72	43.08 ± 0.73
Lymphocytes $\times 10^3/\mu L$	Placebo	1.97 ± 0.14	1.93 ± 0.13
	G1	1.95 ± 0.12	1.93 ± 0.11
	G2	1.83 ± 0.10	2.08 ± 0.09
Platelets $\times 10^3/\mu L$	Placebo	204.19 ± 9.5	199.94 ± 8.07
	G1	213.23 ± 16.7	227.57 ± 19.0
	G2	236.65 ± 13.2 *	250.24 ± 13.6 *

* Statistically significant differences $p < 0.05$.

Table 3. Blood serum biochemical parameters of study population (mean $\pm$ SEM).

Parameters	Group	Before Treatment	After Treatment
Creatinine [mg/dL]	Placebo	0.74 ± 0.02	0.78 ± 0.03
	G1	0.81 ± 0.03	0.83 ± 0.04
	G2	0.76 ± 0.04	0.78 ± 0.04
Bilirubin [mg/dL]	Placebo	0.64 ± 0.08	0.65 ± 0.05
	G1	0.71 ± 0.12	0.64 ± 0.09
	G2	0.65 ± 0.09	0.80 ± 0.13
ALT [U/L]	Placebo	33.30 ± 4.50	32.68 ± 6.70
	G1	24.77 ± 3.00	23.24 ± 3.10
	G2	26.82 ± 1.90	26.59 ± 2.19
AST [U/L]	Placebo	$25.63 \pm 1,48$	26.37 ± 4.00
	G1	28.48 ± 4.69	26.14 ± 3.90
	G2	22.47 ± 1.12	22.82 ± 1.24

3.5. Blood Serum Immunological Parameters

A decreased concentration of CRP in the peripheral blood serum of all patients after 30 days of nutritional intervention was found. These changes were confirmed by the significant ANOVA analysis ($p < 0.02$), but only in the G2 group were these changes statistically significant (Figure 1A). Neither the type of supplement nor the duration of nutritional intervention (ANOVA NS) had a considerable effect on TNF-alpha concentration (Figure 1B).

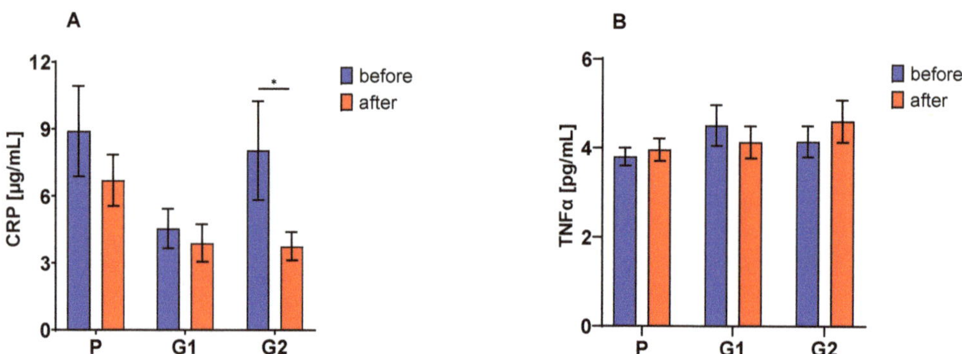

Figure 1. Serum C-reactive protein (CRP) (**A**) and TNFα (**B**) concentration (mean ± SE); The asterisk (*) denotes a statistically significant difference after vs. before dietary intervention, set at *p* < 0.05.

3.6. Peripheral Blood Redox Balance Parameters

The activity of SOD (Figure 2A) after a 30-day nutritional intervention increased (ANOVA, $p = 0.0000$), with significant changes only in the group of patients with placebo treatment ($p < 0.002$). The activity of both GPx (Figure 2B) and GR (Figure 2C) after 30 days of nutritional intervention significantly decreased (ANOVA, in both cases $p < 0.0001$), regardless of the molar mass of the supplement used (in the case of GPx) or only when oat beta-glucan preparations were used (in the case of GR, there was a significant effect of the supplement type $p < 0.04$).

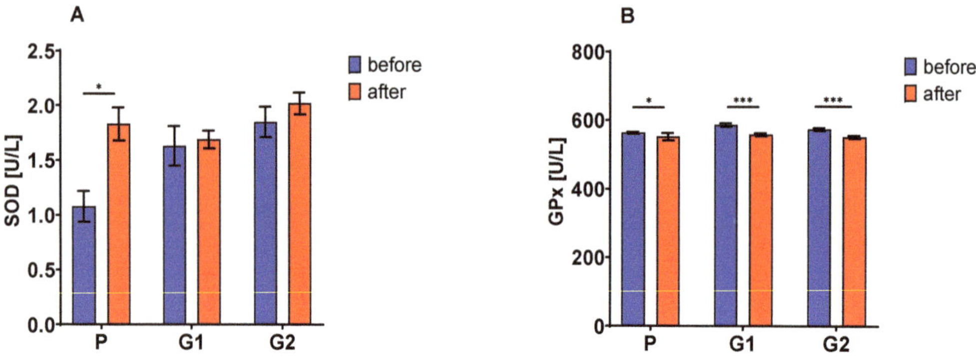

Figure 2. *Cont.*

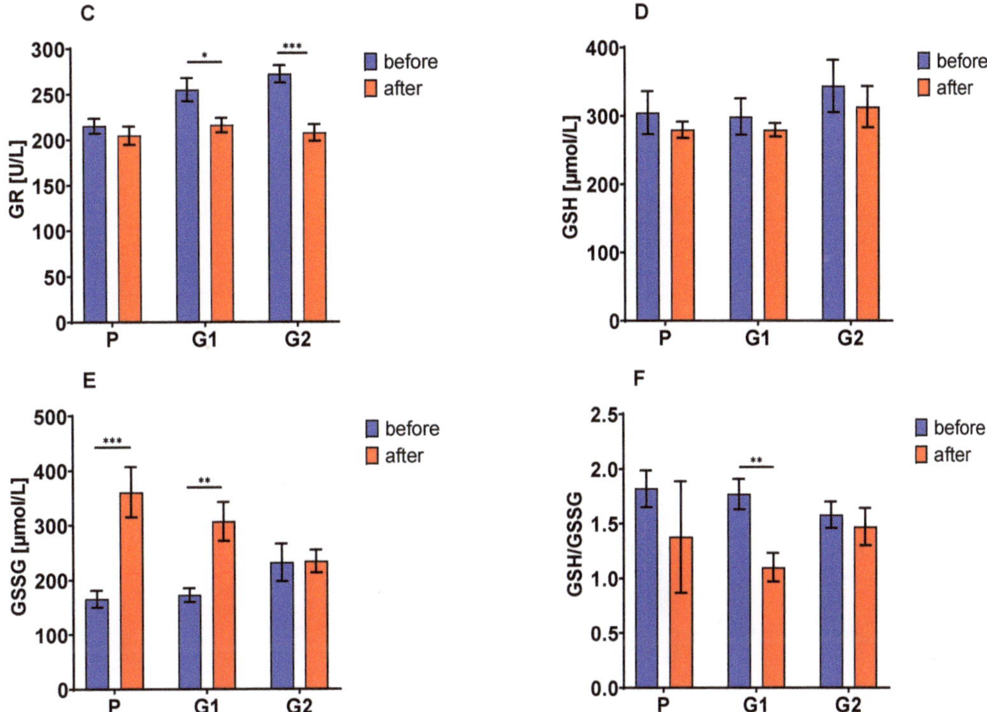

Figure 2. Blood superoxide dismutase (SOD) (**A**), glutathione peroxidase (GPx) (**B**) and reductase (GR) (**C**) activities; serum reduced (GSH) (**D**) and oxidized (GSSG) (**E**) glutathione and GSH to GSSG ratio (**F**). The asterisks (*, **, ***) denote a statistically significant difference after vs. before dietary intervention, set at $p < 0.05$; $p < 0.01$; $p < 0.001$, respectively.

The concentration of the reduced form of glutathione (Figure 2D) did not change as a result of the 30-day nutritional intervention (ANOVA NS). In contrast, the concentration of the oxidized form of this compound (Figure 2E) increased significantly in patients from the placebo and G1 groups (ANOVA, $p = 0.0000$; interaction of type of supplement vs. duration of use $p < 0.03$). In the G1 group the GSH/GSSG ratio (Figure 2F) decreased, which confirmed the significance of the analysis of variance (ANOVA, $p = 0.0000$).

3.7. The Fecal Number of LAB and SCFA Concentration

The number of lactic acid bacteria increased after 30 days of oat beta-glucan with high molar mass consumption, but it was not a statistically significant change. The analysis of variance also did not confirm any substantial changes in this parameter (Figure 3A). Additionally, no significant differences were found due to the 30-day nutritional intervention in the fecal concentration of lactic acid (Figure 3B), while the concentration of acetic acid (Figure 3C), propionic acid (Figure 3D) and hydroxybutyric acid (Figure 3E) increased significantly in patients consuming both oat beta-glucan fractions (ANOVA, in all cases $p = 0.0000$).

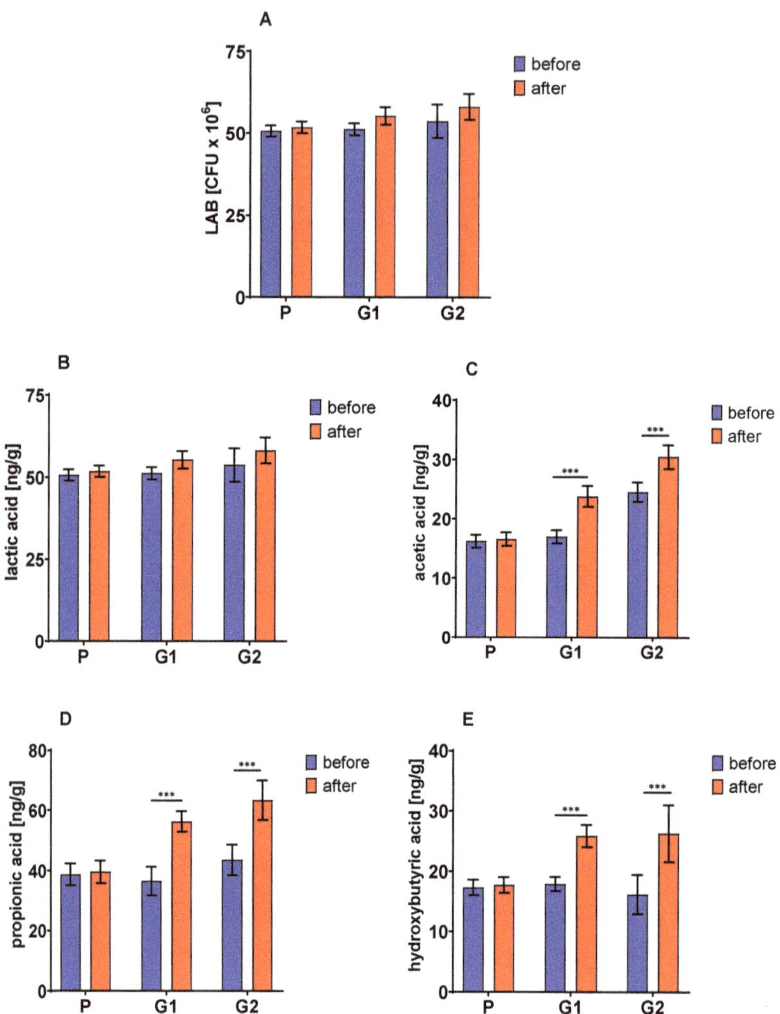

Figure 3. Lactic acid bacteria (**A**) and short-chain fatty acids (lactic, acetic, propionic and hydroxybutyric) (**B**–**E**) in feces. The asterisks (***) denote a statistically significant difference after vs. before dietary intervention, set at $p < 0.001$.

3.8. Fisher's Linear Discriminant Analysis (FLD)

Fisher's linear discriminant analysis (FLD) was performed to explore other differences in the above parameters associated with beta-glucan supplementation. The results of FLD obtained for the experimental data are presented in Figure 4. FLD was used to optimize the linear combination of antioxidative defense and oxidative stress parameters and SCFA level to obtain the best separation of the experimental groups, supplemented with G1 or G2 oat beta-glucan or placebo at the beginning of the treatment and after one month of the supplementation. The FLD test provides the most efficient linear combination of analyzed parameters needed to separate groups treated with both fractions of oat beta-glucan. The results of the linear discriminant analysis at two time points are shown in Figure 4A,C. In each figure, three treatment groups are separated. The data are shown between the linear combinations of parameters (FLDs), marked as FLD1 and FLD2, which separate the best predefined group.

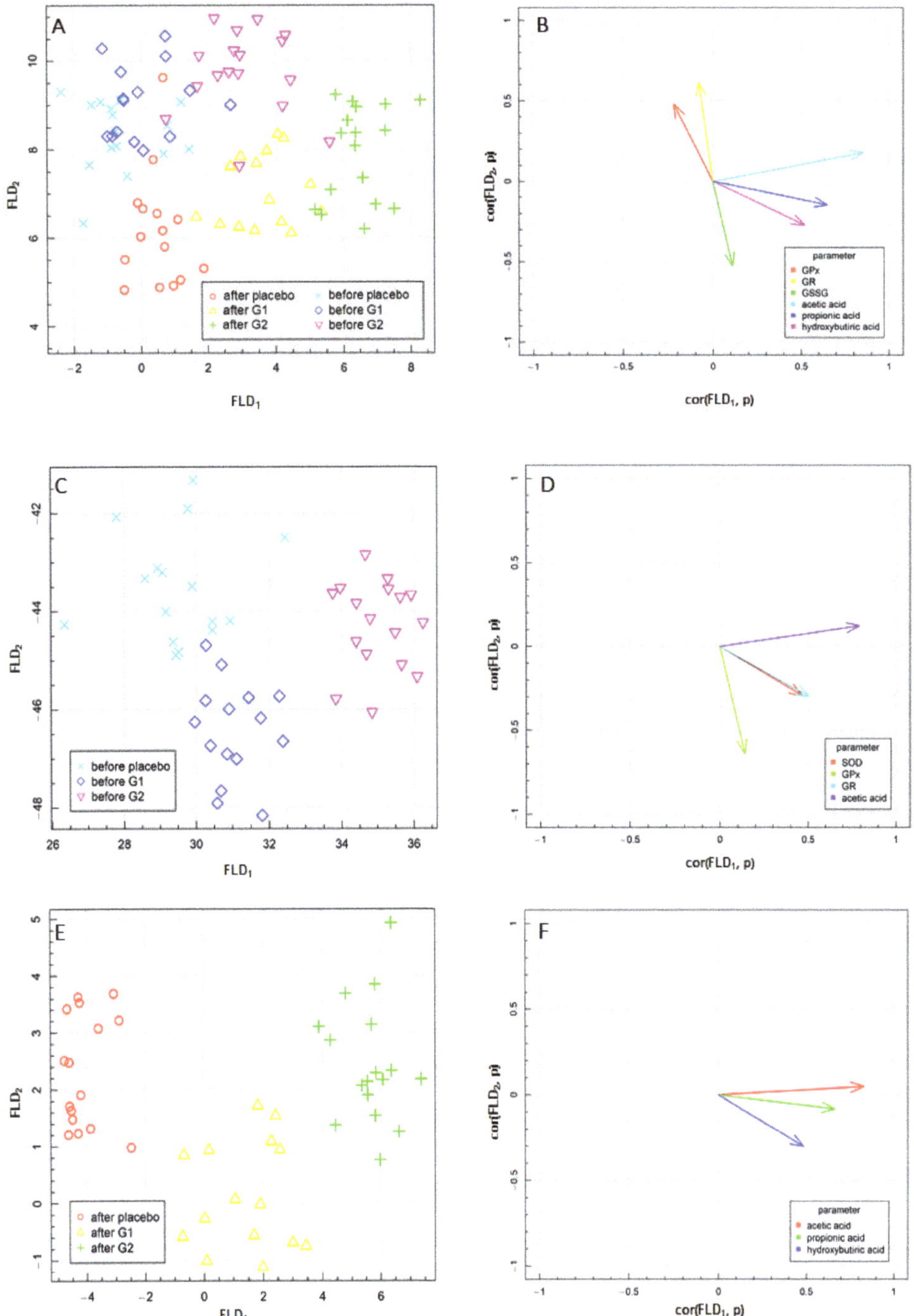

Figure 4. Fisher's Linear Discriminant (FLD) Analysis: (**A,C,E**) experimental data on the plane spanned by two the most data separating FLDs and (**B,D,F**) parameters contributing the most to FLDs. (**A,B**)—before and after the beta-glucans intervention; (**C,D**)—before the beta-glucans intervention; (**E,F**)—after 30 days of the beta-glucans intervention.

The results of the FLD together and at two experimental time points are shown in Figure 4A,C,E. In Figure 4A, six experimental groups are isolated, and in Figure 4C,D, three experimental groups per time point are separated. The data are presented in the space between the linear combinations of parameters (FLDs), marked as FLD1 and FLD2, ensuring the optimal separation of the predefined groups. Figure 4B,D,F show the correlation vectors of the studied parameters with FLD1 and FLD2 predictors, which establish the direction and influence of a given parameter on the separation of the experimental groups: for all aggregated data (Figure 4A,B), before treatment with both fractions of oat beta-glucans or placebo (Figure 4C,D) and after one month of supplementation (Figure 4E,F). The graphs highlight the parameters that proved to have the most critical impact on the separation of data. The biological marker compatible with a particular vector caused the data to move in the direction indicated by the vector. The FLD complemented and confirmed the statistical analysis of ANOVA, which showed the increased concentration of acetic acid, propionic acid and hydroxybutyric acid after 30 days of high molar mass oat beta-glucan consumption.

FLD performed for all data obtained in the present study (Figure 4A,B) confirmed that the factors that most differentiated these experimental groups were SCFAs, in particular acetic and propionic acid, as well as antioxidative defense enzymes including GPx and GR activity in the blood, which was also confirmed by ANOVA results.

The FLD provided information about combining the above parameters, which are helpful to distinguish the groups of data acquired before and after oat beta-glucan supplementation with G1 or G2 beta-glucans in the vertical plane (FLD2). Additionally, as indicated in the horizontal plane (FLD1), they did not enable the separation of the placebo group from the other groups before dietary intervention (Figure 4A) but simultaneously showed that it is possible after one month of nutritional enrichment with oat beta-glucan. Therefore, we decided to perform separate FLDs for each time point of the present study.

At the beginning of the research, the separation of the groups was possible, but the correlation vector power is low. The horizontal plane (FLD1) allowed the separation of the placebo group from the group that took low molecular mass oat beta-glucan (G2), as well as the separation of the low molecular mass beta-glucan group and the group that was supplemented with high molar mass beta-glucans (G1). Blood serum activity of GPx presented the most significant influence on FLD2, while acetic acid had a strong effect on FLD2. Both FLDs were impacted by the blood serum activity of SOD and GR. Based on this, it can be concluded that at the beginning of the experiment, all the groups were characterized by similar propionic and hydroxybutyric acid levels.

After one month of dietary intervention (Figure 4E,F), the analyzed parameters differentiated the experimental groups much more than at the earlier time point. Interestingly, all markers that allowed us to separate the experimental groups belong to carboxylic acids. The placebo group, as well as both groups that consumed oat beta-glucans, were separated by FLD1 and FLD2. The acetic and propionic acid levels in the stool significantly impacted FLD1, whereas hydroxybutyric acid influenced FLD2 (Figure 4F).

4. Discussion

Our study is the first randomized, double-blind experiment to evaluate the effects of oat beta-glucans in chronic gastritis in humans. The obtained results showed that the 30-day dietary supplementation with beta-glucans of different molar masses in a dose of 3 g/day resulted in molar mass-dependent changes in some immunological and redox balance parameters in peripheral blood serum and SCFA concentration in the stool of patients with gastritis. These supplements did not cause any significant changes in the hematological and biochemical blood indices and only a slight normalization of histopathological changes in patients from the G1 group. It should be emphasized that the dietary supplementation of beta-glucans in patients with gastritis did not adversely affect their well-being and did not exacerbate the clinical gastrointestinal symptoms, which proves the harmlessness of these highly purified beta-glucan preparations isolated with the use of alkaline extraction from oat grains.

This inflammation is driven by genetic factors, bacterial pathogenicity, patient age or nutritional factors, among others [20]. Maintaining the integrity of the gastric mucosa also depends on exposure to reactive oxygen species, which can increase lipid peroxidation and cause damage to this mucosal barrier [21–23]. For the endoscopic and histopathological assessment of chronic gastritis, the Sydney System is commonly used, which allows for a standardized approach to biopsy interpretation and allows for a clinically correct diagnosis [24]. Using the Sydney System, chronic gastritis was confirmed in all patients enrolled in the experiment, which was only slightly reduced in those who used oat beta-glucan with a high molar mass as a dietary supplement for 30 days.

Chronic gastritis often does not manifest changes in hematological and biochemical parameters. However, as shown in a few studies, gastritis caused by an increased presence of *Helicobacter pylori* causes changes in the percentage of platelets and the average platelet volume. However, due to the large population variability of these parameters caused by the gender and age of patients, they currently do not play an essential role in carrying out the diagnosis of gastritis [25]. Any chronic inflammation activates platelets, and the mean platelet volume (MPV) reflects this effect well. Still, there are more and more studies showing that MPV and platelet counts do not differ between *H. pylori*-positive and *H. pylori*-negative patients. There were also no differences or correlations between MPV and the severity of gastritis according to the Sydney classification [26]. As was shown in pediatric gastritis, no significant differences in hematological parameters, including platelet count and mean volume, were found between control and stomach inflammation patients, except severe gastritis, which manifested with a decrease in hemoglobin and hematocrit levels [27]. The lack of changes in the hematological and biochemical blood picture found in our patients confirms, on the one hand, mild or moderate gastritis and, on the other hand, good tolerability and safety of the use of both oat beta-glucan preparations. There was also no change in the patients' general well-being during oat beta-glucan supplementation compared to the placebo group, which indicated that these cereal polysaccharides are well tolerated in patients suffering from inflammation of the gastric mucosa.

Maintaining adequate antioxidant protection in stomach cells is possible by antioxidant enzyme activities, such as superoxide dismutase (SOD), peroxidase (GPx) and glutathione reductase (GR). Additionally, glutathione, a non-enzymatic antioxidant that should dominate in its reduced form, is crucial [28]. Glutathione is an endogenous component that plays a crucial role in critical physiological processes, helping to protect against redox balance, reducing oxidative stress, enhancing metabolic detoxification and regulating immune function. The latter aspect of glutathione metabolism is especially essential in the context of inflammatory diseases of individual components of the digestive system. Much evidence indicates that glutathione metabolism is a biomarker and indicator of the effectiveness of nutritional management in gastrointestinal disorders [29,30]. Our study focused on the GSH/GSSG parameter calculated based on the quantitative analysis of glutathione in the reduced (GSH) and oxidized (GSSG) forms and the activity of two critical enzymes in glutathione metabolism: glutathione reductase and glutathione peroxidase. Selenocysteine-containing glutathione peroxidases use glutathione to reduce H_2O_2 or lipid peroxides, generating oxidized glutathione [31].

In turn, regeneration of GSH from GSSG is catalyzed by glutathione reductase (GR) in the GSH redox cycle. The reduction of GSSG occurs at the expense of NADPH, produced by the pentose phosphate pathway from glucose oxidation [32,33]. Our study confirmed that in patients with diagnosed gastritis, the use of beta-glucans for 30 days, regardless of their molar mass, caused changes in glutathione metabolism: blood serum activity of both GPx and GR decreased, and in the case of high molar mass beta-glucan, the GSSG blood serum concentration was increased and the quotient GSH/GSSG ratio was decreased. As the levels of GSH decrease in states with generated oxidative stress, this would suggest increased oxidative stress after beta-glucan diet treatment; however, other parameters of oxidative stress (i.e., SOD, TAS) do not confirm this. In our opinion, the endogenous mechanism of glutathione synthesis is disturbed due to the decrease in the pool of available substrates for glutathione synthesis. Inflammation of the gastrointestinal tract causes a reduction in dietary availability, including amino acids. The primary substrates for the synthesis of glutathione are glutamate, cysteine and glycine, which are used in cell inflammation for high-energy processes. The reduced concentration of substrates for the synthesis of glutathione is responsible for oxidative stress and dietary supplementation, e.g., glutamine and cysteine reduce the effects of oxidative stress by increasing the parameter of the antioxidant potential of the blood [34]. This is confirmed by the fact that glutathione reductase activity is reduced, leading to a decreased amount of GSH from GSSG. As noted in some studies [35,36], pathological conditions such as inflammation and neoplasms lower GSH levels (and thus raise the GSSG/GSH ratio) of glutathione metabolism and it seems to be more folded up than expected. For example, the sustained constant activity of GPx does not guarantee a continuous concentration of both GSH and GSSG [37].

Our previous studies have confirmed the beneficial effect of oat beta-glucan in different gastrointestinal diseases in animal models [38]. In our previous study on rats with experimentally DSS-induced gastritis, we also found a beneficial effect of oat beta-glucan consumption which confirmed the important molar mass-dependent role of these cereal polysaccharides in maintaining the antioxidative defense of stomach tissue undergoing inflammation [39]. Consumption of low molar mass oat beta-glucan decreased SOD activity and concentration of GSH in the stomach tissue of animals with induced gastritis. In contrast, consumption of a high molar mass fraction caused a lowering of the concentration of GSSG but did not affect the concentration of GSH in the group of rats with gastritis. In rats, all parameters were analyzed in the stomach tissue, while in patients they were measured in peripheral blood, which could have resulted in slightly different results. It should be added that in studies on animal models of gastritis, we used the same oat beta-glucans as in patients suffering from this gastric inflammation. Both high and low molar mass beta-glucans were integral rat feed additives, while gastritis patients were given these polysaccharides in the form of a drinking gel. Different forms of beta-glucan preparations consumed by humans and animals could also have contributed to the differences in the obtained results.

The differences in the obtained results also concern the immunological parameters. In rats with gastritis, high beta-glucan consumption resulted in a significant reduction in the concentration of TNF-alpha in the stomach tissue [39], while in humans, the concentration of this immune parameter in the blood did not change.

In the stool of patients diagnosed with gastritis and consuming oat beta-glucans, the number of lactic acid bacteria (LAB) and the level of intestinal microbiota metabolites—short-chain fatty acids (SCFAs)—were tested. Although the subject of the study was not the colon itself, which is to a greater extent a beneficiary of fecal LAB changes, there are reasons to believe that any method influencing the fecal LAB profile has positive effects on the whole organism [40–42]. It is true that we measured only the LAB number without identification and assignment to families of intestinal bacteria. Still, this observation confirmed that LAB changes in the stool are an individual feature, dependent on many factors. Statistical analysis of the LAB results showed no statistically significant differences between the study groups. Still, the exact individual changes between the beginning and end of beta-glucan treatment showed changes within individual patients. These changes are mainly individual differences in the consumption of these food products, which are the primary substrate for LAB biodiversity in the stool. This variability has covered up the effect of beta-glucans. A too-small group of patients did not allow us to show significant differences in our entire study. Individual changes in the number of LAB are consistent with changes in metabolites of the whole intestinal microflora.

In our studies, the concentrations of SCFAs in the stool, acetic, hydroxybutyric and propionic acids, are statistically significantly correlated with both the pathological state and the beta-glucan diet therapy. The lactic acid concentration in stool remained constant, regardless of the type of beta-glucans used. Changes in the concentration of individual SCFAs and their mutual proportions are a characteristic feature of quantitative changes in microbiota populations, especially in pathophysiological states. Still, it seems impossible to clearly state which SCFAs play the most critical role in the pathomechanism of these changes at the present stage. The results of our studies to date clearly indicate that beta-glucans in the model of LPS-induced inflammation cause changes in SCFA profiles in the feces of rats [43]. The direction of these changes is similar in the case of the described use of beta-glucans. Likewise, we observed an increase in propionic, lactic and hydroxybutyric acid concentrations. Their increased presence in the stool determines their increased availability for enterocytes and demonstrates a positive effect on their metabolism.

On the one hand, evidence points to the fact that fecal propionic acid levels are affected by irritable bowel syndrome (IBS) [44]. Still, propionic acid is of great interest due to its substantial immunomodulatory effects. Its supplementation reduces the exacerbation of colitis in murine models, suggesting that its modulation may be a therapeutic intervention in inflammatory bowel disease [45]. On the other hand, studies by Ormsby and co-workers highlight the potential risk of widespread use of propionic acid as an antimicrobial, demonstrating this mechanism through the increased virulence and persistence of the pathotype adherent-invasive *E. coli* (AIEC) strain in a mouse colitis model [46]. The results of our studies indicate only the increased concentrations of SCFAs in the stool as a result of beta-glucan treatment. They are not the basis for concluding their safety.

Our study had several strengths. First, it was conducted in a double-blind, randomized trial. Second, we used highly purified oat beta-glucan with specific molar mass ranges. The duration of the study, 30 days, was sufficient to assess the safety of the systemic use of beta-glucans in patients with gastritis, as well as the therapeutic effect of these cereal polysaccharides. The limitations of our study concern primarily a heterogeneous group in terms of age and gender. Another downside is the lack of biochemical analyses in the stomach tissue, where antioxidative and anti-inflammatory defense processes occurring in chronic inflammation of the gastric mucosa are primarily found.

5. Conclusions

Our study demonstrated a beneficial effect of oat beta-glucans with high molar mass in chronic gastritis in humans, resulting in reduced mucosal damage and healthy changes in SCFA fecal concentrations and peripheral blood serum glutathione metabolism and antioxidant defense parameters as well. This fraction of a highly purified oat beta-glucan is safe for the body. Its action is effective after 30 days of use, which sheds new light on the nutritional treatment of chronic gastritis.

Author Contributions: Conceptualization, J.H. and J.G.-O.; Data curation, S.G., K.D., M.O., M.J. (Małgorzata Juszczak) and J.G.-O.; Formal analysis, M.J. (Małgorzata Juszczak); Funding acquisition, J.G.-O.; Investigation, S.G., J.W. and M.J. (Małgorzata Jałosińska); Methodology, S.G., R.F., J.W., M.J.(Małgorzata Juszczak); and J.G.-O.; Project administration, K.D. and J.G.-O.; Resources, S.G., J.H. and E.L.; Software, K.D. and M.O.; Supervision, R.F. and J.G.-O.; Validation, S.G., R.F. and J.G.-O.; Visualization, K.D. and M.O.; Writing—original draft, S.G., J.W. and J.G.-O.; Writing—review and editing, R.F., J.H., J.W., K.D., E.L. and J.G.-O. All authors have read and agreed to the published version of the manuscript.

Funding: This study was supported by the National Science Centre, Poland through project no. NN312 427440.

Institutional Review Board Statement: The study was conducted according to the guidelines of the Declaration of Helsinki, and approved by the Research Ethics Board of the Institute of Rural Health in Lublin, Poland (Commission Decisions No. 17/2011 and 17/2013).

Informed Consent Statement: Informed consent was obtained from all subjects involved in the study.

Data Availability Statement: The data that support the findings of this study are available on request from the corresponding author (S.G.).

Acknowledgments: The authors wish to thank Katarzyna Błaszczyk and Dominika Suchecka for their help with data collection.

Conflicts of Interest: The authors declare no conflict of interest.

References

1. Du, Y.; Bai, Y.; Xie, P.; Fang, J.; Wang, X.; Hou, X.; Tian, D.; Wang, C.; Liu, Y.; Sha, W.; et al. Chronic Gastritis in China: A National Multi-Center Survey. *BMC Gastroenterol.* **2014**, *14*, 21. [CrossRef] [PubMed]
2. Blaser, M.J.; Atherton, J.C. Helicobacter Pylori Persistence: Biology and Disease. *J. Clin. Investig.* **2004**, *113*, 321–333. [CrossRef] [PubMed]
3. Filip, R.; Huk, J.; Jarosz, B.; Skrzydło-Radomańska, B. Endoscopic and histological verification of upper GI tract side effects after long-term therapy with alendronate and strontium ranelate. *Gastroenterol. Rev. Przegląd Gastroenterol.* **2009**, *4*, 23–30.
4. Wolf, E.-M.; Plieschnegger, W.; Geppert, M.; Wigginghaus, B.; Höss, G.M.; Eherer, A.; Schneider, N.I.; Hauer, A.; Rehak, P.; Vieth, M.; et al. Changing Prevalence Patterns in Endoscopic and Histological Diagnosis of Gastritis? Data from a Cross-Sectional Central European Multicentre Study. *Dig. Liver Dis.* **2014**, *46*, 412–418. [CrossRef]
5. Nagata, N.; Niikura, R.; Aoki, T.; Sakurai, T.; Moriyasu, S.; Shimbo, T.; Sekine, K.; Okubo, H.; Watanabe, K.; Yokoi, C.; et al. Effect of Proton-Pump Inhibitors on the Risk of Lower Gastrointestinal Bleeding Associated with NSAIDs, Aspirin, Clopidogrel, and Warfarin. *J. Gastroenterol.* **2015**, *50*, 1079–1086. [CrossRef] [PubMed]
6. Du, B.; Meenu, M.; Liu, H.; Xu, B. A Concise Review on the Molecular Structure and Function Relationship of β-Glucan. *IJMS* **2019**, *20*, 4032. [CrossRef] [PubMed]
7. Tanaka, K.; Tanaka, Y.; Suzuki, T.; Mizushima, T. Protective Effect of β-(1,3 → 1,6)-D-Glucan against Irritant-Induced Gastric Lesions. *Br. J. Nutr.* **2011**, *106*, 475–485. [CrossRef] [PubMed]
8. Ho, H.V.T.; Sievenpiper, J.L.; Zurbau, A.; Blanco Mejia, S.; Jovanovski, E.; Au-Yeung, F.; Jenkins, A.L.; Vuksan, V. The Effect of Oat β-Glucan on LDL-Cholesterol, Non-HDL-Cholesterol and ApoB for CVD Risk Reduction: A Systematic Review and Meta-Analysis of Randomised-Controlled Trials. *Br. J. Nutr.* **2016**, *116*, 1369–1382. [CrossRef]
9. Wolever, T.M.S.; Jenkins, A.L.; Prudence, K.; Johnson, J.; Duss, R.; Chu, Y.; Steinert, R.E. Effect of Adding Oat Bran to Instant Oatmeal on Glycaemic Response in Humans—A Study to Establish the Minimum Effective Dose of Oat β-Glucan. *Food Funct.* **2018**, *9*, 1692–1700. [CrossRef] [PubMed]
10. Vetvicka, V.; Vannucci, L.; Sima, P. β-glucan as a New Tool in Vaccine Development. *Scand. J. Immunol.* **2020**, *91*, e12833. [CrossRef]
11. Pan, W.; Hao, S.; Zheng, M.; Lin, D.; Jiang, P.; Zhao, J.; Shi, H.; Yang, X.; Li, X.; Yu, Y. Oat-Derived β-Glucans Induced Trained Immunity Through Metabolic Reprogramming. *Inflammation* **2020**, *43*, 1323–1336. [CrossRef] [PubMed]
12. Zheng, X.; Zou, S.; Xu, H.; Liu, Q.; Song, J.; Xu, M.; Xu, X.; Zhang, L. The Linear Structure of β-Glucan from Baker's Yeast and Its Activation of Macrophage-like RAW264.7 Cells. *Carbohydr. Polym.* **2016**, *148*, 61–68. [CrossRef] [PubMed]
13. Laroche, C.; Michaud, P. New developments and prospective applications for beta (1,3) glucans. *Recent Pat. Biotechnol.* **2007**, *1*, 59–73. [CrossRef]
14. Harasym, J.; Suchecka, D.; Gromadzka-Ostrowska, J. Effect of Size Reduction by Freeze-Milling on Processing Properties of Beta-Glucan Oat Bran. *J. Cereal. Sci.* **2015**, *61*, 119–125. [CrossRef]
15. Harasym, J.; Brach, J.; Czarnota, J.L.; Stechman, M.; Słabisz, A.; Kowalska, A.; Chorowski, M.; Winkowski, M.; Madera, A. A Kit and a Method of Producing Beta-Glucan, Insoluble Food Fibre as Well as a Preparation of Oat Proteins. European Patent EP2515672B1, 6 July 2016.
16. Harasym, J.; Gromadzka-Ostrowska, J. PL226915B1 Method for Obtaining Cereal Beta-Glucan with Low Molecular Weight. Method for Obtaining Cereal Beta-Glucan with Low Molecular Weight. PL226915. 2017. Available online: https://grab.uprp.pl (accessed on 25 June 2021).
17. EFSA Panel on Dietetic Products, Nutrition and Allergies (NDA). Scientific Opinion on the Substantiation of a Health Claim Related to Oat Beta Glucan and Lowering Blood Cholesterol and Reduced Risk of (Coronary) Heart Disease Pursuant to Article 14 of Regulation (EC) No 1924/2006. *EFSA J.* **2010**, *8*, 1885. [CrossRef]
18. Dixon, M.F.; Genta, R.M.; Yardley, J.H.; Correa, P. Classification and Grading of Gastritis: The Updated Sydney System. *Am. J. Surg. Pathol.* **1996**, *20*, 1161–1181. [CrossRef]
19. Rebrin, I.; Forster, M.J.; Sohal, R.S. Effects of Age and Caloric Intake on Glutathione Redox State in Different Brain Regions of C57BL/6 and DBA/2 Mice. *Brain Res.* **2007**, *1127*, 10–18. [CrossRef] [PubMed]
20. Sipponen, P.; Maaroos, H.-I. Chronic Gastritis. *Scand. J. Gastroenterol.* **2015**, *50*, 657–667. [CrossRef]
21. Arend, A.; Loime, L.; Roosaar, P.; Soom, M.; Lõivukene, K.; Sepp, E.; Aunapuu, M.; Zilmer, K.; Selstam, G.; Zilmer, M. Helicobacter pylori substantially increases oxidative stress in indomethacin-exposed rat gastric mucosa. *Medicina (Kaunas)* **2005**, *41*, 343–347.
22. Kwiecien, S.; Jasnos, K.; Magierowski, M.; Sliwowski, Z.; Pajdo, R.; Brzozowski, B.; Mach, T.; Wojcik, D.; Brzozowski, T. Lipid peroxidation, reactive oxygen species and antioxidative factorsin the pathogenesis of gastric mucosal lesions and mechanism of protection against oxidative stress—Induced gastric injury. *J. Physiol. Pharmacol.* **2014**, *65*, 613–622.
23. Suzuki, R.B. Different Risk Factors Influence Peptic Ulcer Disease Development in a Brazilian Population. *WJG* **2012**, *18*, 5404. [CrossRef] [PubMed]
24. Garg, B.; Sandhu, V.; Sood, N.; Sood, A.; Malhotra, V. Histopathological Analysis of Chronic Gastritis and Correlation of Pathological Features with Each Other and with Endoscopic Findings. *PJP* **2012**, *3*, 172–178. [CrossRef]

25. Noris, P.; Melazzini, F.; Balduini, C.L. New Roles for Mean Platelet Volume Measurement in the Clinical Practice? *Platelets* **2016**, *27*, 607–612. [CrossRef]
26. Akar, T. Can Mean Platelet Volume Indicate Helicobacter Positivity and Severity of Gastric Inflammation? An Original Study and Review of the Literature. *ACC* **2019**, *58*, 576–582. [CrossRef]
27. Săsăran, M.O.; Meliţ, L.E.; Mocan, S.; Ghiga, D.V.; Dobru, E.D. Pediatric Gastritis and Its Impact on Hematologic Parameters. *Medicine* **2020**, *99*, e21985. [CrossRef] [PubMed]
28. Bhattacharyya, A.; Chattopadhyay, R.; Mitra, S.; Crowe, S.E. Oxidative Stress: An Essential Factor in the Pathogenesis of Gastrointestinal Mucosal Diseases. *Physiol. Rev.* **2014**, *94*, 329–354. [CrossRef] [PubMed]
29. Liu, Y.; Hyde, A.S.; Simpson, M.A.; Barycki, J.J. Emerging Regulatory Paradigms in Glutathione Metabolism. In *Advances in Cancer Research*; Elsevier: Amsterdam, The Netherlands, 2014; Volume 122, pp. 69–101. [CrossRef]
30. Minich, D.M.; Brown, B.I. A Review of Dietary (Phyto)Nutrients for Glutathione Support. *Nutrients* **2019**, *11*, 2073. [CrossRef]
31. Brigelius-Flohé, R.; Maiorino, M. Glutathione Peroxidases. *Biochim. Biophys. Acta (BBA) Gen. Subj.* **2013**, *1830*, 3289–3303. [CrossRef] [PubMed]
32. Couto, N.; Wood, J.; Barber, J. The Role of Glutathione Reductase and Related Enzymes on Cellular Redox Homoeostasis Network. *Free. Radic. Biol. Med.* **2016**, *95*, 27–42. [CrossRef]
33. Silvagno, F.; Vernone, A.; Pescarmona, G.P. The Role of Glutathione in Protecting against the Severe Inflammatory Response Triggered by COVID-19. *Antioxidants* **2020**, *9*, 624. [CrossRef]
34. Kumar, P.; Liu, C.; Hsu, J.W.; Chacko, S.; Minard, C.; Jahoor, F.; Sekhar, R.V. Glycine and N-acetylcysteine (GlyNAC) Supplementation in Older Adults Improves Glutathione Deficiency, Oxidative Stress, Mitochondrial Dysfunction, Inflammation, Insulin Resistance, Endothelial Dysfunction, Genotoxicity, Muscle Strength, and Cognition: Results of a Pilot Clinical Trial. *Clin. Transl. Med.* **2021**, *11*, e372. [CrossRef] [PubMed]
35. Gamcsik, M.P.; Kasibhatla, M.S.; Teeter, S.D.; Colvin, O.M. Glutathione Levels in Human Tumors. *Biomarkers* **2012**, *17*, 671–691. [CrossRef] [PubMed]
36. Sarıkaya, E.; Doğan, S. Glutathione Peroxidase in Health and Diseases. In *Glutathione System and Oxidative Stress in Health and Disease*; Dulce Bagatini, M., Ed.; IntechOpen: London, UK, 2020. [CrossRef]
37. Lu, S.C. Dysregulation of Glutathione Synthesis in Liver Disease. *Liver Res.* **2020**, *4*, 64–73. [CrossRef]
38. Kopiasz, Ł.; Dziendzikowska, K.; Gajewska, M.; Wilczak, J.; Harasym, J.; Żyła, E.; Kamola, D.; Oczkowski, M.; Królikowski, T.; Gromadzka-Ostrowska, J. Time-Dependent Indirect Antioxidative Effects of Oat Beta-Glucans on Peripheral Blood Parameters in the Animal Model of Colon Inflammation. *Antioxidants* **2020**, *9*, 375. [CrossRef] [PubMed]
39. Suchecka, D.; Błaszczyk, K.; Harasym, J.; Gudej, S.; Wilczak, J.; Gromadzka-Ostrowska, J. Impact of Purified Oat 1-3,1-4-β-d-Glucan of Different Molecular Weight on Alleviation of Inflammation Parameters during Gastritis. *J. Funct. Foods* **2017**, *28*, 11–18. [CrossRef]
40. Cano-Garrido, O.; Seras-Franzoso, J.; Garcia-Fruitós, E. Lactic Acid Bacteria: Reviewing the Potential of a Promising Delivery Live Vector for Biomedical Purposes. *Microb. Cell Fact.* **2015**, *14*, 137. [CrossRef]
41. Mathur, H.; Beresford, T.P.; Cotter, P.D. Health Benefits of Lactic Acid Bacteria (LAB) Fermentates. *Nutrients* **2020**, *12*, 1679. [CrossRef]
42. De Filippis, F.; Pasolli, E.; Ercolini, D. The Food-Gut Axis: Lactic Acid Bacteria and Their Link to Food, the Gut Microbiome and Human Health. *FEMS Microbiol. Rev.* **2020**, *44*, 454–489. [CrossRef]
43. Wilczak, J.; Błaszczyk, K.; Kamola, D.; Gajewska, M.; Harasym, J.P.; Jałosińska, M.; Gudej, S.; Suchecka, D.; Oczkowski, M.; Gromadzka-Ostrowska, J. The Effect of Low or High Molecular Weight Oat Beta-Glucans on the Inflammatory and Oxidative Stress Status in the Colon of Rats with LPS-Induced Enteritis. *Food Funct.* **2015**, *6*, 590–603. [CrossRef]
44. Sun, Q.; Jia, Q.; Song, L.; Duan, L. Alterations in Fecal Short-Chain Fatty Acids in Patients with Irritable Bowel Syndrome: A Systematic Review and Meta-Analysis. *Medicine* **2019**, *98*, e14513. [CrossRef]
45. Smith, P.M.; Howitt, M.R.; Panikov, N.; Michaud, M.; Gallini, C.A.; Bohlooly-Y, M.; Glickman, J.N.; Garrett, W.S. The Microbial Metabolites, Short-Chain Fatty Acids, Regulate Colonic Treg Cell Homeostasis. *Science* **2013**, *341*, 569–573. [CrossRef] [PubMed]
46. Ormsby, M.J.; Johnson, S.A.; Carpena, N.; Meikle, L.M.; Goldstone, R.J.; McIntosh, A.; Wessel, H.M.; Hulme, H.E.; McConnachie, C.C.; Connolly, J.P.R.; et al. Propionic Acid Promotes the Virulent Phenotype of Crohn's Disease-Associated Adherent-Invasive Escherichia Coli. *Cell Rep.* **2020**, *30*, 2297–2305.e5. [CrossRef] [PubMed]

 *nutrients*

Article

Effects of β-glucan Rich Barley Flour on Glucose and Lipid Metabolism in the Ileum, Liver, and Adipose Tissues of High-Fat Diet Induced-Obesity Model Male Mice Analyzed by DNA Microarray

Kento Mio [1,2], Chiemi Yamanaka [3], Tsubasa Matsuoka [2], Toshiki Kobayashi [2] and Seiichiro Aoe [1,3,*]

1. Studies in Human Life Sciences, Graduate School of Studies in Human Culture, Otsuma Women's University, Chiyoda-ku, Tokyo 102-8357, Japan; mio.kento@hakubaku.co.jp
2. Research and Development Department, Hakubaku Co. Ltd., Chuo-City, Yamanashi 409-3843, Japan; matsuoka.tsubasa@hakubaku.co.jp (T.M.); k.toshiki@hakubaku.co.jp (T.K.)
3. The Institute of Human Culture Studies, Otsuma Women's University Chiyoda-ku, Tokyo 102-8357, Japan; chiemiy@gmail.com
* Correspondence: s-aoe@otsuma.ac.jp; Tel.: +81-3-5275-6048

Received: 17 October 2020; Accepted: 17 November 2020; Published: 19 November 2020

Abstract: We evaluated whether intake of β-glucan-rich barley flour affects expression levels of genes related to glucose and lipid metabolism in the ileum, liver, and adipose tissues of mice fed a high-fat diet. C57BL/6J male mice were fed a high-fat diet supplemented with high β-glucan barley, for 92 days. We measured the expression levels of genes involved in glucose and lipid metabolism in the ileum, liver, and adipose tissues using DNA microarray and q-PCR. The concentration of short-chain fatty acids (SCFAs) in the cecum was analyzed by GC/MS. The metabolic syndrome indices were improved by barley flour intake. Microarray analysis showed that the expression of genes related to steroid synthesis was consistently decreased in the liver and adipose tissues. The expression of genes involved in glucose metabolism did not change in these organs. In liver, a negative correlation was showed between some SCFAs and the expression levels of mRNA related to lipid synthesis and degradation. Barley flour affects lipid metabolism at the gene expression levels in both liver and adipose tissues. We suggest that SCFAs are associated with changes in the expression levels of genes related to lipid metabolism in the liver and adipose tissues, which affect lipid accumulation.

Keywords: barley; β-glucan; dietary fiber; microarray; short chain fatty acids; lipid metabolism.

---

1. Introduction

Metabolic syndrome is widely known as a multi-factorial disorder with symptoms such as hypertension, hyperglycemia, insulin resistance, and visceral fat obesity. This syndrome is promoted by irregular eating, lack of exercise, excessive drinking of alcohol and stress, which may cause risk factors such as cardiovascular disease, coronary disease, and type 2 diabetes [1]. In particular, quality of diet is closely associated with the development of metabolic syndrome. In recent years, the beneficial effects of whole-grains have become apparent: a meta-analysis of cohort studies showed that a high consumption of whole-grains is associated with reduced incidence of type 2 diabetes [2]. Another report showed that whole-grain consumption (48–80 g/3–5 serving/day) reduced the incidence of obesity, type 2 diabetes, and concentrations of total and LDL-cholesterol in serum [3]. A systematic review also reported that intake of a whole-grain and fiber diet reduced the risk of type 2 diabetes, the incidence of being overweight and obesity in Japanese men and women [4]. Whole-grain and grains are good sources of fiber, and the effect of dietary fiber on lipid metabolism has been studied over a long period [5–7].

Barley, a cereal grain, contains higher amounts of β-glucan compared with other grains such as oats, rye, and wheat. Since barley β-glucan is distributed throughout the kernel, the amount of β-glucan in refined barley is unchanged. Barley β-glucan is a polysaccharide polymer in which glucose is linked by β-glycoside bonds (β-1,3–1,4), and is almost soluble in water [8]. β-Glucan increases water solubilization in the stomach and small intestine and delays the absorption of other nutrients [9]. These effects suppress increases in postprandial blood glucose, and maintains satiety, thereby improving insulin resistance and suppressing visceral fat accumulation by reduced excessive insulin secretion [10]. It has also been reported that blood cholesterol levels were reduced by barley intake in vivo and human studies [11–13]. Studies have indicated that the mechanism of barley β-glucan involves inhibition of cholesterol synthesis in the liver, inhibition of bile acid reabsorption and alteration of cholesterol metabolism due to decreased insulin secretion and promotion of cholesterol excretion [14,15].

Water-soluble dietary fibers such as barley β-glucan are fermented by bacteria in the cecum to the distal colon. Fermentation produces short-chain fatty acids (SCFAs) such as acetic acid, propionic acid, and butyric acid as metabolites which serve as energy sources for the host. It was recently reported that SCFAs affect lipid metabolism in several organs such as liver, muscle, and brown adipose tissue [16]. Acetic acid is mainly metabolized in the liver and is involved in fat and cholesterol synthesis in peripheral tissues [17]. Propionic acid is mainly metabolized in the liver and intestinal tract, and butyric acid is involved in cell differentiation and maintenance of intestinal barrier formation in the intestinal tract [18]. These SCFAs also contribute to reducing intestinal pH, which inhibits the absorption of toxic products such as *p*-cresol and phenols [19]. Moreover, recent studies have indicated the SCFAs act in the host metabolism as signal molecules, regulating intestinal hormone secretion and insulin signaling via SCFA receptors [20].

The findings of the above studies support the beneficial effects of barley β-glucan in normalizing cholesterol concentration, improving glucose tolerance, and reducing visceral fat. These effects are considered to be related to the physiological functions of delaying digestion and absorption of nutrients in the diet from the digestive tract, and physiological functions of host energy by gut microbiota-derived SCFAs. However, the mechanism of barley β-glucan affecting lipid metabolism is still not understood at the gene level. There are very few reports which confirm the effects of barley β-glucan on lipid metabolism in several organs, such as intestinal tract, adipose, and liver tissues; therefore, it is still debatable whether the intake of β-glucan-rich barley flour is involved in lipid metabolism at the gene level.

To investigate the effect of β-glucan-rich barley on glucose and lipid metabolism, we measured gene expression levels using DNA microarrays, which are a tool to comprehensively detect total gene expression. In recent years, several studies have clarified the physiological effects of dietary fiber materials using DNA microarrays. Drew et al. reported that genes involved in the intestinal barrier function and metabolism of the intestinal epithelium were upregulated in the cecum of mice fed inulin or β-glucan extract of barley [21]. Other studies have reported that the expression of genes involved in fatty acid oxidation and lipid transport in skeletal muscle are upregulated in mice fed psyllium [22]. The purpose of this study was to investigate whether the intake of β-glucan-rich barley flour affects the expression levels of genes related to glucose and lipid metabolism in the ileum, liver, and adipose tissues of mice fed a high-fat diet using a DNA microarray and subsequent analysis by q-PCR. We also investigated the relationship between gene expression levels in each organ and SCFAs in the cecum.

2. Materials and Methods

2.1. Sample Preparation and Chemical Analysis

A new hulless barley cultivar, "Beau Fiber" (BF), was used in our experiments. BF flour pearled to 70% yield was obtained from the National Agriculture and Food Research Organization (Tsukuba, Japan). Total dietary fiber was analyzed using the AOAC 991.43 method [23]. β-Glucan content was analyzed using the McCleary method (AOAC 995.16) [24]. Protein and lipid content in BF were analyzed by the Kjeldahl and acid hydrolysis method, respectively. The nutritional content of BF is shown in Table 1.

Table 1. Nutritional components of Beau fiber (BF).

Beau Fiber (BF)	(g/100 g)
Moisture	8.1
Fat	3.0
Protein	8.8
Ash	0.1
Available carbohydrate	63.5
Total dietary fiber	16.5
β-Glucan	8.0

Available carbohydrate: (100 − ("Moisture" + "Fat" + "Protein" + "Ash" + "Total dietary fiber")).

2.2. Animals and Study Design

Four-week-old male C57BL/6J mice were purchased from Charles River Laboratories Japan, Inc. (Yokohama, Japan). Mice were housed on a 12 h light/dark cycle (light on at 07:30 h) under the conditions of constant air exchange at a temperature of 22 ± 1 °C and humidity of 50 ± 5%. Mice were acclimatized for one week on commercial chow (NMF, Oriental Yeast Co., Ltd., Shiga, Japan), before they were randomly assigned into two groups according to body weight (n = 10 per group). Mice were individually housed in plastic cages, and given the experimental powdered diets (Table 2). A high fat diet was prepared by the addition of lard to the AIN-93G diet (fat energy ratio 50%). The control and BF diets were supplemented with cellulose and BF flour, respectively, to give 5% total dietary fiber (Table 1). Mice were given free access to water and experimental diets during the whole experimental period of 92 days. Food intake and body weights were monitored three times a week during the study. Oral glucose tolerance test (OGTT) were performed in mice after 8 h fasting during the 11th week of the experimental diets. Mice were orally administered 20% glucose (1.5 g/kg body) and blood samples were collected from the tail at 0, 15, 30, 60, 120 min, before glucose levels were analyzed using electrode method (Glutest Neo Super, Sanwa Kagaku Kenkyusho Co., Ltd., Aichi, Japan). At the end of the study, mice were fasted for 8 h and sacrificed by isoflurane/CO_2 anesthesia. Blood samples were then collected from the postcaval vein, centrifuged to obtain serum which was stored at −80 °C until biochemical analysis. The weights of liver, cecum, and adipose (epididymal fat, retroperitoneal fat, mesenteric fat) tissues were measured and then ileum, liver, and adipose (epididymal fat) tissues were immediately soaked in RNA protect Tissue Reagent (RNAlater, Qiagen, Hilden, Germany), while samples of liver and cecum contents were stored at −30 °C. The animal protocol was approved by the Animal Research Committee of Otsuma Women's University (Tokyo, Japan) and was implemented in accordance with their regulations (No. 17012, February 2018).

Table 2. Composition of the control and BF experimental diets (g/kg).

	Control	BF
Dextrinized corn starch	329.5	114.8
Casein	200.0	200.0
Sucrose	100.0	100.0
Soybean oil	70.0	70.0
Lard	200.0	190.9
Cellulose powder	50.0	-
BF *		303.8
AIN-93G mineral mixture	35.0	35.0
AIN-93 vitamin mixture	10.0	10.0
L-Cystine	3.0	3.0
Choline bitartrate	2.5	2.5
t-Butylhydroquinone	0.014	0.014
The amounts of β-glucan (%)	0.0	2.4

* BF obtained from the National Agriculture and Food Research Organization (Tsukuba, Japan).

2.3. Concentration of Liver Lipid and Biochemical Analysis in Serum

Liver lipid concentration was measured using Folch's method [25]. Chloroform-methanol solution (2:1 *v/v*) was used to extract lipids from the liver, and isopropanol containing 10% polyoxyethylene octylphenyl ether (Triton X-100, FUJIFILM Wako Pure Chemical Corporation, Osaka, Japan) was added to dissolve the lipids. Triglyceride and cholesterol concentrations in the extracts were measured enzymatically using the Cholesterol E-test and Triglyceride E-test, respectively (FUJIFILM Wako Pure Chemical Corporation, Osaka, Japan).

Total cholesterol (TC), low-density lipoprotein (LDL), high-density lipoprotein (HDL)-cholesterol, triglycerides (TG) and non-esterified fatty acids (NEFA) were measured in mice serum using Hitachi 7180 auto-analyzers at the Nagahama Research Institute (Oriental Yeast Co., Ltd., Shiga, Japan). Enzyme-linked immunosorbent assays (ELISAs) were used to measure serum insulin (mouse insulin ELISA kit, Shibayagi Co., Ltd., Gunma, Japan) and leptin (mouse leptin immunoassay kit R & D Systems, Inc., Minneapolis, MN, USA) concentrations.

2.4. Short-Chain Fatty Acids Analysis in Cecum Contents

SCFA concentration in the cecum contents was analyzed by gas chromatography-mass spectrometry (GC/MS) based on a previous report [26]. Twenty mg of cecum contents was added to a 2 mL microtube containing 100 µL of internal standard (100 µM crotonic acid), 50 µL of HCl, 300 µL of diethyl ether, and 5 mm stainless beads (AS ONE Corp., Osaka, Japan), and homogenized using a Tissue Lyser II (Qiagen, Hilden, Germany) at 2000 rpm for 2 min twice. After homogenates were centrifuged (3000 rpm, at 25 °C, for 15 min), 80 µL of supernatant (ether layer) was collected into a glass vial (Agilent Technologies Japan, Ltd., Tokyo, Japan), and mixed with 16 µL of N-tert-butyldimethylsilyl-N-methyltrifluoroacetamide (derivatization reagent). Vials were then sealed with a cap, and heated at 80 °C for 20 min. Vials were then placed at room temperature for 48 h for derivatization. The derivatized samples were analyzed using the 7890B GC system equipped with a 5977A mass selective detector (Agilent Technologies Japan, Ltd., Tokyo, Japan) and DB-5MS column (30 m × 0.53 mm). The oven temperature was initially kept at 60 °C, then ramped up to 120 °C at a rate of 5 °C/min. The oven was then ramped to 300 °C at a rate of 20 °C/min, and finally maintained at 300 °C for 2 min. Helium was used as the carrier gas at 1.2 mL/min. The temperature of the front inlet, transfer line, and electron impact ion source were set at 250, 260, and 230 °C, respectively. The mass spectral data were collected in a selective ion monitoring mode. Concentrations of SCFA were calculated by comparing the peak area with the internal standard.

2.5. DNA Microarray Analysis

The RNeasy Mini kit (Qiagen, Hilden, Germany) was used to extract total RNA in ileum, liver, and adipose (epididymal fat) tissues. The RNA integrity number (RIN) of all total RNA samples was checked by a visual inspection of the Bioanalyzer electropherograms (Agilent technologies Japan, Ltd., Tokyo, Japan). Based on a previous report [27], samples with RIN values higher than 6.5 were used for the analysis. Each RNA was mixed equally to obtain pooled RNA for each group (n = 10, per group). Total RNA (100 ng) was processed for use on the microarray using the GeneChip WT PLUS Reagent Kit (Thermo Fisher Scientific, Inc., Waltham, MA, USA) according to the manufacturer's instructions. The resultant single-strand cDNA was fragmented and labeled with biotin, then hybridized to the GeneArray Mouse 2.0ST Array. The arrays were washed, stained and scanned using the Affymetrix 450 Fluidics Station and GeneChip Scanner 3000 7G (Thermo Fisher Scientific, Inc., Waltham, MA, USA) according to the manufacturer's recommendations (Thermo Fisher Scientific, Inc., Waltham, MA, USA). Expression values were generated using Expression Console software, version 1.3 (Thermo Fisher Scientific, Inc.) with default robust multichip analysis parameters. These analyzes were performed by Kurabo Industries Ltd. (Osaka, Japan)

Raw and standardized (logarithmic transformation) microarray data were registered in Gene Expression Omnibus (GEO) at the National Center for Biotechnology Information (NCBI). GEO accession number was GSE157828.

2.6. mRNAs Expression Analysis in Ileum, Liver, and Adipose Tissues

mRNA expression levels were analyzed by Applied Biosystems Quant3 Real-Time polymerase chain reaction (PCR) system. Synthesized cDNA was used to measure the mRNA expression level using the $2^{-\Delta\Delta CT}$ method and Power-up SYBR®® Green PCR Master Mix (Thermo Fisher Scientific, Waltham, MA, USA). We used threshold cycle (CT) for data analysis, which indicates the fractional cycle number as the amount of amplified target reaches a fixed threshold. The ΔCT is the difference in threshold cycles for target genes compared to the reference gene, 36B4. The $\Delta\Delta$CT is the difference between the ΔCT for the control group and the ΔCT for the BF group. Relative expression levels are showed as fold changes to the control group (arbitrary unit). Primer sequences are shown in Table S1.

2.7. Statistical Analysis

All statistical analyses were performed using R software (ver. 3.6.3, R Foundation for Statistical Computing, Vienna, Austria). Data are presented as mean ± standard deviation (SD) of the mean. Significant differences between the control group and BF group were analyzed by Student's *t*-test if homoscedasticity and normality were confirmed, and by Wilcoxon test if not confirmed. Difference were assessed with two-side test with an α level of 0.05. The relationships among the expression levels of mRNA related to lipid metabolism and SCFAs were assessed using Spearman's rank correlation coefficient.

3. Results

3.1. Changes in Body Weight, Food Intake, and Organ Weights

Body weight, food intake, and organ weights in mice fed the experimental diets are shown in Table 3. There were no significant differences in final weight, body weight gain, food intake, and food efficiency ratio between the two experimental groups. Organ weights in mice fed the experimental diets are shown in Table 4. Liver weight and the weights of retroperitoneal and mesenteric fat were significantly lower in the BF group compared with the control group ($p < 0.05$). The weight of cecum with digesta was significantly higher in the BF group compared with the control group ($p < 0.05$).

Table 3. Final weight, body weight gain, food intake, and food efficiency ratio in mice fed the control and BF diets.

	Control	BF	*p* Value
Initial weight (g)	20.3 ± 0.9	20.3 ± 0.8	N.S.
Final weight (g)	41.5 ± 2.8	38.5 ± 4.0	N.S.
Body weight gain (g/day)	0.23 ± 0.03	0.20 ± 0.04	N.S.
Food intake (g/day)	3.3 ± 0.2	3.2 ± 0.1	N.S.
Food intake (g/day/10 g final weight)	0.79 ± 0.04	0.82 ± 0.05	N.S.
Food efficiency ratio (%)	7.0 ± 0.7	6.3 ± 1.0	N.S.

Values are mean ± SD, Food efficiency ratio (%) = Body weight gain/Food intake × 100, N.S.: not significant difference. BF: Beau fiber group.

Table 4. Organ weights in mice fed the control BF and diets.

	Control	BF	*p* Value
Liver (g)	1.50 ± 0.23	1.25 ± 0.11	$p < 0.05$
Epididymal fat (g)	2.45 ± 0.40	2.24 ± 0.54	N.S.
Retroperitoneal fat (g)	0.93 ± 0.15	0.74 ± 0.19	$p < 0.05$
Mesenteric fat (g)	0.95 ± 0.26	0.67 ± 0.18	$p < 0.05$
Cecum with digesta (g)	0.23 ± 0.05	0.42 ± 0.07	$p < 0.05$

Values are mean ± SD, N.S.: not significant difference.

3.2. Concentrations of the Liver and Serum Lipid

The concentration of liver lipids, and serum biochemical markers are shown in Table 5. Accumulation of liver cholesterol and triglyceride was significantly lower in the BF group compared with the control group ($p < 0.05$). Serum TC, LDL and leptin concentrations were significantly lower in the BF group compared with the control group ($p < 0.05$). Serum TG concentration was significantly higher in the BF group compared with the control group ($p < 0.05$). There were no significant differences in serum HDL, NEFA, and insulin concentrations between the experimental groups.

Table 5. Concentrations of liver and serum lipids and hormones in mice fed the control and BF diets.

	Control	BF	*p* Value
Concentration of liver lipids			
Cholesterol (mmol/liver)	0.7 ± 0.1	0.4 ± 0.1	$p < 0.05$
(mmol/g liver)	0.5 ± 0.1	0.3 ± 0.1	$p < 0.05$
Triglyceride (mmol/liver)	8.0 ± 4.3	3.8 ± 1.5	$p < 0.05$
(mmol/g liver)	5.1 ± 1.9	3.0 ± 1.0	$p < 0.05$
Biochemical levels in serum			
Total cholesterol (mmol/L)	10.6 ± 0.7	8.8 ± 1.2	$p < 0.05$
LDL-cholesterol (mmol/L)	0.5 ± 0.1	0.4 ± 0.1	$p < 0.05$
HDL-cholesterol (mmol/L)	4.7 ± 0.2	4.6 ± 0.4	N.S.
Non-esterified fatty acid (NEFA) (μEq/L)	661.9 ± 63.7	666.5 ± 44.3	N.S.
Triglyceride (mmol/L)	2.4 ± 0.4	3.3 ± 0.8	$p < 0.05$
Insulin (ng/mL)	6.0 ± 3.3	3.9 ± 1.5	N.S.
Leptin (ng/mL)	92.8 ± 19.5	54.9 ± 22.1	$p < 0.05$

Values are mean ± SD, N.S.: not significant difference.

3.3. Oral Glucose Tolerance Test (OGTT)

The OGTT results are shown in Figure 1. Blood glucose levels after glucose administration were significantly lower in the BF group compared with the control group at 15 and 60 min ($p < 0.05$). No significant differences were observed at other times.

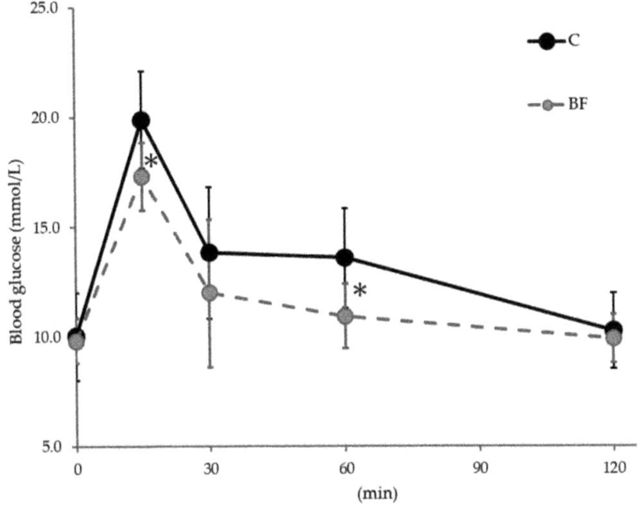

Figure 1. Blood glucose levels in the oral glucose tolerance test (OGTT). Data are shown as mean ± SD. * Significantly different from the control group ($p < 0.05$). C: Control group, BF: Beau fiber group.

3.4. Concentrations of SCFAs in Cecum Contents

Concentrations of SCFAs and organic acids in the cecum are shown in Table 6. The amounts of acetic acid, propionic acid, lactic acid, succinic acid, and total SCFAs were higher in the BF group compared with the control group ($p < 0.05$). No significant difference in the amount of butyric acid was observed between the experimental groups.

Table 6. Concentration of short-chain fatty acids (SCFAs) and organic acids in cecum contents.

(μmol/cecum)	Control	BF	*p* Value
Acetic acid	2.2 ± 0.9	3.7 ± 1.0	$p < 0.05$
Propionic acid	0.7 ± 0.3	1.4 ± 0.4	$p < 0.05$
Butyric acid	0.7 ± 0.2	0.9 ± 0.5	N.S.
Lactic acid	0.2 ± 0.0	0.4 ± 0.1	$p < 0.05$
Succinic acid	0.1 ± 0.1	0.3 ± 0.1	$p < 0.05$
Total short-chain fatty acids (SCFAs)	4.5 ± 1.5	7.7 ± 2.0	$p < 0.05$

Values are mean ± SD, N.S.: not significant difference.

3.5. Gene Expression Profiles of the Ileum, Liver, and Adipose Tissues from Microarray

3.5.1. Differential Expressed Genes (DEGs) by Microarray Analysis

The microarray analyses of each organ was compared by logarithmically converting values (Log-Ratio) of gene expression levels between the experimental groups, excluding non-coding genes. In the ileum of mice in the BF group, the expression levels of 1536 genes had increased >1.3 fold and 1529 genes had decreased <0.77 fold when compared with the control group (Table S2). In the liver of mice in the BF group, the expression levels of 2197 genes had increased >1.3 fold and 2054 genes had decreased <0.77 fold when compared with the control group (Table S3). In the adipose tissue of mice in the BF group, the expression levels of 2111 genes had increased >1.3 fold and 1878 genes had decreased <0.77 fold when compared with the control group (Table S4). The DEGs were annotated with gene symbols using Transcriptome Viewer (Kurabo Industries Ltd., Osaka, Japan) and subjected to further analyses.

3.5.2. DEG Profiles Related to Lipid Metabolism

Figures 2 and 3 show the ratio of DEGs to all genes involved in the lipid metabolism pathway, as described in the Kyoto Encyclopedia of Gene and Genome (KEGG) Pathway Database [28]. More than 10% of genes involved in steroid biosynthesis (mmu00140) were downregulated in all organ tissues. In the liver, more than 10% of genes involved in fatty acid elongation (mmu00062), glycerolipid metabolism (mmu00561), primary bile acid biosynthesis (mmu00120), biosynthesis of unsaturated fatty acids (mmu01040), and alpha-linolenic acid metabolism (mmu00592) were downregulated, while more than 10% of genes involved in steroid hormone biosynthesis (mmu00140), linoleic acid metabolism (mmu00591), and fatty acid biosynthesis (mmu00061) were upregulated. In adipose tissue, more than 10% of genes involved in fatty acid degradation (mmu00071), glycerolipid metabolism, ether lipid metabolism (mmu00565), and alpha-linolenic acid metabolism were downregulated, while more than 10% genes of genes involved in fatty acid biosynthesis were upregulated. In the ileum, more than 10% of genes involved in steroid hormone biosynthesis, ether lipid metabolism, arachidonic acid metabolism (mmu00590), linoleic acid metabolism, and alpha-linolenic acid metabolism were upregulated.

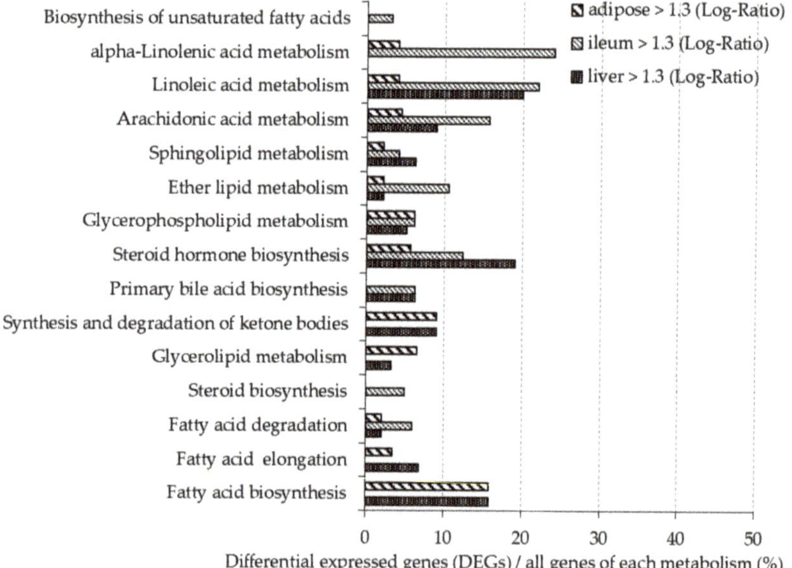

Figure 2. In mice fed the BF diet, the ratio of Differential Expressed Genes (DEGs) upregulated (gene expression > 1.3-fold difference compared with the control group) in each organ is shown. The *y*-axis shows the lipid metabolic pathway described in Kyoto Encyclopedia of Gene and Genome (KEGG). The *x*-axis shows the percentage of the DEGs of all genes involved in each metabolic pathway.

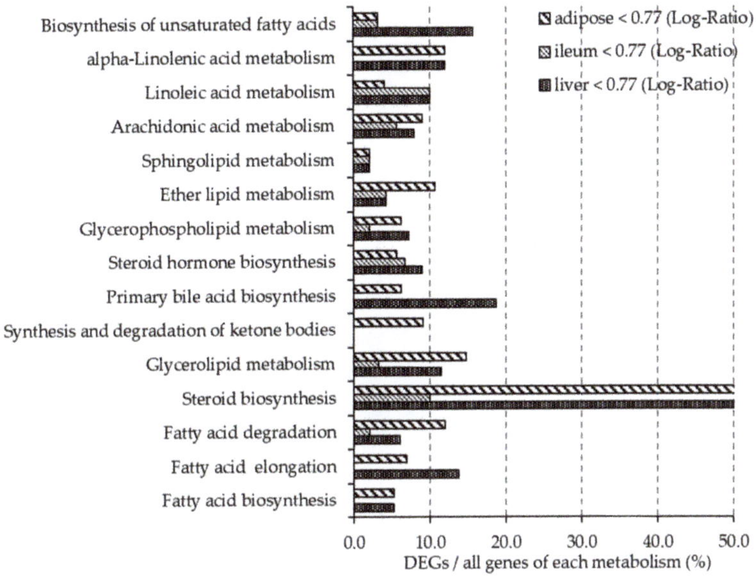

Figure 3. In mice fed the BF diet, the ratio of DEGs downregulated (gene expression < 0.77-fold difference compared with the control group) in each organ is shown. The *y*-axis shows the lipid metabolic pathways described in KEGG. The *x*-axis shows the percentage of DEGs of all genes involved in each metabolic pathway.

3.5.3. DEGs Profiles Related to Carbohydrate Metabolism

Figures 4 and 5 show the ratio of DEGs to all genes involved each carbohydrate metabolism pathway, as described in the KEGG Pathway Database. In the liver, more than 10% of genes involved in ascorbate and aldarate metabolism (mmu00053) were upregulated, and more than 10% of genes involved in galactose metabolism (mmu00052) and fructose and mannose metabolism (mmu00051) were downregulated. Less than 10% of the genes involved in carbohydrate pathways in the ileum were differentially expressed. In adipose tissue, more than 10% of the genes involved in inositol phosphate metabolism, starch and sucrose metabolism (mmu00500), galactose metabolism and pentose phosphate metabolism (mmu00040) were downregulated.

3.6. mRNA Expression Levels in Ileum, Liver, and Adipose Tissues Using q-PCR

mRNA expression levels in the ileum, liver, and adipose tissues are shown in Figure 6. No significant differences in mRNA expression levels were observed in the ileum (Figure 6A). In liver, the mRNA expression levels of fatty acid synthase (FAS), acyl-CoA oxidase (ACOX), diacylglycerol o-acyltransferase-1 (DGAT1), stearoyl-CoA-desaturase-1 (SCD-1), carnitine palmitoyltransferase-1 (CPT-1), 3-hydroxy-3-methyl-glutaryl-coenzyme A reductase (HMG-CoA r), and cytochrome P450 7A1 (CYP7a1) were significantly lower in the BF group compared with the control group ($p < 0.05$) (Figure 6B). In adipose tissue, the mRNA expression levels of hormone-sensitive lipase (HSL) was higher in the BF group compared with the control group ($p < 0.05$), and the mRNA expression of monocyte chemotactic protein 1 (MCP-1), F4/80, and NADPH oxidase subunit p67 phox (p67[phox]) were significantly lower in the BF group compared with the control group ($p < 0.05$) (Figure 6C).

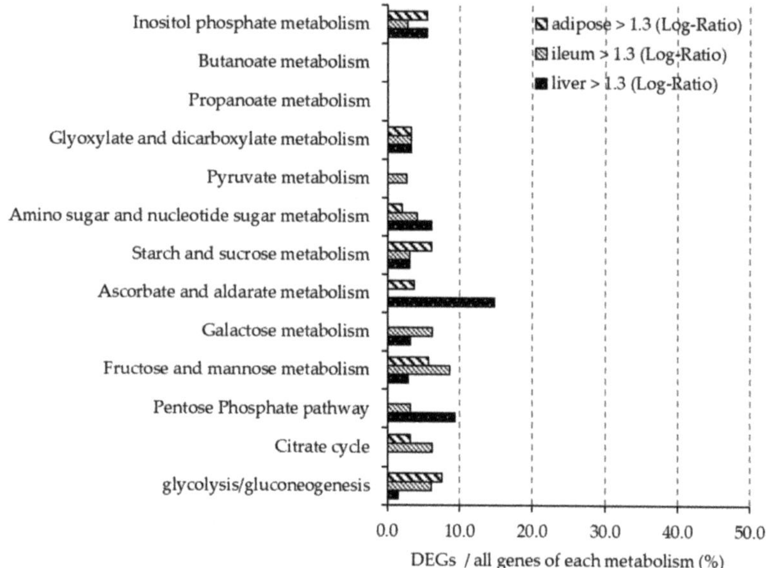

Figure 4. In mice fed the BF diet, the ratio of DEGs upregulated (gene expression > 1.3-fold difference compared with the control group) in each organ is shown. The *y*-axis shows the carbohydrate metabolic pathway described in KEGG. The *x*-axis shows the percentage of DEGs of all genes involved in each metabolic pathway.

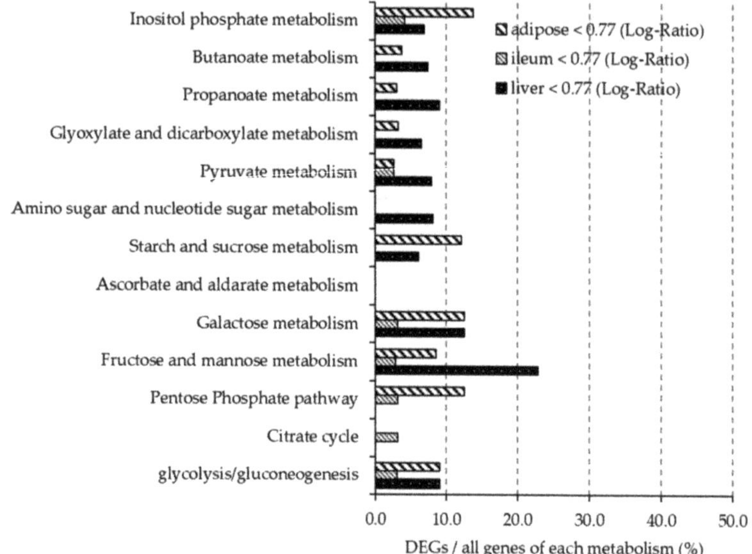

Figure 5. In mice fed the BF diet, the ratio of DEGs downregulated (gene expression < 0.77-fold difference compared with the control group) in each organ is shown. The *y*-axis shows the carbohydrate metabolic pathway described in KEGG. The *x*-axis shows the percentage of DEGs of all genes involved in each metabolic pathway.

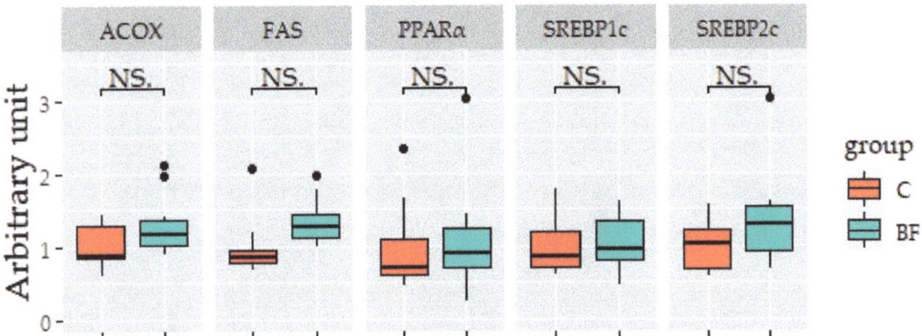

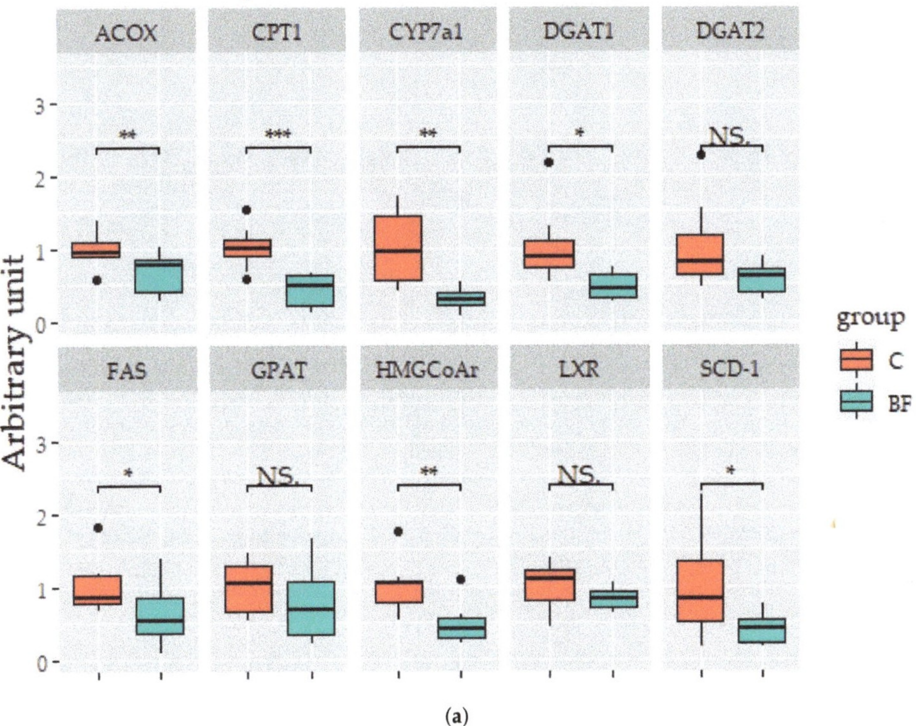

(a)

Figure 6. *Cont.*

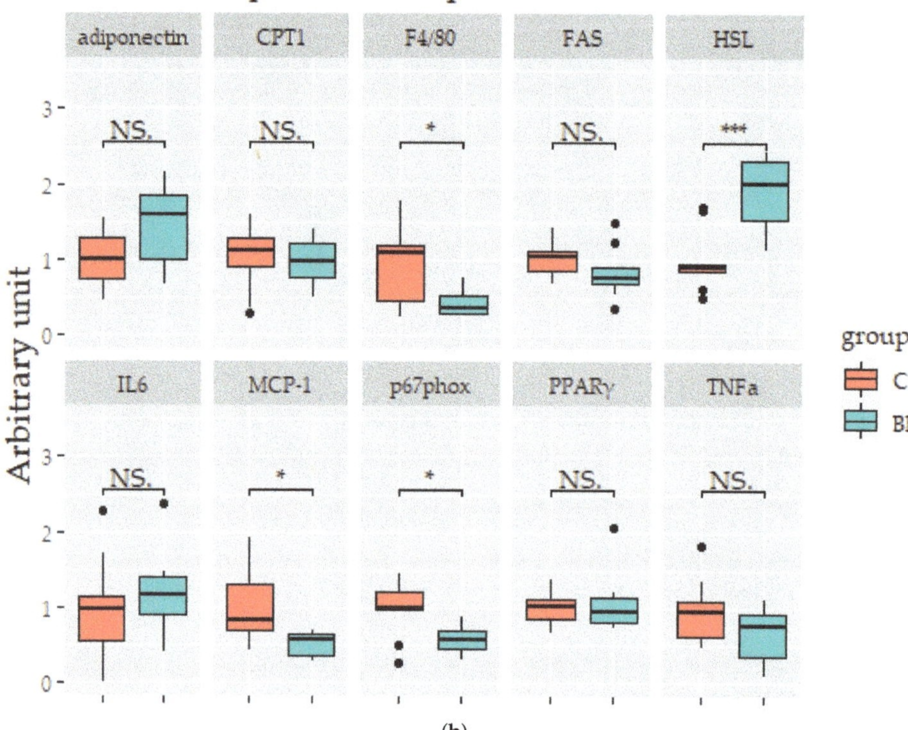

(b)

Figure 6. The expression levels of mRNA in the ileum (**A**), liver (**B**), and adipose (**C**) tissues. Values are means ± SD. * $p < 0.05$, ** $p < 0.01$, *** $p < 0.001$ significantly different from the control group. NS.: not significant. ACOX, Acyl-CoA oxidase; FAS, Fatty acid synthase; PPARα, Peroxisome proliferator-activated receptor-α; SREBP1c, Sterol regulatory element-binding protein-1c; SREBP2, Sterol regulatory element-binding protein-2; CPT1, Carnitine palmitoyl transferase 1; CYP7a1, Cholesterol 7alpha-hydroxylase 1; DGAT1, Diacylglycerol acyltransferase-1; DGAT2, Diacylglycerol acyltransferase-2; GPAT, glycerol-3-phosphate acyltransferase; HMGCoAr, Hydroxymethylglutaryl-CoA reductase, reductase; LXR, Liver X receptor; SCD-1, Stearoyl-CoA desaturase 1; HSL, Hormone-Sensitive Lipase; IL6, Interleukin-6; MCP-1, Monocyte chemotactic protein-1; p67phox, NADPH oxidase subunit p67 phox; PPARγ, Peroxisome proliferator-activated receptor-γ; TNF-α, Tumor necrosis factor-α. C: Control group, BF: Beau fiber group.

3.7. Correlation Analysis between SCFAs and Expression of mRNA Related to Lipid Metabolism in Liver and Adipose Tissues

Figure 7 shows the correlation coefficients by Spearman's rank correlation analysis between the concentration of SCFAs in the cecum and mRNA expression in the liver. The concentration of acetic acid in the cecum negatively correlated with mRNA expression of ACOX, HMG-CoA reductase, and CYP7a1 in the liver ($p < 0.05$). The concentration of propionic acid in the cecum negatively correlated with mRNA expression of ACOX, CPT1, DGAT1, DGAT2, HMG-CoA reductase, and CYP7a1 ($p < 0.05$). The concentration of lactic acid in the cecum was negatively correlated with mRNA expression of CYP7a1 ($p < 0.05$). The concentration of succinic acid in the cecum negatively correlated with mRNA expression of CYP7a1 ($p < 0.05$). There was no relationship between the concentration of butyric acid in cecum and mRNA expression in the liver.

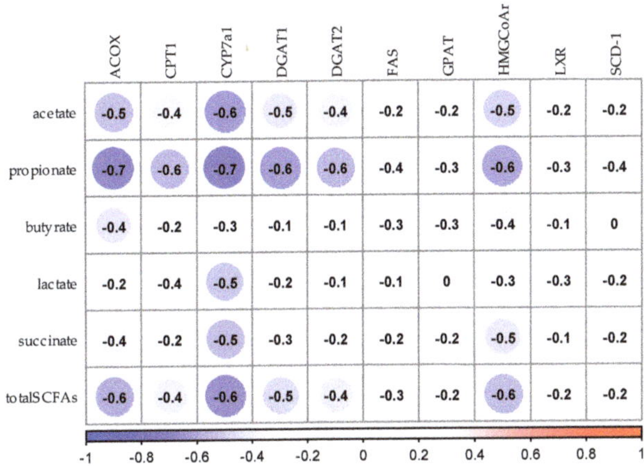

Figure 7. Spearman's rank correlation analysis between SCFAs and expression of mRNA related to lipid metabolism in the liver. The values in the figure shows the correlation coeffcient. Blue circles show negative correlation and red circles show positive correlation. ACOX, Acyl-CoA oxidase; CPT1, Carnitine palmitoyl Table 1. CYP7a1, Cholesterol 7alpha-hydroxylase 1; DGAT1, Diacylglycerol acyltransferase-1; DGAT2, Diacylglycerol acyltransferase-2; FAS, Fatty acid synthase; GPAT, glycerol-3-phosphate acyltransferase; HMGCoAr, Hydroxymethylglutaryl-CoA reductase, reductase; LXR, Liver X receptor; SCD-1, Stearoyl-CoA desaturase 1; total SCFAs, total short-chain fatty acids.

The relationship between mRNA expression in adipose tissue and the concentration of SCFAs in the cecum is showed in Figure 8. The concentrations of propionic acid and lactic acid in cecum positively correlated with mRNA expression of HSL in adipose tissue ($p < 0.05$). The concentration of lactic acid in the cecum negatively correlated with mRNA expression of MCP-1 ($p < 0.05$).

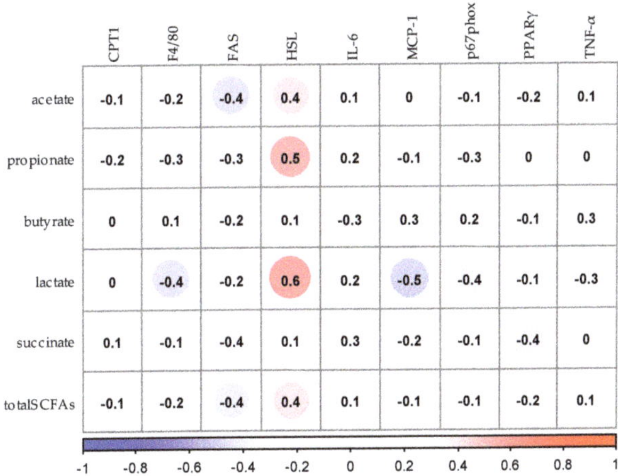

Figure 8. Spearman's rank correlation analysis between SCFAs and expression of mRNA related to lipid metabolism in adipose tissues. The values in the figure shows the correlation coeffcient. Blue circles show negative correlation and red circles show positive correlation. CPT1, Carnitine palmitoyl transferase 1; FAS, Fatty acid synthase; HSL, Hormone-Sensitive Lipase; IL6, Interleukin-6; MCP-1, Monocyte chemotactic protein-1; p67phox, NADPH oxidase subunit p67 phox; PPARγ, Peroxisome proliferator-activated receptor-γ; TNF-α, Tumor necrosis factor-α; total SCFAs, total short-chain fatty acids.

4. Discussion

We investigated whether β-glucan-rich barley flour affected glucose and lipid metabolism at the gene expression level in high-fat obesity model mice by comparing them with a control group. Firstly, we performed OGTTs and found that barley flour intake reduced postprandial blood glucose levels in BF mice compared with the control mice. Barley β-glucan is known to moderate the digestion and absorption of carbohydrates in the intestine, which results in suppressing excessive elevation of postprandial blood glucose. Previous studies have shown that long-term barley intake suppressed elevated blood glucose levels in mice [29]. Brockman et al. reported that consumption of high concentrations of barley β-glucan improved glucose control during glucose tolerance tests [30]. Our results were consistent with these previous studies and suggested that barley intake improves glucose tolerance. Moreover, this effect may be regulated by SCFAs. A previous study speculated that barley suppresses food intake and improves insulin sensitivity through SCFAs, which promote intestinal hormone secretion, such as glucagon-like peptide 1 (GLP-1) from intestinal enteroendocrine cells [31]. GLP-1 has the ability to decrease blood glucose levels by modulating glucose-dependent insulin secretion [32]. We confirmed that barley intake increased SCFAs in the cecum in this study (Table 4); however, further studies are needed to elucidate the effect of gut hormone secretion on glucose metabolism. DNA microarray analysis showed that there was only a slight change in the expression level of genes related to glucose metabolism in ileum, liver and adipose tissues after barley intake. Therefore, we speculated that barley flour intake reduced postprandial blood glucose levels and was mainly influenced by retardation of glucose absorption and secretion of intestinal hormones, but not through changes in the expression of genes related to glucose metabolism in ileum, liver, and adipose tissues.

Intake of a diet rich in β-glucan barley improved lipid metabolism in mice. DNA microarray analysis showed that the expression of 50% of the genes related to steroid biosynthesis (mmu00100) pathways were greatly downregulated in the BF group compared with the control group in liver and adipose tissues. In particular, the mRNA expression level of HMG-CoA reductase, the rate-limiting enzyme in cholesterol synthesis, was lower in the BF group compared with the control group in the liver. Xia et al. also reported that intake of whole-grain significantly lowered the expression of HMG-CoA reductase in the liver and plasma and liver lipid levels when compared with the no barley group in high-fat diet model rats [33]. These reports are similar to the results obtained in our study. We found that intake of barley flour lowered liver and serum TC concentrations by consistently reducing the expression of genes involved in cholesterol synthesis. Moreover, it was previously suggested that improvements in lipid metabolism after barley intake is due to the promotion of cholesterol excretion in the intestine and inhibition of bile acid re-absorption. It is known that CYP7a1 is the rate-limiting enzyme for bile acid synthesis in the liver. A previous study reported that intake of barley β-glucan and waxy barley promotes excretion of bile acids in the feces; concomitantly increasing CYP7a1 expression and bile acid synthesis, thereby suppressing cholesterol synthesis [34]. However, in our study, the mRNA expression levels of CYP7a1 were significantly lower in the BF group compared with the control group. We also found that the fecal lipid levels did not change between the experimental and control groups (data not shown). The differing results may be due to differences in the source of barley, the experimental period and/or the animal model. Further studies are needed to elucidate how changes in the intake of barley affects bile acid metabolism.

The metabolic syndrome indices (abdominal fat organs, liver lipids, serum TC and LDL) is improved in mice fed barley flour: we previously reported that barley intake reduced fat accumulation in mice [35] and Choi et al. reported that consumption of barley β-glucan extract is involved in preventing obesity in mice fed a high-fat diet [36]. Another study reported that barley intake has an improving effect, both histologically and biochemically, on the liver of diabetic-model rats [37]. Similar effects have been shown in human studies; Japanese patients with hyperlipidemia had decreased abdominal fat and serum LDL after barley intake [38] and abdominal fat was decreased by whole-grain cereal intake in healthy humans [39]. Our results support previous studies. A DNA microarray showed that more than 10% of genes involved in glycerolipid metabolism (mmu00561) and the fatty acid elongation pathway (mmu00062) were downregulated in the BF group compared with the control group in the liver. q-PCR analyses also showed the expression of genes involved in lipid synthesis, such as DGAT1, SCD-1, FAS, and degradation, ACOX, CPT-1, were lowered after β-glucan-rich barley flour intake. The data suggest that gene expression levels related to lipid metabolism were lower overall in the liver, and affected an improvement in the obesity index.

Intake of β-glucan-rich barley flour affected the expression levels of mRNA related to lipid metabolism in adipocytes in a different way to the control group. Adipocytes are not only energy storage organs but also secrete some adipocytokines to regulate energy metabolism. In the present study, intake of barley flour downregulated genes related to several lipid metabolism pathways, such as steroid biosynthesis (mmu00100), and glycerolipid metabolism (mmu00561), in adipose tissue. It is considered that lipids flowing into adipose tissue were lower in the BF group compared with the control group due to lower lipid synthesis in the liver. The mRNA expression levels of HSL were significantly higher in the BF group compared with the control group. It is known that HSL catalyzes the degradation of triacylglycerol and decreases in the expression of HSL are associated with insulin concentration and obesity [40,41]. In this study, serum insulin concentration was reduced by barley flour intake (BF: 3.9 ± 1.5 ng/mL, control 6.0 ± 3.3 ng/mL); therefore, it was suggested that increased HSL expression occurs by the lower concentration of insulin accelerating fat degradation. Positive correlations between several SCFAs and HSL expression in adipose tissue were detected; however, further studies are needed to evaluate the relationship between HSL and SCFAs.

The expression levels of genes related to glucose and lipid metabolism were unchanged in the ileum. A previous study using fructo-oligosaccharides reported that succinic acid induced intestinal gluconeogenesis in vivo [42]; however, our study produced different results and this may be due to the differences in soluble fiber material. Therefore, our results suggest that the main organs in which barley flour is involved in lipid metabolism at the gene expression level are the liver and adipose tissues, but not the ileum.

Some inflammatory markers (MCP-1, p67^{phox}, F4/80) were significantly decreased in the adipose tissue of barley-supplemented mice. It has been reported that adipocytes induce early inflammation by producing active oxygen via activation of NADPH oxidase in mice on a high-fat diet [43]. These results suggest that a decrease in adipocyte hypertrophy reduces fat synthesis in the liver, thus lowering visceral fat weight and inflammation in adipocytes.

Previous studies have described SCFAs as energy sources for the intestine and liver [44,45]. We found that several SCFAs negatively correlated with the expression levels of mRNA involved in hepatic lipid synthesis and degradation. Our results are supported by previous studies in which SCFAs suppressed lipid synthesis in the liver [46] and the expression levels of FAS and carbohydrate-response element-binding protein (Chrebp) were lower in mice fed SCFAs compared with mice fed cellulose [47]. Based on these results, we suggest that SCFAs alter the expression of genes related to lipid metabolism in the cecum. A recent study revealed that SCFAs influence the metabolism of lipids and glucose to regulate GLP-1 and transcription factor activity involved in lipid metabolism via intestinal SCFA receptors [48]. It is known that G protein-coupled receptors (GPCRs) are activated by SCFAs; notably, GPR43 and GPR41 regulate gut hormone secretion [49]. Further studies using mice deficient in these SCFA receptors will clarify the relationship between barley intake, SCFAs and the expression of genes related to lipid metabolism.

5. Conclusions

Our study provides evidence that β-glucan-rich barley flour affects lipid metabolism in the liver and adipose tissues at the gene expression level in high-fat obesity model mice. Using a DNA microarray, we observed that the intake of barley flour consistently reduced the expression levels of genes related to steroid biosynthesis in the liver and adipose tissue; however, in the ileum, barley flour is suggested to affect lipid metabolism through digestive absorption and intestinal fermentation, rather than changes in gene expression. The expression of genes involved in glucose metabolism did not change in each organ; therefore, we suggest that the improving effect of barley flour on postprandial blood glucose levels is caused by the modification of the gut events such as the retardation of glucose absorption and intestinal fermentation. In conclusion, microbiota-derived SCFAs affect changes in the expression levels of genes related to lipid metabolism in the liver and adipose tissues, thus affecting lipid accumulation and abdominal fat.

Supplementary Materials: The following are available online at http://www.mdpi.com/2072-6643/12/11/3546/s1, Table S1: Primers used in the real-time reverse transcription polymerase chain reaction. Table S2: Differentially expressed genes in the ileum by microarray analysis. Table S3: Differentially expressed genes in the liver by microarray analysis., Table S4: Differentially expressed genes in the adipose tissue by microarray analysis.

Author Contributions: Conceptualization, K.M. and S.A.; methodology, K.M. and S.A.; formal analysis, K.M., C.Y. and S.A.; resources, T.M. and T.K.; original draft preparation and writing, K.M.; review and editing, K.M., C.Y., T.M., T.K. and S.A.; supervision, S.A. All authors have read and agreed to the published version of the manuscript.

Funding: This research received no external funding.

Acknowledgments: This study was funded by Hakubaku Co., Ltd. (Yamanashi, Japan).

Conflicts of Interest: The author belongs to Hakubaku Co., Ltd., and part of this study included funding from Hakubaku. Co., Ltd.

References

1. Grundy, S.M. Metabolic syndrome update. *Trends Cardiovasc. Med.* **2016**, *26*, 364–373. [CrossRef] [PubMed]
2. Aune, D.; Norat, T.; Romundstad, P.; Vatten, L.J. Whole grain and refined grain consumption and the risk of type 2 diabetes: A systematic review and dose-response meta-analysis of cohort studies. *Eur. J. Epidemiol.* **2013**, *28*, 845–858. [CrossRef] [PubMed]
3. Ye, E.Q.; Chacko, S.A.; Chou, E.L.; Kugizaki, M.; Liu, S. Greater whole-grain intake is associated with lower risk of type 2 diabetes, cardiovascular disease, and weight gain. *J. Nutr.* **2012**, *142*, 1304–1313. [CrossRef] [PubMed]
4. JACC Study Group. Dietary fiber intake is associated with reduced risk of mortality from cardiovascular disease among Japanese men and women. *J. Nutr.* **2010**, *140*, 1445–1453. [CrossRef]
5. Fuller, S.; Beck, E.; Salman, H.; Tapsell, L. New horizons for the study of dietary fiber and health: A review. *Plant Foods Hum. Nutr.* **2016**, *71*, 1–12. [CrossRef]
6. Cho, S.S.; Qi, L.; Fahey, G.C., Jr.; Klurfeld, D.M. Consumption of cereal fiber, mixtures of whole grains and bran, and whole grains and risk reduction in type 2 diabetes, obesity, and cardiovascular disease. *American J. Clin. Nutr.* **2013**, *98*, 594–619. [CrossRef]
7. Anderson, J.W.; Story, L.; Sieling, B.; Chen, W.J.; Petro, M.S.; Story, J. Hypocholesterolemic effects of oat-bran or bean intake for hypercholesterolemic men. *Am. J. Clin. Nutr.* **1984**, *40*, 1146–1155. [CrossRef]
8. Nakashima, A.; Yamada, K.; Iwata, O.; Sugimoto, R.; Atsuji, K.; Ogawa, T.; Ishibashi-Ohgo, N.; Suzuki, K. β-Glucan in foods and its physiological functions. *J. Nutr. Sci. Vitaminol.* **2018**, *64*, 8–17. [CrossRef]
9. Chutkan, R.; Fahey, G.; Wright, W.L.; McRorie, J. Viscous versus nonviscous soluble fiber supplements: Mechanisms and evidence for fiber-specific health benefits. *J. Am. Acad. Nurse Pract.* **2012**, *24*, 476–487. [CrossRef]
10. Weickert, M.O.; Pfeiffer, A.F.H. Metabolic effects of dietary fiber consumption and prevention of diabetes. *J. Nutr.* **2008**, *138*, 439–442. [CrossRef]
11. Son, B.K.; Kim, J.Y.; Lee, S.S. Effect of adlay, buckwheat and barley on lipid metabolism and aorta histopathology in rats fed an obesogenic diet. *Ann. Nutr. Metab.* **2008**, *52*, 181–187. [CrossRef] [PubMed]
12. Åman, P. Cholesterol-lowering effects of barley dietary fibre in humans: Scientific support for a generic health claim. *Scand. J. Food Nutr.* **2006**, *50*, 173–176. [CrossRef]
13. Keenan, J.M.; Goulson, M.; Shamliyan, T.; Knutson, N.; Kolberg, L.; Curry, L. The effects of concentrated barley b-glucan on blood lipids in a population of hypercholesterolaemic men and women. *Br. J. Nutr.* **2007**, *97*, 1162–1168. [CrossRef] [PubMed]
14. Wang, L.; Newman, R.K.; Newman, C.W.; Hofer, P.J. Barley beta-glucans alter intestinal viscosity and reduce plasma cholesterol concentrations in chicks. *J. Nutr.* **1992**, *122*, 2292–2297. [CrossRef] [PubMed]

15. Qureshi, A.A.; Burger, W.C.; Herbert, N.P.; Bird, R.; Sunde, M.L. Regulation of lipid metabolism in chicken liver by dietary cereals. *J. Nutr.* **1980**, *110*, 388–393. [CrossRef] [PubMed]

16. Den Besten, G.; van Eunen, K.; Groen, A.K.; Venema, K.; Reijngoud, D.J.; Baller, B.M. The role of short-chain fatty acids in the interplay between diet, gut microbiota, and host energy metabolism. *J. Lipid Res.* **2013**, *54*, 2325–2340. [CrossRef] [PubMed]

17. Cummings, J.H.; Branch, W.J. Fermentation and the production of short chain fatty acids in the human large intestine. In *Dietary Fiber: Basic and Clinical Aspect*; Vahouny, G.V., Kritchevsky, D., Eds.; Springer: Boston, MA, USA, 1986; pp. 131–149.

18. Hague, A.; Butt, A.J.; Paraskeva, C. The role of butyrate in human colonic epithelial cells: An energy source or inducer of differentiation and apoptosis? *Proc. Nutr. Soc.* **1996**, *55*, 937–943. [CrossRef]

19. Joanne, S. Fiber and prebiotics: Mechanisms and health benefits. *Nutrients* **2013**, *5*, 1415–1435.

20. Kimura, I.; Ichimura, A.; Ohue-Kitano, R.; Igarashi, M. Free fatty acid receptors in health and disease. *Physiol. Rev.* **2019**, *100*, 171–210. [CrossRef]

21. Drew, J.E.; Reichardt, N.; Williams, L.M.; Mayer, C.-D.; Walker, A.W.; Farquharson, A.J.; Kastora, S.; Farquharson, F.; Milligan, G.; Morrison, D.J.; et al. Dietary fibers inhibit obesity in mice, but host responses in the cecum and liver appear unrelated to fiber-specific changes in cecal bacterial taxonomic composition. *Sci. Rep.* **2018**, *8*, 15566. [CrossRef]

22. Togawa, N.; Takahashi, R.; Hirai, S.; Fukushima, T.; Egashira, Y. Gene expression analysis of the liver and skeletal muscle of psyllium-treated mice. *Br. J. Nutr.* **2013**, *109*, 383–393. [CrossRef] [PubMed]

23. Lee, S.C.; Rodriguez, F.; Storey, M.; Farmkalidis, E.; Prosky, L. Determination of soluble and insoluble dietary fiber in psyllium containing cereal products. *J. AOAC Int.* **1995**, *78*, 724–729. [CrossRef] [PubMed]

24. McCleary, B.V.; Codd, R. Measurement of (1-3), (1-4)-β-D-glucan in barley and oats: A streamlined enzymic procedure. *J. Sci. Food Agric.* **1991**, *55*, 303–312. [CrossRef]

25. Folch, J.; Lee, M.; Stanley, G.H. A simple method for the isolation and purification of total lipids from animal tissues. *J. Biol. Chem.* **1957**, *226*, 497–509. [PubMed]

26. Atarashi, K.; Tanoue, T.; Oshima, K.; Suda, W.; Nagano, Y.; Nishikawa, H.; Fukuda, S.; Saito, T.; Narushima, S.; Hase, K.; et al. Treg induction by a rationally selected mixture of clostridia strains from the human microbiota. *Nature* **2013**, *500*, 232–236. [CrossRef]

27. Steffen, M. *Application Note: Optimizing Real-Time Quantitative PCR Experiments with the Agilent 2100 Bioanalyzer*; Agilent Technologies, Inc.: Santa Clara, CA, USA, 2008.

28. KEGG. Kyoto Encyclopedia of Genes and Genomes. Available online: https://www.genome.jp/kegg/pathway.html (accessed on 9 April 2020).

29. Kato, M.; Tsubaki, K.; Kuge, T.; Aoe, S. Effect of a high-fat diet with barley β-glucan on glucose tolerance and abdominal fat-liver lipid accumulation in mice. *Jpn. J. Nutr. Diet.* **2016**, *74*, 60–68. [CrossRef]

30. Brockman, D.A.; Chen, X.; Gallaher, D.D. Consumption of a high β-glucan barley flour improves glucose control and fatty liver and increases muscle acylcarnitines in the zucker diabetic fatty rat. *Eur. J. Nutr.* **2013**, *52*, 1743–1753. [CrossRef]

31. Miyamoto, J.; Watanabe, K.; Taira, S.; Kasubuchi, M.; Li, X.; Irie, J.; Kimura, I. Barley β-glucan improves metabolic condition via short-chain fatty acids produced by gut microbial fermentation in high fat diet fed mice. *PLoS ONE* **2018**, *13*, e0196579. [CrossRef]

32. Silva, A.D.; Salem, V.; Long, C.J.; Makwana, A.; Newbould, R.D.; Rabiner, E.A.; Ghatei, M.A.; Bloom, S.R.; Matthews, P.M.; Beaver, J.D.; et al. The gut hormones PYY3-36 and GLP-17-36 amide reduce food intake and modulate brain activity in appetite centers in humans. *Cell Metab.* **2011**, *14*, 700–706.

33. Xia, X.; Li, G.; Song, J.; Zheng, J.; Kan, J. Hypocholesterolaemic effect of whole-grain highland hull-less barley in rats fed a high-fat diet. *Br. J. Nutr.* **2018**, *119*, 1102–1110. [CrossRef]

34. Yang, J.L.; Kim, Y.H.; Lee, H.S.; Lee, M.S.; Moon, Y.K. Barley β-glucan lowers serum cholesterol based on the up-regulation of cholesterol 7α-hydroxylase activity and mRNA abundance in cholesterol-fed rats. *J. Nutr. Sci. Vitaminol.* **2003**, *49*, 381–387. [CrossRef] [PubMed]

35. Aoe, S.; Ichinose, Y.; Kohyama, N.; Komae, K.; Takahashi, A.; Yoshioka, T.; Yanagisawa, T. Effects of β-glucan content and pearling of barley in diet-induced obese mice. *Cereal Chem.* **2017**, *94*, 956–962. [CrossRef]

36. Choi, J.S.; Kim, H.; Jung, M.H.; Hong, S.; Song, J. Consumption of barley beta-glucan ameliorates fatty liver and insulin resistance in mice fed a high-fat diet. *Mol. Nutr. Food Res.* **2010**, *54*, 1004–1013. [CrossRef] [PubMed]

Nutrients **2020**, *12*, 3546

37. Gehan, K.; Abdel-Gabbar, M. Effect of barley (Hordeum Vulgare) on the liver of diabetic rats: Histological and biochemical study. *Egypt. J. Histol.* **2008**, *31*, 245–255.

38. Shimizu, C.; Kihara, M.; Aoe, S.; Araki, S.; Ito, K.; Hayashi, K.; Watari, J.; Sakata, Y.; Ikegami, S. Effect of high β-glucan barley on serum cholesterol concentrations and visceral fat area in Japanese men—A randomized, double-blinded, placebo-controlled trial. *Plant Foods Hum. Nutr.* **2008**, *63*, 21–25. [CrossRef]

39. Alminger, M.; Eklund Jonsson, C. Whole-grain cereal products based on a high-fiber barley or oat genotype lower post-prandial glucose and insulin responses in healthy humans. *Eur. J. Nutr.* **2008**, *47*, 294–300. [CrossRef]

40. Gaidhu, M.P.; Anthony, N.M.; Patel, P.; Hawke, T.J.; Ceddia, R.B. Dysregulation of lipolysis and lipid metabolism in visceral and subcutaneous adipocytes by high-fat diet: Role of ATGL, HSL, and AMPK. *Am. J. Physiol. Cell Physiol.* **2010**, *298*, C961–C971. [CrossRef]

41. Lan, Y.L.; Lou, J.C.; Lyu, W.; Zhang, B. Update on the synergistic effect of HSL and insulin in the treatment of metabolic disorders. *Ther. Adv. Endocrinol. Metab.* **2019**, *10*, 31565213. [CrossRef]

42. Vadder, F.D.; Kovatcheva-Datchary, P.; Zitoun, C.; Duchamp, A.; Bäckhed, F.; Mithieux, G. Microbiota-produced succinate improves glucose homeostasis via intestinal gluconeogenesis. *Cell Metab.* **2016**, *24*, 151–157. [CrossRef]

43. Furukawa, S.; Fujita, T.; Shimabukuro, M.; Iwaki, M.; Yamada, Y.; Nakajima, Y.; Nakayama, O.; Makishima, M.; Matsuda, M.; Shimomura, I. Increased oxidative stress in obesity and its impact on metabolic syndrome. *J. Clin. Investig.* **2004**, *114*, 1752–1761. [CrossRef]

44. Bergman, E.N. Energy contributions of volatile fatty acids from the gastrointestinal tract in various species. *Physiol. Rev.* **1990**, *70*, 567–590. [CrossRef]

45. Flint, H.J.; Bayer, E.A.; Rincon, M.T.; Lamed, R.; White, B.A. Polysaccharide utilization by gut bacteria: Potential for new insights from genomic analysis. *Nat. Rev. Microbiol.* **2008**, *6*, 121–131. [CrossRef] [PubMed]

46. Li, X.; Shimizu, Y.; Kimura, I. Gut microbial metabolite short-chain fatty acids and obesity. *Biosci. Microb. Food Health* **2017**, *36*, 135–140. [CrossRef] [PubMed]

47. Shimizu, H.; Masujima, Y.; Ushiroda, C.; Mizushima, R.; Taira, S.; Ohue-Kitano, R.; Kimura, I. Dietary short-chain fatty acid intake improves the hepatic metabolic condition via FFAR3. *Sci. Rep.* **2019**, *9*, 16574. [CrossRef] [PubMed]

48. den Besten, G.; Bleeker, A.; Gerding, A.; van Eunen, K.; Havinga, R.; van Dijk, T.H.; Oosterveer, M.H.; Jonker, J.W.; Groen, A.K.; Reijngoud, D.J.; et al. Short-chain fatty acids protect against high-fat diet-induced obesity via a PPARγ-dependent switch from lipogenesis to fat oxidation. *Diabetes* **2015**, *64*, 2398–2408. [CrossRef]

49. Tolhurst, G.; Heffron, H.; Lam, Y.S.; Parker, H.E.; Habib, A.M.; Diakogiannaki, E.; Cameron, J.; Grosse, J.; Reimann, F.; Gribble, F.M. Short-chain fatty acids stimulate glucagon-like peptide-1 secretion via the G-protein–coupled receptor FFAR2. *Diabetes* **2012**, *61*, 364–371. [CrossRef]

Publisher's Note: MDPI stays neutral with regard to jurisdictional claims in published maps and institutional affiliations.

Article

High β-Glucan Barley Supplementation Improves Glucose Tolerance by Increasing GLP-1 Secretion in Diet-Induced Obesity Mice

Sachina Suzuki [1] and Seiichiro Aoe [1,2,*]

1   The Institute of Human Culture Studies, Otsuma Women's University, Chiyoda-ku, Tokyo 102-8357, Japan; suzusachi0421@gmail.com

2   Studies in Human Life Sciences, Graduate School of Studies in Human Culture, Otsuma Women's University, Chiyoda-ku, Tokyo 102-8357, Japan

*   Correspondence: s-aoe@otsuma.ac.jp; Tel.: +81-3-5275-6048

Abstract: The aim of this study was to investigate the underlying mechanism for the improvement of glucose tolerance following intake of high β-glucan barley (HGB) in terms of intestinal metabolism. C57BL/6J male mice were fed a fatty diet supplemented with HGB corresponding to 5% of dietary fiber for 83 days. An oral glucose tolerance test was performed at the end of the experimental period. The concentration of short-chain fatty acids (SCFAs) in the cecum was analyzed by GC–MS (gas chromatography–mass spectrometry). The mRNA expression levels related to L cell function in the ileum were measured by real-time PCR. Glucagon-like peptide-1 (GLP-1) levels in the portal vein and cecal content were assessed by enzyme-linked immunosorbent assay. GLP-1-producing L cells of the ileum were quantified by immunohistochemistry. HGB intake improved glucose tolerance and increased the cecal levels of SCFAs, acetate, and propionate. The number of GLP-1-positive L cells in the HGB group was significantly higher than in the control group. GLP-1 levels in the portal vein and cecal GLP-1 pool size in the HGB group were significantly higher than the control group. In conclusion, we report improved glucose tolerance after HGB intake induced by an increase in L cell number and subsequent rise in GLP-1 secretion.

Keywords: barley; β-glucan; L cell; glucagon-like peptide 1 (GLP-1); glucose tolerance

Citation: Suzuki, S.; Aoe, S. High β-Glucan Barley Supplementation Improves Glucose Tolerance by Increasing GLP-1 Secretion in Diet-Induced Obesity Mice. *Nutrients* **2021**, *13*, 527. https://doi.org/10.3390/nu13020527

Academic Editor: Lindsay Brown

Received: 28 December 2020
Accepted: 4 February 2021
Published: 6 February 2021

Publisher's Note: MDPI stays neutral with regard to jurisdictional claims in published maps and institutional affiliations.

1. Introduction

A diet that includes barley has several beneficial effects, such as lowering the postprandial glucose rise [1–3] and improved blood cholesterol concentration [4–6]. Barley is rich in β-glucan, which comprises linear homopolysaccharides of β-(1→4) and β-(1→3) linkages and high molecular weight viscous polysaccharides [6]. Barley β-glucan is a soluble dietary fiber that is fermented in human and animal colon to produce short-chain fatty acids (SCFAs) [7–9]. Thus, the fermentability of β-glucans modifies the balance and diversity of the microbiota. These prebiotic actions may be the main mechanism for the beneficial effects of barley intake. However, the mechanism for the observed improvement of glucose tolerance following barley intake remains controversial.

In a previous report, we showed that high β-glucan barley (HGB) reduced postprandial glucose rise in healthy subjects [10,11], serum cholesterol levels in hypercholesterolemic subjects [12], and abdominal fat area in moderately obese subjects [13], and regulated the calorie intake and satiety index of healthy subjects [14]. It is speculated these beneficial effects were caused by either the viscosity of β-glucan or prebiotic effects such as SCFA production.

With regard to the improvement of glucose tolerance, barley intake is reported to reduce postprandial glucose rise and the second meal effect [15]. Moreover, it is suggested that the second meal effect is the result of colonic fermentation [16]. The second meal effect [17,18] refers to the ability of foods to reduce postprandial glucose rise not only after the first meal but also after the next meal of the day. It is reported that boiled barley kernels as an evening meal decreased incremental blood glucose area at the following breakfast and increased plasma glucagon-like peptide-1 (GLP-1) at fasting and during the day by comparison with the consumption of white wheat bread [15].

G protein-coupled receptors, which sense SCFAs, are thought to contribute to the regulation of glucose metabolism [19]. Indeed, rats fed a high fiber diet had a higher plasma GLP-1 level after oral glucose administration than those fed a low fiber diet. GLP-1 is secreted from L cells, which are predominantly located in the ileum and colon [20,21]. A previous study involving rats reported that fermentable dietary fibers increase the contents of the distal digestive tract, augment GLP-1 secretion, and boost proglucagon mRNA expression [22–24].

Our previous report showed that barley intake increased the biosynthesis of short-chain fatty acids (SCFAs) by the gut microbiota, but we did not measure gut hormone levels [25]. Given that high β-glucan barley stimulates SCFA production, we hypothesized that a high β-glucan barley diet might result in greater GLP-1 secretion by comparison to a cellulose based diet. A recent study showed that the secretion of GLP-1 increased in mice fed high β-glucan barley and improved insulin sensitivity through modification to the gut microbiota and elevated levels of SCFAs [26]. However, another report showed that barley β-glucan promoted fermentation in cecum but did not alter glucose tolerance or insulin secretion [27].

The main aim of this study was to examine the mechanism for improvement of glucose tolerance by increases in GLP-1 secretion in mice given a diet rich in β-glucan barley. We quantified the number of L cells, mRNA expression related to L cell functions in the ileum, and GLP-1 levels in the portal vein and cecum contents of mice fed a fatty diet with and without supplementation of high β-glucan barley. An improvement of glucose tolerance mediated by GLP-1 secretion in mice fed a diet of high β-glucan barley might provide useful information for a therapeutic approach to diabetes and obesity.

2. Materials and Methods

2.1. Chemical Composition of High β-Glucan Barley

These experiments were performed using a new two-rowed, waxy, hull-less high β-glucan barley cultivar Kirari-mochi. HGB flour pearled to 60% yield was obtained from NARO (Tsukuba, Japan). Total dietary fiber was quantified using a previously published protocol [28]. The β-glucan content was determined by the McCleary procedure [29]. Protein and lipid content in HGB were analyzed by the Kjeldahl and acid hydrolysis method, respectively. The overall nutritional make-up of HGB is summarized in Table 1.

Table 1. Nutritional components of high β-glucan barley (HGB).

	(g/100 g)
Moisture	9.1
Fat	1.8
Protein	12.8
Ash	0.7
Available carbohydrate	65.7
Total dietary fiber	9.9
β-glucan	5.4

Available carbohydrate: (100 − ("Moisture" + "Fat" + "Protein" + "Ash" + "Total dietary fiber")).

2.2. Experimental Design

Male C57BL/6J mice at 5 weeks of age were obtained from Charles River Laboratories Japan, Inc. (Yokohama, Japan). The mice were kept in the same holding room under a 12 h light/12 h dark regime with continual air exchange at a temperature of 22 °C (±1 °C) and humidity of 50% (± 5%). The mice were given commercial chow (NMF; Oriental

Yeast Co., Ltd., Shiga, Japan) for 1 week before being divided into 2 groups matched for body mass (n = 8 per group). Each mouse was individually held in a plastic cage and fed the experimental diet (Table 2). The modified fatty AIN-93G diet (fat energy ratio 50%) was prepared by adding lard. Cellulose and HGB flour were included in the control and HGB diets, respectively, to give 5% total dietary fiber (Table 2). Mice were allowed ad libitum access to water and food throughout the study (83 days). Animal experiments were performed twice; the first experiment analyzed cecal GLP-1 pool size and the second experiment (n = 8 for the new control and experimental groups) evaluated cecal short-chain fatty acid (SCFA) pool size. Both food intake and body weight for each mouse were recorded 3 times per week. Body weight measurements were always performed at the same time of the day. On the final week, an oral glucose tolerance test (OGTT) was carried out. Mice were fasted for 6 h and then glucose (1.5 g/kg body) was orally administered. Blood glucose concentrations were analyzed from the tail using a blood glucose meter (Sanwa Kagaku Kenkyusho Co., Ltd., Aichi, Japan) at 5 timepoints (0, 15, 30, 60, and 120 min). At the end of the study, the mice were sacrificed by isoflurane/CO_2 anesthesia after fasting for 6 h. Blood was withdrawn from portal vein (the first experiment only) and the heart and, following centrifugation, the resultant serum was stored at -80 °C. Liver, cecum, and adipose tissues were weighed. Samples of ileum were quickly saturated in RNAprotect Tissue Reagent (Qiagen, Hilden, Germany) to facilitate stabilization of RNA. The samples of cecum were stored at -40 °C until further analysis. All animal experiments were conducted with the approval of the Animal Research Committee of Otsuma Women's University (Tokyo, Japan) (no. 12001, 16006).

Table 2. Composition of the experimental diets.

	(g/kg Diet)	
	Control	**HGB**
Casein	200	120.6
L-cystine	3	3
Corn starch	197.486	-
Dextrinized corn starch	132	-
Sucrose	100	62.886
Soybean oil	70	70
Lard	200	190.9
Cellulose	50	-
HGB flour	-	505.1 *
AIN-93G mineral mixture	35	35
AIN-93 vitamin mixture	10	10
Choline bitartrate	2.5	2.5
t-Butylhydroquinone	0.014	0.014

* β-glucan content in the HGB diet was 27.3 g/kg diet. Total energy in the control and HGB diets were 4.96 kcal/g diet: carbohydrate 35 en %, protein 14 en %, fat 50 en %. HGB; high β-glucan barley.

2.3. Biochemical Analyses of the Serum

Serum cholesterol (TC), triglyceride (TG), and non-esterified fatty acids (NEFAs) were determined using a Hitachi 7180 automatic analyzer (Hitachi Ltd., Tokyo, Japan). Enzyme-linked immunosorbent assays (ELISAs) were performed to quantify serum insulin and leptin levels. GLP-1 levels in plasma derived from the portal vein were determined. Dipeptidyl peptidase IV inhibitor (Millipore, Billerica, MA, USA) was immediately added to the blood to prevent degradation of GLP-1. Quantification of GLP-1 was performed using a commercial ELISA kit (GLP-1 Active; Shibayagi Corp., Gunma, Japan).

2.4. Analysis of Short-Chain Fatty Acids in Cecal Digesta

Cecal SCFA content was measured using a previously described method [30]. A 7890B GC system (Agilent, Tokyo, Japan) equipped with a 5977A MSD (Agilent) was used for analysis of SCFAs. A DB-5MS capillary column (30 m × 0.53 mm) (Agilent) was used to separate the SCFAs. SCFA concentrations (expressed as μmol/cecum) were calculated by comparing their peak areas with an internal standard (crotonic acid).

2.5. Measurement of Total GLP-1 Level in the Cecum

Cecal tissues were extracted using Kenny's method [31]. In brief, frozen tissue samples were extracted using an acidic ethanol solution (5 mL/g wet weight tissue). The tissue samples were subsequently homogenized and then incubated for 24 h prior to clarification by centrifuging. All extraction procedures were performed at 4 °C. The supernatant was decanted, and total GLP-1 level was measured using a total GLP-1 ELISA kit (FUJIFILM Wako Pure Chemical Corporation, Osaka, Japan).

2.6. mRNA Expression Analysis in the Ileum

Oligonucleotide primer sequences are shown in Table S1. mRNA levels were quantified by real-time PCR (QuantStudio3 RT-PCR system; Applied Biosystems, Foster City, CA, USA). Complementary DNA was used to quantify the mRNA expression levels according to the $2^{-\Delta\Delta CT}$ method with a commercial PCR Master Mix (Thermo Fisher Scientific, Waltham, MA, USA). Data were assessed from threshold cycle (Ct) values, which indicate the cycle number at which the fluorescence signal reaches a fixed threshold that is well above background. ΔCT is the difference in Ct values for genes of interest compared to transcription factor II B (TFII B), which served as a reference. The $\Delta\Delta CT$ is the difference between the ΔCT for the control group and the ΔCT for the HGB group. Relative expression levels are shown as fold differences compared to the control group (arbitrary units).

2.7. Measurement of the Number of L Cells in the Ileum

After collecting the ileum, samples were immersed in 4% paraformaldehyde/phosphate buffer and fixed, then dehydrated, degreased, paraffin-penetrated, paraffin-embedded, sliced, spread, and dried at the Institute of Nutrition and Pathology, Inc. (Kyoto, Japan). Deparaffinization and rehydration were performed with xylene, anhydrous ethanol, 99% ethanol, and 70% ethanol. Endogenous peroxidase inhibition was achieved with 30% (*v/v*) H_2O_2 and methanol. Then, each ileal section was blocked using diluted 5% (*v/v*) goat serum (Normal) (DakoCytomation, Glostrup, Denmark) and 0.5% (*w/v*) BSA/PBS (bovine serum albumin/phosphate buffered saline). Anti-GLP-1 (7–36) -NH2; rabbit Ab (Yanaihara Institute Inc., Shizuoka, Japan) was used as the primary antibody. For the secondary antibody, Histofine Simple Stain Mouse MAX-PO (Rabbit) (Nichirei Bioscience Inc., Tokyo, Japan) with 1% (*v/v*) mouse serum (Normal) (DakoCytomation, Glostrup, Denmark) was used. After the antigen–antibody reaction, samples were colored with the ENVISION kit/HRP (horseradish peroxidase) (DAB; 3,3′-Diaminobenzidine) (DakoCytomation, Glostrup, Denmark). In addition, nuclear staining was performed with hematoxylin and the samples were enclosed with Mount Quick (Daido Sangyo Corporation, Saitama, Japan). Each ileal section was photographed under a light microscope. From images of the ileum, we measured the number of cells per unit area (1 mm^2) using WinROOF (Version 6.0.1; Mitani Corporation, Fukui, Japan).

2.8. Statistical Analyses

Data are given as the mean ± standard error of the mean (SE). Comparisons of data between two groups were performed by unpaired Student's *t*-test (JMP Version 14.2.0; SAS Institute Inc., Cary, NC, USA). Time-dependent changes in the blood glucose levels during the OGTT were analyzed by two-way ANOVA (diet × time). A *p*-value of <0.05 was considered statistically significant.

3. Results

3.1. Gross Changes to Mice during the Study Period

No significant differences were identified between the two experimental groups in terms of weight gain, food intake, and food efficiency ratio (Table 3). Similar results were obtained for the second experiment (Table S2). Organ weights are shown in Table 4. There were no significant differences in liver weight and the weights of retroperitoneal, epididymal, and mesenteric fat between the two experimental groups. However, the weight of cecum with digesta was significantly higher in the HGB group compared to the control group. Similar results were obtained for the second experiment (Table S3).

Table 3. Weight gain, food intake, and food efficiency ratio.

	Control	HGB
Initial weight (g)	18.5 ± 0.2	18.5 ± 0.2
Final weight (g)	41.2 ± 0.7	42.4 ± 0.9
Body weight gain (g/d)	0.27 ± 0.01	0.30 ± 0.01
Food intake (g/d)	2.8 ± 0.0	3.0 ± 0.1
Food efficiency ratio (%)	9.73 ± 0.41	9.92 ± 1.06

Values are means $\pm$ standard error of the mean (SE), $n = 8$. HGB; high β-glucan barley.

Table 4. Weight of organs.

	Control	HGB
Liver (g)	1.47 ± 0.06	1.56 ± 0.09
Cecum with digesta (g)	0.28 ± 0.02	0.38 ± 0.02 *
Total abdominal fat	4.53 ± 0.20	4.38 ± 0.16
Retroperitoneal fat (g)	0.95 ± 0.05	0.91 ± 0.02
Epididymal fat (g)	2.61 ± 0.11	2.48 ± 0.01
Mesenteric fat (g)	0.97 ± 0.10	0.99 ± 0.10

Values are means $\pm$ standard error of the mean (SE), $n = 8$. Asterisk indicates a significant difference (Student's t-test, * $p < 0.05$). HGB; high β-glucan barley.

3.2. Biochemical Markers in the Serum

The concentrations of serum biochemical markers are shown in Table 5. There were no significant differences in serum total cholesterol, triglyceride, NEFA, glucose, insulin, and leptin concentrations between the two experimental groups.

Table 5. Serum biochemical concentrations.

	Control	HGB
Cholesterol (mmol/L)	5.60 ± 0.34	5.96 ± 0.55
Triglyceride (mmol/L)	0.84 ± 0.11	0.67 ± 0.04
NEFA (μmol/L)	639.8 ± 45.5	728.0 ± 33.0
Glucose (mmol/L)	16.61 ± 0.86	14.88 ± 0.55
Insulin (ng/mL)	6.10 ± 1.03	8.36 ± 1.22
Leptin (ng/mL)	71.35 ± 7.22	73.48 ± 6.14

Values are means $\pm$ standard error of the mean (SE), $n = 8$. NEFA; non-esterified fatty acids. HGB; high β-glucan barley.

3.3. Assessment of Glucose Tolerance

The OGTT and area under the curve of blood glucose (AUC) results are shown in Figure 1. Significant interaction (diet $\times$ time) in the blood glucose levels was not observed, but blood glucose levels in the HGB group were significantly lower than in the control group (main effect was significant), as shown by two-way ANOVA. Blood glucose levels after glucose administration were significantly lower in the HGB group compared with the control group at 15 and 60 min in the first experiment. In the second experiment, significant differences between the control and HGB groups were also observed at 30 and 60 min. AUC was significantly lower in the HGB group compared with the control group in both experiments.

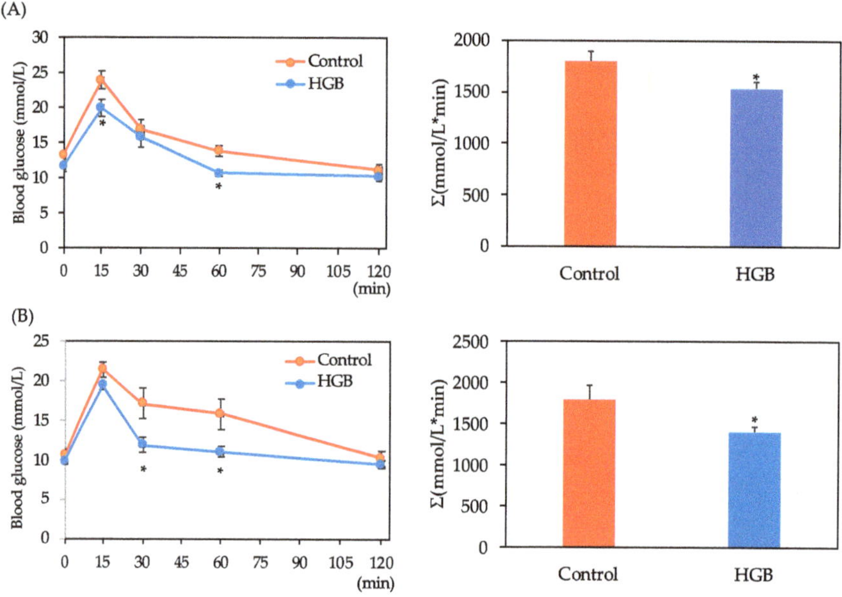

Figure 1. Blood glucose levels in the oral glucose tolerance test (OGTT) and area under the curve of blood glucose for (**A**) the first experiment for cecal glucagon-like peptide-1 (GLP-1) analysis and (**B**) the second experiment for cecal short-chain fatty acid (SCFA) analysis. Data are shown as mean ± standard error of the mean (SE). Means marked by an asterisk differ significantly (Student's *t*-test, * $p < 0.05$). HGB; high β-glucan barley.

3.4. Analysis of Short-Chain Fatty Acids in Cecal Digesta

The cecal pool size of SCFAs is shown in Figure 2. Total SCFA concentrations, as well as acetate and propionate concentrations, were significantly higher in the HGB group than the control group.

3.5. Concentrations of GLP-1 in the Portal Vein and Cecum

The concentration of GLP-1 in the portal vein and cecum are shown in Table 6. Portal vein and cecal GLP-1 levels were significantly higher in the HGB group compared with the control group.

Table 6. Portal vein and cecal GLP-1 levels.

	Control	HGB
Portal vein active GLP-1 (7-36) levels (pg/mL)	50.0 ± 9.5	81.4 ± 10.0 *
Cecal total GLP-1 pool size (ng/cecum)	31.1 ± 7.4	64.5 ± 8.4 *

Values are means ± standard error of the mean (SE), $n = 8$. Means marked with an asterisk differ significantly (Student's *t*-test, * $p < 0.05$). HGB; high β-glucan barley.

3.6. mRNA Expression Levels in the Ileum Determined by q-PCR

The experimentally determined mRNA levels in the ileum are given in Figure 3. No significant differences were found in peroxisome proliferator-activated receptor β/δ (PPARβ/δ), proglucagon (PGCG), prohormone convertase 1/3 (PC1/3), G-protein-coupled bile acid receptor 1 (GPBAR1), and G-protein-coupled receptor 43 (GPR43) between the two experimental groups. Likewise, there were no significant differences in neurogenin 3 (NGN3). By contrast, the mRNA expression level of NeuroD was elevated in the HGB group in comparison with the control group.

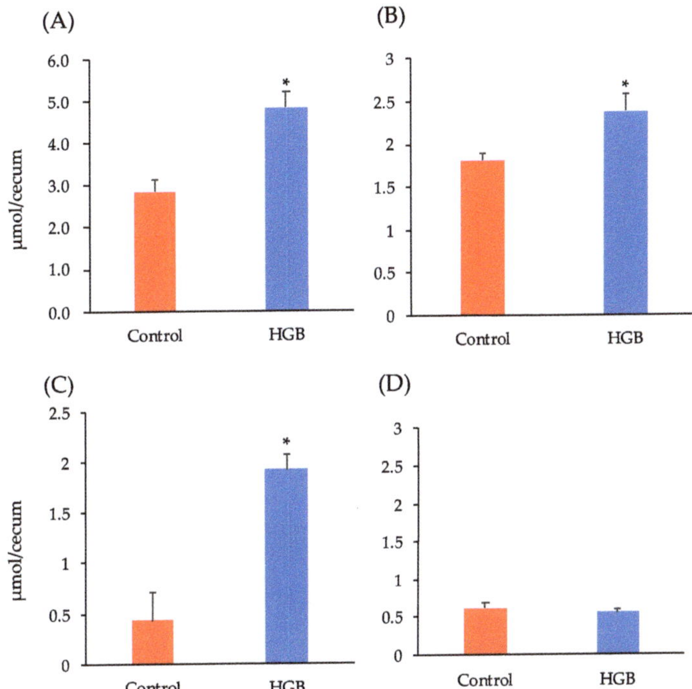

Figure 2. Short-chain fatty acid (SCFA) content in the cecal digesta of mice fed the test diets. (**A**) Total SCFA, (**B**) Acetate, (**C**) Propionate, (**D**) Other SCFA (the sum of the concentrations of formate, *n*-butyrate, *iso*-butyrate, *iso*-valerate, and valerate). Bars represent means and standard error of the mean (SE), $n = 8$. Means marked by an asterisk differ significantly (Student's *t*-test, * $p < 0.05$). HGB; high β-glucan barley.

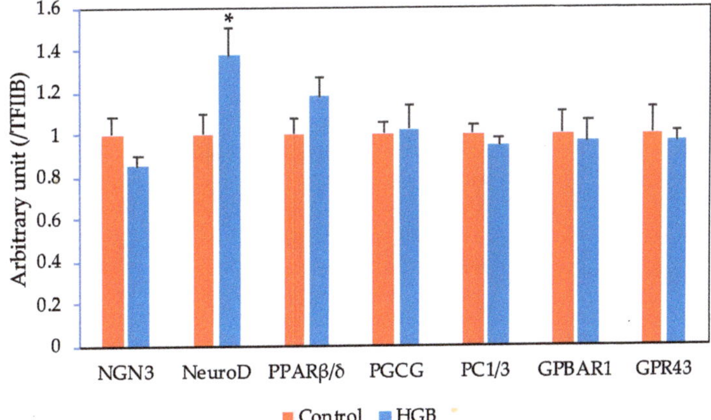

Figure 3. Expression of mRNAs related to L cell function in the ileum bars represent means and standard error of the mean (SE), $n = 8$. Means marked with an asterisk differ significantly (Student's *t*-test, * $p < 0.05$). NGN3, neurogenin 3; NeuroD, neurogenic differentiation factor; PPARβ/δ, peroxisome proliferator-activated receptor β/δ; PGCG, proglucagon; PC1/3, prohormone convertase 1/3; GPBAR1, G-protein-coupled bile acid receptor 1; GPR43, G-protein-coupled receptor 43; TFIIB, transcription factor II B. HGB; high β-glucan barley.

3.7. Number of L Cells in the Ileum

A representative image of GLP-1 positively stained cells of the ileum is shown in Supplementary Figure S1. The number of L cells in the ileum are shown in Figure 4. L cell numbers were significantly higher in the HGB group compared with the control group.

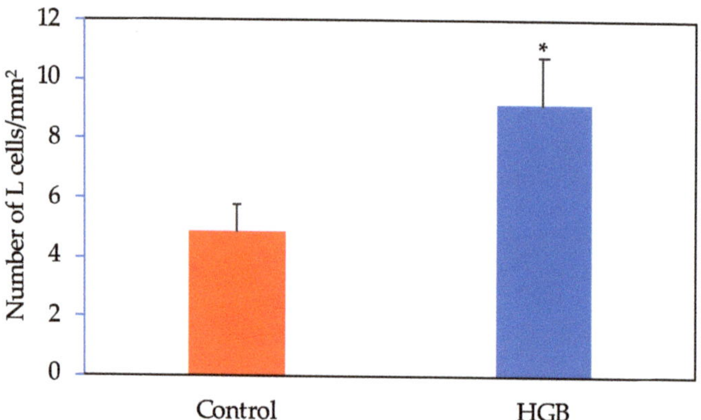

Figure 4. Number of L cells in the ileum. Data are shown as mean ± standard error of the mean (SE). * Significantly different from the control group ($p < 0.05$). HGB; high β-glucan barley.

4. Discussion

Herein, we investigated the mechanism for improvement of glucose tolerance in mice fed a high-fat diet containing HGB. Results from this study indicated that HGB intake improved glucose tolerance in OGTT. GLP-1 levels in the portal vein and cecal pool size were significantly increased in the HGB group. These findings suggest that the effect of HGB on glucose metabolism was mediated by increased GLP-1 secretion. Previous studies in human [15] and animal experiments [26] showed that the serum GLP-1 concentrations were increased by the barley intake. However, effects of HGB intake on the L cell function related to the GLP-1 secretion were not observed in the previous experiments. This is the first evidence to show an increase in the number of ileal L cells causing an increase in intestinal GLP-1 pool size and GLP-1 secretion without the alteration of mRNA expression related to GLP-1 secretion.

These results, together with those from our previous report [25], suggest that augmentation in the number of L cells and the subsequent elevation in the level of GLP-1 were induced by production of SCFAs. Thus, a new mechanism for increasing GLP-1 secretion by HGB intake is proposed. It has been generally recognized that the preventive effect of barley intake on postprandial blood glucose rise is triggered by delayed gastric emptying time due to the viscous properties of barley β-glucan and the suppression of digestion and absorption of starch in barley [32,33]. However, our previous report showed that improvement of glucose tolerance became more pronounced when HGB with partially hydrolyzed β-glucan was supplemented into the test diet [34]. It is possible that a prebiotic effect, such as intestinal fermentation, contributed more to the improved glucose tolerance than the physicochemical effect, such as increased viscosity. Furthermore, HGB imparts a second meal effect, which is induced through intestinal fermentation [15]. In a human intervention study, it was reported that the postprandial blood glucose rise was suppressed when a standard diet without barley was consumed at lunch after the intake of a HGB diet at breakfast [35,36].

Our experiments showed no difference in serum insulin concentration between the two groups, but the active GLP-1 level in the portal vein of the HGB group was significantly elevated by comparison to the control group. The total GLP-1 pool size of the cecal contents in the HGB group was also significantly elevated compared to the control group. GLP-1 is produced by prohormone convertase 1/3 from the precursor proglucagon in L cells. However, changes in the expression levels of peroxisome proliferator-activated receptor β/δ (PPARβ/δ), proglucagon, and prohormone convertase 1/3 related to GLP-1 secretion in the ileum were not observed. The expression levels of NeuroD, which is thought to reflect the number of L cells, increased significantly in the HGB group. These results indicated that the improvement of glucose tolerance following HGB intake is fostered by an increase in GLP-1 secretion accompanying the increased number of L cells.

Several studies propose secretion of GLP-1 and gastric inhibitory peptide (GIP) may be coupled with changes to the microflora of the lower gut as well as glucose metabolism [37–40]. In 1996, the first published report appeared, suggesting fermentation in the lower digestive tract promoted an increase of GLP-1 secretion [22]. It was also reported that a prebiotic oligofructose led to an increase in total cecal GLP-1 concentration [41]. Similar effects were observed with resistant starch, which increased the concentration of plasma GLP-1 [42,43]. Our study indicated prebiotic barley β-glucan is one of the dietary fibers that promotes GLP-1 secretion.

Previous studies report that inert dietary fibers did not enhance colonic proliferation of epithelial cells, although fermentable soluble fibers were able to promote proliferation in the lower gut, which is associated with increased enteroglucagon secretion [44,45]. Enteroglucagon, which regulates intestinal epithelial cell proliferation and acts as a trophic factor for the intestinal mucosa, is synthesized in the L cells of the ileum and colon. It is speculated that increases in the number of L cells is mediated by the trophic action of enteroglucagon. Luminal infusion of acetate and butyrate was reported to significantly increase colonic GLP-1 secretion, whereas propionate had no such effect [46]. Moreover, SCFAs may increase the release of GLP-1 via the SCFA receptor GPR43, which is expressed on the L cells [47].

A major limitation of this study was the lack of data regarding serum insulin and GLP-1 concentrations during the OGTT. Future studies are needed to further elucidate the relationship between GLP-1 release and insulin response by HGB intake.

5. Conclusions

In conclusion, we confirm that the observed improvement of glucose tolerance by HGB intake was induced by an increase in L cell number and the subsequent enhancement in GLP-1 secretion. We demonstrated for the first time a relationship between the level of GLP-1, the number of L cells, and the level of NeuroD mRNA in the lower digestive tract. On the basis of these findings, we propose that the presence of barley β-glucan in the lower gut increases L cell differentiation by upregulation of NeuroD gene expression. The present results are consistent with the notion that events occurring in the lower gut, such as fermentation and alteration of gut microbiota, exert a key mechanism on glucose metabolism.

Supplementary Materials: The following are available online at https://www.mdpi.com/2072-664 3/13/2/527/s1: Table S1: Primers used for real-time reverse transcription polymerase chain reaction. Table S2: Body weight gain, food intake, and food efficiency ratio (2nd Exp). Table S3: Weight of organs (2nd Exp). Supplementary Figure S1: Representative GLP-1 staining of cells from the ileum: (a) control and (b) HGB staining of GLP-1-positive cells.

Author Contributions: Conceptualization, S.A.; data curation, S.A. and S.S.; formal analysis, S.S.; investigation, S.A. and S.S.; methodology, S.S.; project administration, S.A.; supervision, S.A.; writing—original draft, S.A. and S.S.; writing—review and editing, S.A. and S.S. All authors have read and agreed to the published version of the manuscript.

Funding: This research received no external funding.

Institutional Review Board Statement: The studies were approved by the Otsuma Women's University Animal Research Committee (Tokyo, Japan) and were performed in accordance with the Regulation on Animal Experimentation at Otsuma Women's University (no. 12001,16006, 27 July 2016).

Informed Consent Statement: Not applicable.

Data Availability Statement: Data are contained within the article or supplementary material.

Acknowledgments: This investigation was funded by the Council of Japan Barley Foods Promotion.

Conflicts of Interest: The authors (S.A. and S.S.) have no conflict of interest to disclose.

References

1. Behall, K.M.; Scholfield, D.J.; Hallfrisch, J. Comparison of hormone and glucose responses of overweight women to barley and oats. *J. Am. Coll. Nutr.* **2005**, *24*, 182–188. [CrossRef]
2. Casiraghi, M.C.; Garsetti, M.; Testolin, G.; Brighenti, F. Post-prandial responses to cereal products enriched with barley β-glucan. *J. Am. Coll. Nutr.* **2006**, *25*, 313–320. [CrossRef]
3. Aldughpassi, A.; Abdel-Aal, E.-S.M.; Wolever, T.M.S. Barley cultivar, kernel composition, and processing affect the glycemic index. *J. Nutr.* **2012**, *142*, 1666–1671. [CrossRef]
4. AbuMweis, S.S.; Jew, S.; Ames, N.P. Beta-glucan from barley and its lipid-lowering capacity, a meta-analysis of randomized, controlled trials. *Eur. J. Clin. Nutr.* **2010**, *64*, 1472–1480. [CrossRef]
5. Behall, K.M.; Scholfield, D.J.; Hallfrisch, J. Diets containing barley significantly reduce lipids in mildly hypercholesterolemic men and women. *Am. J. Clin. Nutr.* **2004**, *80*, 1185–1193. [CrossRef]
6. Wang, Y.; Harding, S.V.; Eck, P.; Thandapilly, S.J.; Gamel, T.H.; Abdel-Aal, E.-S.M.; Crow, G.H.; Tosh, S.M.; Jones, P.J.; Ames, N.P. High-Molecular-Weight beta-Glucan Decreases Serum Cholesterol Differentially Based on the CYP7A1 rs3808607 Polymorphism in Mildly Hypercholesterolemic Adults. *J. Nutr.* **2016**, *146*, 720–727. [CrossRef] [PubMed]
7. Hughes, S.A.; Shewry, P.R.; Gibson, G.R.; McCleary, B.V.; Rastall, R.A. In vitro fermentation of oat and barley derived beta-glucans by human faecal microbiota. *FEMS Microbiol. Ecol.* **2008**, *64*, 482–493. [CrossRef] [PubMed]
8. Bindelle, J.; Pieper, R.; Montoya, C.A.; Van Kessel, A.G.; Leterme, P. Nonstarch polysaccharide-degrading enzymes alter the microbial community and the fermentation patterns of barley cultivars and wheat products in an in vitro model of the porcine gastrointestinal tract. *FEMS Microbiol. Ecol.* **2011**, *76*, 553–563. [CrossRef] [PubMed]
9. Aoe, S.; Yamanaka, C.; Fuwa, M.; Tamiya, T.; Nakayama, Y.; Miyoshi, T.; Kitazono, E. Effects of BARLEYmax and high-beta-glucan barley line on short-chain fatty acids production and microbiota from the cecum to the distal colon in rats. *PLoS ONE* **2019**, *14*, e0218118. [CrossRef] [PubMed]
10. Kanamoto, I.; Koyama, S.; Murata, I.; Inoue, Y.; Kohyama, N.; Ichinose, Y.; Komae, K.; Yauagisaya, T.; Aoe, S. Effect of ingestion of high β-glucan barley breads with different particle size on postprandial blood glucose levels. *Luminacoids Res.* **2017**, *21*, 19–23.
11. Aoe, S.; Komae, K.; Inoue, Y.; Murata, I.; Minegishi, Y.; Kanmotom, I.; Kohyama, N.; Ichinose, Y.; Yoshioka, T.; Yanagisawa, T. Effects of Various Blending Ratios of Rice and Waxy Barley on Postprandial Blood Glucose Levels. *J. Jpn. Soc. Nutr. Food Sci.* **2018**, *71*, 283–288. [CrossRef]
12. Shimizu, C.; Kihara, M.; Aoe, S.; Araki, S.; Ito, K.; Hayashi, K.; Watari, J.; Sakata, Y.; Ikegami, S. Effect of high beta-glucan barley on serum cholesterol concentrations and visceral fat area in Japanese men—A randomized, double-blinded, placebo-controlled trial. *Plant Foods Hum. Nutr.* **2008**, *63*, 21–25. [CrossRef] [PubMed]
13. Aoe, S.; Ichinose, Y.; Kohyama, N.; Komae, K.; Takahashi, A.; Abe, D.; Yoshioka, T.; Yanagisawa, T. Effects of high beta-glucan barley on visceral fat obesity in Japanese individuals: A randomized, double-blind study. *Nutrition* **2017**, *42*, 1–6. [CrossRef]
14. Aoe, S.; Ikenaga, T.; Noguchi, H.; Kohashi, C.; Kakumoto, K.; Kohda, N. Effect of cooked white rice with high β-glucan barley on appetite and energy intake in healthy Japanese subjects: A randomized controlled trial. *Plant Foods Hum. Nutr.* **2014**, *69*, 325–330. [CrossRef]
15. Johansson, E.V.; Nilsson, A.C.; Östman, E.M.; Björck, I.M.E. Effects of indigestible carbohydrates in barley on glucose metabolism, appetite and voluntary food intake over 16 h in healthy adults. *Nutr. J.* **2013**, *12*, 46. [CrossRef]
16. Brighenti, F.; Benini, L.; Del Rio, D.; Casiraghi, C.; Pellegrini, N.; Scazzina, F.; Jenkins, D.J.; Vantini, I. Colonic fermentation of indigestible carbohydrates contributes to the second-meal effect. *Am. J. Clin. Nutr.* **2006**, *83*, 817–822. [CrossRef] [PubMed]
17. Jenkins, D.J.; Wolever, T.M.; Taylor, R.H.; Griffiths, C.; Krzeminska, K.; Lawrie, J.A.; Bennett, C.M.; Goff, D.V.; Sarson, D.L.; Bloom, S.R. Slow release dietary carbohydrate improves second meal tolerance. *Am. J. Clin. Nutr.* **1982**, *35*, 1339–1346. [CrossRef]
18. Wolever, T.M.; Jenkins, D.J.; Ocana, A.M.; Rao, V.A.; Collier, G.R. Secondmeal effect: Low-glycemic-index foods eaten at dinner improve subsequent breakfast glycemic response. *Am. J. Clin. Nutr.* **1988**, *48*, 1041–1047. [CrossRef] [PubMed]
19. Priyadarshini, M.; Villa, S.R.; Fuller, M.; Wicksteed, B.; Mackay, C.R.; Alquier, T.; Poitout, V.; Mancebo, H.; Mirmira, R.G.; Gilchrist, A.; et al. An Acetate-Specific GPCR, FFAR2, Regulates Insulin Secretion. *Mol. Endocrinol.* **2015**, *29*, 1055–1066. [CrossRef]
20. Holst, J.J. Glucagon like peptide 1: A newly discovered gastrointestinal hormone. *Gastroenterology* **1994**, *107*, 1848–1855. [CrossRef]
21. Orskov, C.; Rabenhoj, L.; Wettergren, A.; Kofod, H.; Holst, J.J. Tissue and plasma concentrations of amidated and glycine-extended glucagonlike peptide I in humans. *Diabetes* **1994**, *43*, 535–539. [CrossRef] [PubMed]
22. Reimer, R.A.; McBurney, M.I. Dietary fiber modulates intestinal proglucagon messenger ribonucleic acid and postprandial secretion of glucagon-like peptide-1 and insulin in rats. *Endocrinology* **1996**, *137*, 3948–3956. [CrossRef]
23. Reimer, R.A.; Thomson, A.B.; Rajotte, R.V.; Basu, T.K.; Ooraikul, B.; McBurney, M.I. A physiological level of rhubarb fiber increases proglucagon gene expression and modulates intestinal glucose uptake in rats. *J. Nutr.* **1997**, *127*, 1923–1928. [CrossRef]

24. Reimer, R.A.; Thomson, A.B.R.; Rajotte, R.V.; Basu, T.K.; Ooraikul, B.; McBurney, M.I. Proglucagon messenger ribonucleic acid and intestinal glucose uptake are modulated by fermentable fiber and food intake in diabetic rats. *Nutr. Res.* **2000**, *20*, 851–864. [CrossRef]

25. Mio, K.; Yamanaka, C.; Matsuoka, T.; Kobayashi, T.; Aoe, S. Effects of beta-glucan rich barley flour on glucose and lipid metabolism in the ileum, liver, and adipose tissues of high-fat diet induced-obesity model male mice analyzed by DNA microarray. *Nutrients* **2020**, *12*, 3546. [CrossRef] [PubMed]

26. Miyamoto, J.; Watanabe, K.; Taira, S.; Kasubuchi, M.; Li, X.; Irie, J.; Itoh, H.; Kimura, I. Barley beta-glucan improves metabolic condition via short-chain fatty acids produced by gut microbial fermentation in high fat diet fed mice. *PLoS ONE* **2018**, *13*, e0196579. [CrossRef]

27. Belobrajdic, D.P.; Jobling, S.A.; Morell, M.K.; Taketa, S.; Bird, A.R. Wholegrain barley beta-glucan fermentation does not improve glucose tolerance in rats fed a high-fat diet. *Nutr. Res.* **2015**, *35*, 162–168. [CrossRef] [PubMed]

28. Lee, S.C.; Rodriguez, F.; Storey, M.; Farmakalidis, E.; Prosky, L. Determination of soluble and insoluble dietary fiber in psyllium containing cereal products. *J. AOAC Int.* **1995**, *78*, 724–729. [CrossRef]

29. McCleary, B.V.; Codd, R. Measurement of (1–3), (1–4)-β-D-glucan in barley and oats: A streamlined enzymic procedure. *J. Sci. Food Agric.* **1991**, *55*, 303–312. [CrossRef]

30. Atarashi, K.; Tanoue, T.; Oshima, K.; Suda, W.; Nagano, Y.; Nishikawa, H.; Fukuda, S.; Saito, T.; Narushima, T.; Hase, K.; et al. Treg induction by a rationally selected mixture of Clostridia strains from the human microbiota. *Nature* **2013**, *500*, 232–236. [CrossRef]

31. Knapper, J.M.; Morgan, L.M.; Fletcher, J.M.; Marks, V. Plasma and intestinal concentrations of GIP and GLP-1 (7-36) amide during suckling and after weaning in pigs. *Horm. Metab. Res.* **1995**, *27*, 485–490. [CrossRef] [PubMed]

32. Thondre, P.S.; Shafat, A.; Clegg, M.E. Molecular weight of barley beta-glucan influences energy expenditure, gastric emptying and glycaemic response in human subjects. *Br. J. Nutr.* **2013**, *110*, 2173–2179. [CrossRef]

33. El Khoury, D.; Cuda, C.; Luhovy, B.L.; Anderson, G.H. Beta glucan: Health benefits in obesity and metabolic syndrome. *J. Nutr. Metab.* **2012**, *2012*, 851362. [CrossRef]

34. Mio, K.; Yamanaka, C.; Ichinose, Y.; Kohyama, N.; Yanagisawa, T.; Aoe, S. Effects of barley β-glucan with various molecular weights partially hydrolyzed by endogenous β-glucanase on glucose tolerance and lipid metabolism in mice. *Cereal Chem.* **2020**, *97*, 1056–1065. [CrossRef]

35. Nilsson, A.C.; Ostman, E.M.; Granfeldt, Y.; Björck, I.M. Effect of cereal test breakfasts differing in glycemic index and content of indigestible carbohydrates on daylong glucose tolerance in healthy subjects. *Am. J. Clin. Nutr.* **2008**, *87*, 645–654. [CrossRef]

36. Matsuoka, T.; Tsuchida, A.; Yamaji, A.; Kurosawa, C.; Shinohara, M.; Takayama, I.; Nakagomi, H.; Izumi, K.; Ichikawa, Y.; Hariya, N.; et al. Consumption of a meal containing refined barley flour bread is associated with a lower postprandial blood glucose concentration after a second meal compared with one containing refined wheat flour bread in healthy Japanese, A randomized control trial. *Nutrition* **2020**, *72*, 110637. [CrossRef] [PubMed]

37. Burcelin, R.; Cani, P.D.; Knauf, C. Glucagon-like peptide-1 and energy homeostasis. *J. Nutr.* **2007**, *137*, 2534S–2538S. [CrossRef]

38. Cani, P.D.; Holst, J.J.; Drucker, D.J.; Delzenne, N.M.; Thorens, B.; Burcelin, R.; Knauf, C. GLUT2 and the incretin receptors are involved in glucoseinduced incretin secretion. *Mol. Cell. Endocrinol.* **2007**, *276*, 18–23. [CrossRef] [PubMed]

39. Knauf, C.; Cani, P.D.; Ait-Belgnaoui, A.; Benani, A.; Dray, C.; Cabou, C.; Colom, A.; Uldry, M.; Rastrelli, S.; Sabatier, E.; et al. Brain glucagon-like peptide 1 signaling controls the onset of high-fat diet-induced insulin resistance and reduces energy expenditure. *Endocrinology* **2008**, *149*, 4768–4777. [CrossRef] [PubMed]

40. Knauf, C.; Cani, P.D.; Kim, D.H.; Iglesias, M.A.; Chabo, C.; Waget, A.; Colom, A.; Rastrelli, S.; Delzenne, N.M.; Drucker, D.J.; et al. Role of central nervous system glucagon-like Peptide-1 receptors in enteric glucose sensing. *Diabetes* **2008**, *57*, 2603–2612. [CrossRef]

41. Kok, N.N.; Morgan, L.M.; Williams, C.M.; Roberfroid, M.B.; Thissen, J.P.; Delzenne, N.M. Insulin, glucagon-like peptide 1, glucose-dependent insulinotropic polypeptide and insulin-like growth factor I as putative mediators of the hypolipidemic effect of oligofructose in rats. *J. Nutr.* **1998**, *128*, 1099–1103. [CrossRef]

42. Keenan, M.J.; Zhou, J.; McCutcheon, K.L.; Raggio, A.M.; Bateman, H.G.; Todd, E.; Jones, C.K.; Tulley, R.T.; Melton, S.; Martin, R.J.; et al. Effects of resistant starch, a non-digestible fermentable fiber, on reducing body fat. *Obesity* **2006**, *14*, 1523–1534. [CrossRef] [PubMed]

43. Zhou, J.; Hegsted, M.; McCutcheon, K.L.; Keenan, M.J.; Xi, X.; Raggio, A.M.; Martin, R.J. Peptide YY and proglucagon mRNA expression patterns and regulation in the gut. *Obesity* **2006**, *14*, 683–689. [CrossRef] [PubMed]

44. Goodlad, R.A.; Lenton, W.; Ghatei, M.A.; Adrian, T.E.; Bloom, S.R.; Wright, N.A. Effects of an elemental diet, inert bulk and different types of dietary fibre on the response of the intestinal epithelium to refeeding in the rat and relationship to plasma gastrin, enteroglucagon, and PYY concentrations. *Gut* **1987**, *28*, 171–180. [CrossRef] [PubMed]

45. Goodlad, R.A.; Lenton, W.; Ghatei, M.A.; Adrian, T.E.; Bloom, S.R.; Wright, N.A. Proliferative effects of 'fibre' on the intestinal epithelium epithelium: Relationship to gastrin, enteroglucagon and PYY. *Gut* **1987**, *28*, 221–226. [CrossRef]

46. Roberfroid, M.B.; Delzenne, N.M. Dietary fructans. *Annu. Rev. Nutr.* **1998**, *18*, 117–143. [CrossRef] [PubMed]

47. Karaki, S.; Mitsui, R.; Hayashi, H.; Kato, I.; Sugiya, H.; Iwanaga, T.; Furness, J.B.; Kuwahara, A. Short-chain fatty acid receptor, GPR43, is expressed by enteroendocrine cells and mucosal mast cells in rat intestine. *Cell Tissue Res.* **2006**, *324*, 353–360. [CrossRef] [PubMed]

Article

Ingestion of High β-Glucan Barley Flour Enhances the Intestinal Immune System of Diet-Induced Obese Mice by Prebiotic Effects

Kento Mio [1,2], Nami Otake [1], Satoko Nakashima [2], Tsubasa Matsuoka [2] and Seiichiro Aoe [1,*]

1   Studies in Human Life Sciences, Graduate School of Studies in Human Culture, Otsuma Women's University, Chiyoda-ku, Tokyo 102-8357, Japan; mio.kento@hakubaku.co.jp (K.M.); otake.nami@gmail.com (N.O.)
2   Research and Development Department, Hakubaku Co. Ltd., Chuo-City, Yamanashi 409-3843, Japan; nakashima.satoko@hakubaku.co.jp (S.N.); matsuoka.tsubasa@hakubaku.co.jp (T.M.)
*   Correspondence: s-aoe@otsuma.ac.jp; Tel.: +81-3-5275-6048

Abstract: The prebiotic effect of high β-glucan barley (HGB) flour on the innate immune system of high-fat model mice was investigated. C57BL/6J male mice were fed a high-fat diet supplemented with HGB flour for 90 days. Secretory immunoglobulin A (sIgA) in the cecum and serum were analyzed by enzyme-linked immunosorbent assays (ELISA). Real-time PCR was used to determine mRNA expression levels of pro- and anti-inflammatory cytokines such as interleukin (IL)-10 and IL-6 in the ileum as well as the composition of the microbiota in the cecum. Concentrations of short-chain fatty acids (SCFAs) and organic acids were analyzed by GC/MS. Concentrations of sIgA in the cecum and serum were increased in the HGB group compared to the control. Gene expression levels of *IL-10* and polymeric immunoglobulin receptor (*pIgR*) significantly increased in the HGB group. HGB intake increased the bacterial count of microbiota, such as *Bifidobacterium* and *Lactobacillus*. Concentrations of propionate and lactate in the cecum were increased in the HGB group, and a positive correlation was found between these organic acids and the *IL-10* expression level. Our findings showed that HGB flour enhanced immune function such as IgA secretion and *IL-10* expression, even when the immune system was deteriorated by a high-fat diet. Moreover, we found that HGB flour modulated the gut microbiota, which increased the concentration of SCFAs, thereby stimulating the immune system.

Keywords: barley; β-glucan; short-chain fatty acids; immune system; sIgA; prebiotics; microbiota

Citation: Mio, K.; Otake, N.; Nakashima, S.; Matsuoka, T.; Aoe, S. Ingestion of High β-Glucan Barley Flour Enhances the Intestinal Immune System of Diet-Induced Obese Mice by Prebiotic Effects. *Nutrients* **2021**, *13*, 907. https://doi.org/10.3390/nu13030907

Academic Editor: Michael Conlon

Received: 29 December 2020
Accepted: 9 March 2021
Published: 11 March 2021

Publisher's Note: MDPI stays neutral with regard to jurisdictional claims in published maps and institutional affiliations.

1. Introduction

The gastrointestinal tract has to tolerate the presence of the luminal microbiota, but the immune system must protect the intestinal mucosa against potentially harmful dietary antigens and pathogenic agents [1]. Immunoglobulin A (IgA) is an important antibody of this system, which eliminates pathogens and neutralizes toxins. It is known that this system is aggravated by stress, obesity, and disordered eating habits [2]. In particular, foods and food ingredients have the potential to affect the intestinal immune system. Previous studies showed that several food components, such as lactic acid bacteria and vitamin A, promote IgA secretion by different mechanisms [3–5]. Additionally, it is reported that functional foods stimulate IgA secretion by inducing changes in the gut microbiota [6,7]. Indeed, homeostasis of gut microbiota has been shown to be regulated by T cell-dependent IgA [8]. Moreover, a human study indicated that IgA-deficient subjects have different microbiota profiles compared with healthy subjects [9]. Therefore, it is important to identify foods that enhance intestinal immune functions, such as IgA production, by improving gut microbiota.

In recent years, there is increasing interest in utilizing indigestible carbohydrates, to modulate the metabolic function of the microbiota, which are known as prebiotics [10,11]. Grains such as barley contain substantial amounts of dietary fiber and mediate several physiological functions. Beta-glucan in barley and oats is a water-soluble dietary fiber comprising β-glycosidic bonds (β-1-3,1-4 bonds). This fiber is fermented by gut bacteria, potentially leading to health benefits, and thereby acting as a prebiotic [12,13]. A previous study showed that a mixture of barley β-glucan and probiotic microorganisms modulate the transcriptional levels of immune-related genes in vitro [14]. Additionally, whole-grain barley pasta containing barley β-glucan was found to be effective in modulating the composition and metabolism of the gut microbiota in 26 healthy subjects [15]. Another clinical study showed that intake of 60 g/day of whole barley for four weeks increased microbial diversity and the abundance of many genera of gut bacteria [16]. Moreover, β-glucan from oats and barley contribute to the production of intestinal metabolites, such as short-chain fatty acids (SCFAs), via fermentation mediated by gut microbiota [17]. SCFAs are thought to stimulate the immune system and orchestrate an anti-inflammatory effect [18,19]. Park et al. suggested that SCFAs, such as butyrate, propionate, and acetate, activate the naive T cell polarization to regulatory T cells (Tregs) [20]. Another in vitro study showed that butyrate suppressed the induction of T cell apoptosis and interferon gamma (IFN-γ)-mediated inflammation in colonic epithelial cells [21]. Therefore, changes in the gut microbiota and levels of SCFAs following intake of barley are expected to enhance the immune system in the lower gut.

We speculated that barley β-glucan may have an effect on the innate immune function via intestinal fermentation. To date, few studies have focused on changes to the immune system following barley intake. One such study involved patients who had previously undergone a proctocolectomy with ileostomy [22]. The subjects were fed oat β-glucan, and then, digestive waste was collected, freeze-dried, and dissolved in PBS. The resulting solution was incubated with different intestinal cell lines and found to increase parameters related to immune modulation in vitro. Volman et al. suggested that mice fed oat β-glucan activated the gut leukocytes and enterocytes compared to placebo mice [23]. However, most of the studies were conducted using β-glucan extracts, and there have been no studies evaluating barley as a food. Furthermore, there are no reports that have clarified the effects of barley on models of impaired immune function due to poor dietary habits such as obesity. Diet-induced obesity generally causes a low-grade inflammatory state. Several lines of evidence indicate that altered immune function is associated with the etiology of obesity [24–26]. For example, intake of a high-fat diet causes excess lipids and secreted bile acids to flow into the digestive tract, which adversely affects the intestinal environment and immune function. A recent study showed that serum IgA levels were decreased in mice fed a high-fat diet, which was mediated via high-fat-induced changes to the composition of microbiota and gut metabolite production [27]. Therefore, it is important to investigate the effect of food containing prebiotics, such as barley, on inflammation in the intestinal tract caused by obesity.

Here, we focused on the gut fermentability of barley flour and confirmed its effect on the innate immune response under high-fat conditions. Firstly, we investigated gene expression to clarify whether the intake of barley affects the immune system in the ileum using previously published DNA microarray data [28]. Next, as the main aim of this study, we investigated whether or not IgA secretion and mRNA expression levels of cytokines change following the ingestion of high β-glucan barley (HGB) flour in diet-induced obese mice. We also investigated the relationship between gut microbiota and the production of SCFAs.

2. Materials and Methods

2.1. Animals and Study Design

The animal protocol used in this study was approved by the Otsuma Women's University Animal Research Committee (Tokyo, Japan, No.19013, 13 December 2019) and implemented in accordance with animal experimentation according to their regulations. The flowchart of the study design in this animal experiment is shown in Figure S1. Male four-week-old C57BL/6J mice were purchased from Charles River Laboratories Japan,

Inc. (Yokohama, Japan). Each mouse was individually housed in a polycarbonate cage kept in a holding room maintained on a 12 h light/dark cycle (light on at 07:30 h) at a temperature of $22 \pm 1\ ^{\circ}C$ and humidity of $50 \pm 5\%$. After the mice had acclimatized for 11 days on commercial chow (NMF, Oriental Yeast Co., Ltd., Shiga, Japan), they were randomized into 2 groups according to body weight ($n = 8$ per group). Mice were given the experimental diet (powdered diet) over a 90-day period. The experimental diet was a 50% fat energy diet supplemented with cellulose (Control (C) group) or flour of waxy hulled barley "White Fiber" (high β-glucan barley (HGB) group). The total dietary fiber of each diet was adjusted to 5% (Table S1). "White Fiber" refers to barley flour rich in β-glucan (Table S1). "White Fiber" flour which was pearled to 70% and powdered was obtained from the Hakubaku Co. Ltd. (Yamanashi, Japan). Food intake and body weights were monitored 2 or 3 times per week throughout the study period. Feces were collected for 5 days in the 11th week. At the end of the study, mice were fasted for 8 h and sacrificed by isoflurane/CO_2 anesthesia. Then, the cecum with digesta, adipose tissues (epididymal, retroperitoneal, mesenteric fats), and liver were dissected and weighed. Blood samples were collected from the postcaval vein and centrifuged to obtain serum, which was stored at $-30\ ^{\circ}C$ until enzyme-linked immunosorbent assay. Cecum with digesta was stored at $-30\ ^{\circ}C$ until analysis of SCFAs, microbiota, and secretory immunoglobulin A (sIgA). Ileum tissue was soaked in RNA® protect Tissue Reagent (Qiagen, Hilden, Germany) and stored at $-30\ ^{\circ}C$ until extraction of total RNA.

2.2. Gene Expression Analysis of the Immune System Using DNA Microarray Data

The microarray data used in this study have been registered at the National Center for Biotechnology Information (NCBI) for Gene Expression Omnibus (GEO) and GEO series with accession number GSE157828. Male mice were fed a high-fat diet (fat energy ratio of 50%) supplemented with β-glucan rich barley flour before extracting total RNA from the ileum, liver, and adipose tissue and performing DNA microarray analysis. The procedure was also performed for mice fed a diet not supplemented with β-glucan rich barley flour as a control (supplemented with cellulose). We identified differentially expressed genes (DEGs) using cut-off criteria (Log-ratio > 1.3 fold and < 0.77 fold in the barley group compared with the control group) based on our previous study [28]. Kyoto Encyclopedia of Genes and Genomes (KEGG) enrichment analyses of DEGs were performed using the DAVID database (accessed on 12 December 2020). False discovery rate (FDR) was calculated using the Benjamini–Hochberg algorithm. Pathways were extracted with FDR where $p < 0.05$.

2.3. The Levels of Secretory Immunoglobulin A, IL-6, and IL-10 Determined by Enzyme-Linked Immunosorbent Assays

The levels of sIgA in the cecum and serum were determined using an enzyme-linked immunosorbent assay kit for secretory immunoglobulin A (Cloud-Clone Corp., Katy, TX, USA). Serum was analyzed according to the manufacturer's instructions. Cecum was lyophilized, and 10 mg samples were added to 100 μL phosphate buffered saline (PBS) (10 mM, pH 7.0) and then homogenized. After centrifugation (8000 rpm × 20 min, 4 °C), the supernatant was diluted 500 times in PBS and then used for measurement. The concentrations of interleukin (IL)-6 and IL-10 in serum were analyzed using Mouse IL-6 ELISA Kit (RayBiotech, Inc., Norcross, GA, USA) and Mouse IL-10 ELISA Kit (Proteintech Group, Inc., Chicago, IL, USA), respectively.

2.4. Analysis of Short-Chain Fatty Acids in the Cecum and Feces

The concentration of cecum and feces SCFAs was determined as described in a previous report using gas chromatography-mass spectrometry (GC/MS) [29]. First, 10–20 mg of cecum digesta was added to 100 μL of internal standard (100 μM crotonic acid), 300 μL of diethyl ether, and 50 μL of HCl, and then homogenized (Tissue Lyser II; Qiagen) twice at $2000\times g$ rpm for 2 min each. After centrifugation (3000 rpm, at 24 °C, for 10 min), 80 μL of supernatant was added to 16 μL of *N-tert*-butyldimethylsilyl-*N*-methyltrifluoroacetamide in a vial, and derivatization was performed at 80 °C for 20 min. Samples were stored at room temperature for 48 h and then analyzed for SCFAs by GC/MS (7890B GC system equipped with a 5977A MSD; Agilent, Tokyo, Japan). A DB-5MS capillary column (30 m × 0.53 mm) (Agilent) was used to separate the SCFAs. The oven temperature was initially kept at 60 °C and then ramped up to 120 °C at a rate of 5 °C/min. Then, the oven was ramped up to 300 °C at a rate of 20 °C/min and finally maintained at 300 °C for 2 min. Helium was used as the carrier gas at a flow rate of 1.2 mL/min. The temperature of the front inlet, transfer line, and electron impact ion source were set at 250, 260, and 230 °C, respectively. Mass spectral data were collected in selective ion monitoring mode. The concentration of SCFAs was calculated by comparing their peak areas with that of the internal standard.

2.5. Analysis of Counts of Predominant Bacterial Groups in the Cecum Digesta

The gut microbiota in the cecum was analyzed by real-time PCR according to previous studies [30,31]. DNA of cecum digesta was extracted using QIAamp® Fast DNA Stool Mini kit (Qiagen) according to the manufacturer's protocol. DNA was mixed with PowerUp SYBR Green Master Mix (Thermo Fisher Scientific, Waltham, MA, USA) and oligonucleotide primers for predominant bacterial groups (Table S2). Amplification of the DNA was performed using an Applied Biosystems Quant3 Real-Time PCR System (Thermo Fisher Scientific). From the obtained threshold cycle (Ct) values, colony-forming units (CFU) in the cecum were determined from a calibration curve prepared using Ct values after serially diluting a DNA solution extracted from each standard bacterial strain (Table S2).

2.6. Expression Analysis of mRNA Related to Immunity in the Ileum

Total RNAs in the ileum were extracted using an RNeasy Mini Kit (Qiagen). mRNA expression related to intestinal immunity and cytokines was analyzed by real-time PCR using an Applied Biosystems Quant3 Real-Time PCR System and PowerUp SYBR Green Master Mix (Thermo Fisher Scientific) with cDNA synthesized from RNA. Primer sequences are given in Table S3. The $2^{-\Delta\Delta CT}$ method was used for mRNA expression analysis. We used 36B4 as a reference gene and calculated ΔCT compared to the 36B4. Next, we calculated $\Delta\Delta CT$ as the difference between ΔCT for the C group and HGB group in terms of cDNA solution added to each primer. Relative expression levels are presented as fold changes to the C group (arbitrary units).

2.7. Statistical Analysis

All statistical analyses were performed using R Studio (ver. 1.3.1093, R-Tools Technology Inc., Richmond Hill, ON, Canada). Data are presented as mean ± standard error (SE) of the mean. For each parameter, Student's *t*-test was used when the data were based on a normal distribution. If a normal distribution was not confirmed, the Wilcoxon test was used. Significant differences were appraised using a two-side test with an α level of 0.05. The relationships between the sIgA and parameters related to intestinal immunity were analyzed by Spearman's rank correlation coefficient.

3. Results

3.1. KEGG Enrichment Analyses of DEGs by Using DNA Microarray Data

We performed KEGG enrichment analysis of DEGs using DNA microarray data of the ileum of mice fed a high-fat diet containing barley flour. The expression levels of 3065 genes were determined as DEGs. The DEGs up-regulated in the barley group showed an enrichment in pathways related to intestinal immunity, such as "Cytokine–cytokine receptor interaction (mmu04060)" and "B cell receptor signaling pathway (mmu04662)" (Table S4, Figure S2). By contrast, only one pathway (mmu04740: Olfactory transduction) was identified from DEGs down-regulated in the barley group (data not shown).

3.2. Food Intake, Body Weight, and Organ Weight

In the animal study, intake of the experimental diet showed no significant differences with the control group. Therefore, mice in both groups were determined to have been fed a similar level of energy during the study period (Table S5). However, body weight gain and final weight in the HGB group were significantly lower than the control group ($p < 0.05$). As a result, food efficiency ratio in the HGB group was lower than the C group ($p < 0.05$). The organ weight of liver, retroperitoneal, and mesenteric fats were lower in the HGB group than the C group ($p < 0.05$), while the cecum digesta was significantly higher ($p < 0.05$).

3.3. Secretory Immunoglobulin A (sIgA) Concentration in the Cecum and Serum

The concentrations of sIgA in the cecum and serum are shown in Figure 1. Cecum (Figure 1a) and serum (Figure 1b) sIgA levels in the HGB group were significantly higher than in the C group ($p < 0.05$). The concentrations of serum IL-10 and IL-6 were not statistically different between the two groups (Figure S3).

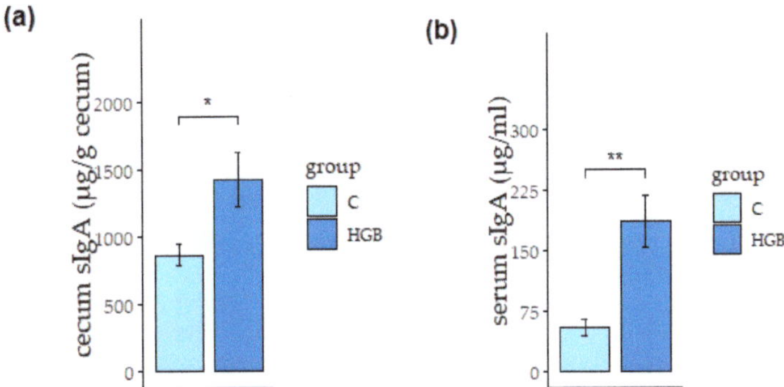

Figure 1. The concentration of sIgA in the cecum (**a**) and serum (**b**). Values are means ± SE, $n = 8$. * $p < 0.05$, ** $p < 0.01$ indicates a significant difference between each group, C: control group, HGB: high β-glucan barley group. sIgA, secretory immunoglobulin A.

3.4. SCFA and Organic Acid Concentration in the Cecum Digesta and Feces

The concentration of SCFAs and organic acids in the cecum digesta are shown in Figure 2. The total level of SCFAs, propionate, isobutyrate, isovalerate, lactate, and succinate concentrations in the HGB group were significantly higher than the C group ($p < 0.05$). The level of acetate also displayed a slight increase in the HGB group compared with the C group, although this was not statistically significant. The concentration of butyrate, propionate, isobutyrate, valerate, isovalerate, lactate, and succinate in the feces were significantly higher in the HGB group than the C group ($p < 0.05$) (Figure S4).

3.5. Counts of Predominant Bacterial Groups in the Cecum Digesta

Bacterial counts of microbiota at the phylum and genus levels in the cecum digesta are shown in Table 1. At the phylum level, the bacterial count of *Bacteroidetes*, *Firmicutes*, and total bacteria were significantly higher in the HGB group than the C group ($p < 0.05$). At the genus level, the bacterial counts of the *Bacteroides*, *Bifidobacterium*, *Lactobacillus*, and *Atopobium* cluster were significantly higher in the HGB group than the C group ($p < 0.05$).

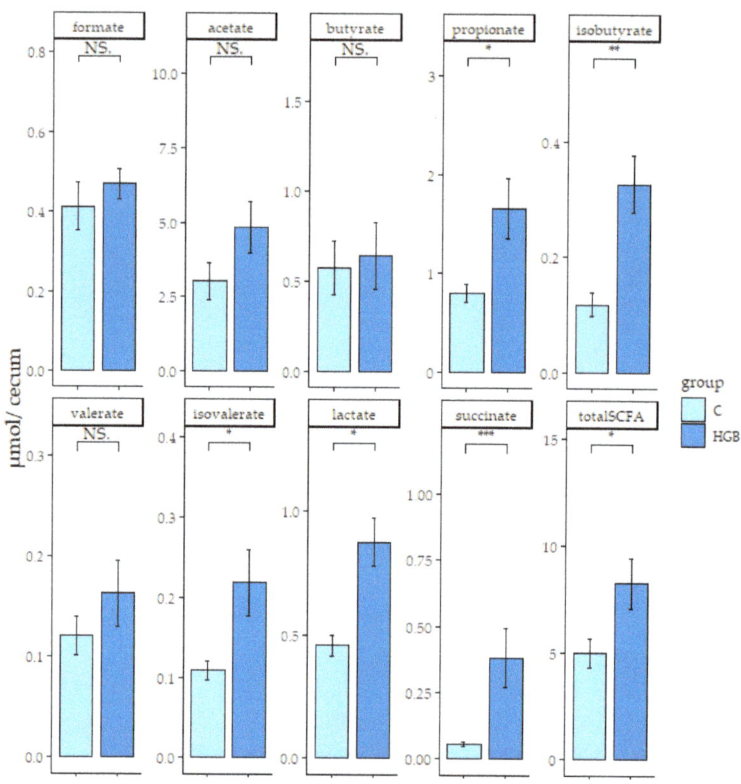

Figure 2. The concentration of short-chain fatty acids (SCFAs), lactate, and succinate in the cecum digesta. Values are means ± SE, $n = 8$. * $p < 0.05$, ** $p < 0.01$, *** $p < 0.001$ showed a significant difference between each group, "NS" is not significant. C: control group, HGB: high β-glucan barley group.

Table 1. Counts of the predominant bacterial groups in the cecum digesta between the control and HGB groups.

LogCFU/g (Cecum Digesta)	Control	HGB
Phylum		
Total bacteria	13.3 ± 0.3	14.9 ± 0.4 *
Bacteroidetes	10.0 ± 0.2	11.2 ± 0.2 *
Firmicutes	12.7 ± 0.1	12.9 ± 0.1 *
Actinobacteria	9.6 ± 0.4	10.1 ± 0.1
Genus		
Bacteroides fragilis group	9.7 ± 0.1	10.3 ± 0.2 *
Bifidobacterium	10.3 ± 0.3	11.1 ± 0.1 *
Lactobacillus	10.2 ± 0.2	11.3 ± 0.1 *
Prevotella	7.4 ± 0.1	7.6 ± 0.1
Clostridium coccoides group	9.8 ± 0.2	10.3 ± 0.1
Clostridium leptum subgroup	11.2 ± 0.2	11.7 ± 0.1
Atopobium cluster	7.8 ± 0.2	9.1 ± 0.2 *

* $p < 0.05$ significantly different between the control and HGB group. Values are means ± SE ($n = 8$). HGB: High β-glucan barley group.

3.6. Expression of mRNA Related to the Immune System in the Ileum

Ileum mRNA expression levels related to the immune system are shown in Figure 3. In the HGB group, mRNA expression levels of *IL-10* were significantly higher than the C group ($p < 0.05$). mRNA expression levels of other cytokines (*IFN-γ, IL-12, IL-1β, IL-4, IL-5, IL-6, IL-33*, transforming growth factor beta (*TGF-β*), tumor necrosis factor-α (*TNF-α*), and *IL-17* were not statistically different between the two groups. mRNA expression level of the polymeric immunoglobulin receptor (*pIgR*) in the HGB group was significantly higher than the C group ($p < 0.05$).

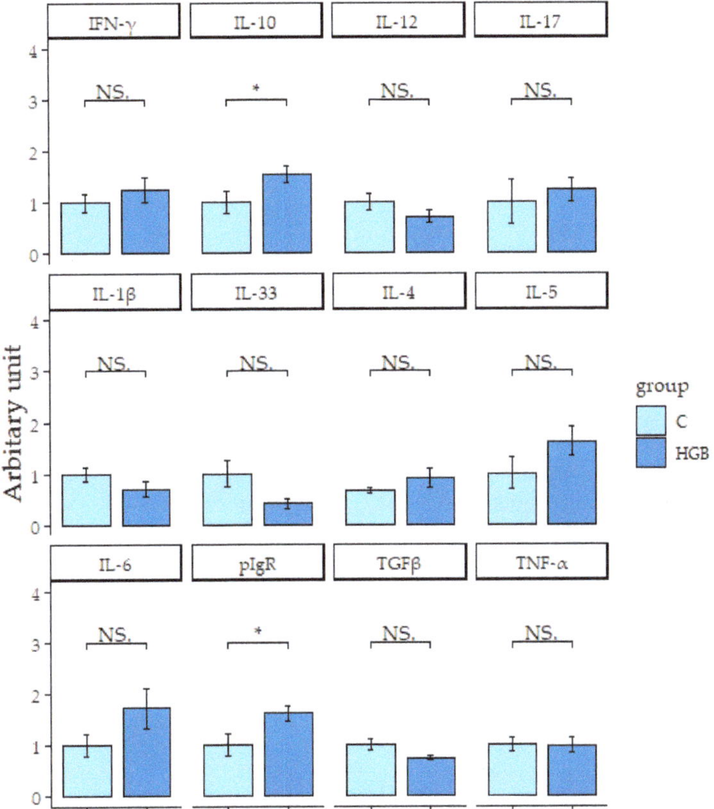

Figure 3. The expression levels of mRNA in the ileum. Values are means $\pm$ SE, $n = 8$. * $p < 0.05$ indicates a significant difference between each group, "NS" is not significant. C: control group, HGB: high β-glucan barley group. *IFN-γ*, interferon gamma; *IL-10*, interleukin 10; *IL-12*, interleukin 12; *IL-17*, interleukin 17; *IL-1β*, interleukin 1 beta; *IL-33*, interleukin 33; *IL-4*, interleukin 4; *IL-5*, interleukin 5; *IL-6*, interleukin 6; *pIgR*, polymeric immunoglobulin receptor; *TGFβ*, transforming growth factor beta; *TNF-α*, tumor necrosis factor-α.

3.7. Correlation Analysis between sIgA Concentration and Parameters Related to the Immune System in the Ileum and Cecum

The correlation coefficients between the concentration of sIgA and parameters related to immunity response are shown in Figure 4. Parameters that were significantly different in the HGB group compared to the C group were used in the analysis. A positive correlation was identified between sIgA in the cecum and isobutyrate, lactate, succinate, *Lactobacillus*, and *pIgR* ($p < 0.05$). A strong positive correlation was identified between sIgA in the serum and most gut bacteria ($p < 0.05$).

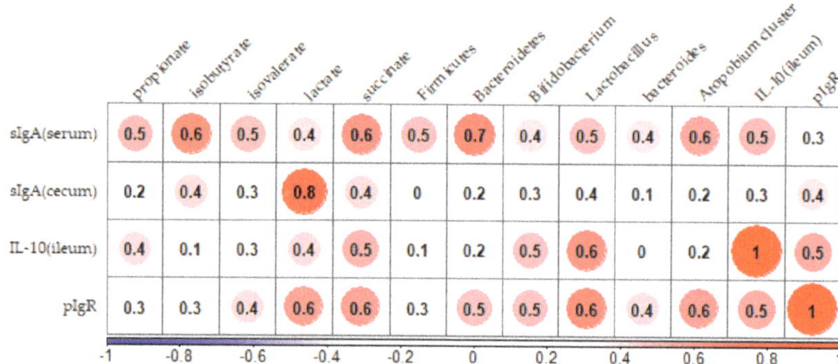

	propionate	isobutyrate	isovalerate	lactate	succinate	Firmicutes	Bacteroidetes	Bifidobacterium	Lactobacillus	bacteroides	Atopobium cluster	IL-10 (ileum)	pIgR
sIgA(serum)	0.5	0.6	0.5	0.4	0.6	0.5	0.7	0.4	0.5	0.4	0.6	0.5	0.3
sIgA(cecum)	0.2	0.4	0.3	0.8	0.4	0	0.2	0.3	0.4	0.1	0.2	0.3	0.4
IL-10(ileum)	0.4	0.1	0.3	0.4	0.5	0.1	0.2	0.5	0.6	0	0.2	1	0.5
pIgR	0.3	0.3	0.4	0.6	0.6	0.3	0.5	0.5	0.6	0.4	0.6	0.5	1

Figure 4. Result of the Spearman's rank correlation coefficients analysis between the concentration of serum and cecum sIgA and parameters related to intestinal immunity. The values in the figure show the correlation, and red circles highlight a significantly positive correlation. sIgA, secretory immunoglobulin A; *IL-10*, interleukin 10; *pIgR*, polymeric immunoglobulin receptor.

4. Discussion

In this study, we investigated the effect of HGB flour on the immune system in diet-induced obese mice. Intake of HGB flour with a high-fat diet increased the concentration of sIgA in the cecum and serum. A positive correlation was confirmed between the intestinal flora and intestinal fermentation metabolites, which are related to an increase in the level of IgA. In addition, the increase in mRNA expression of *IL-10* by HGB intake suggests that SCFAs and lactic acid may regulate the Tregs response and affect the immune response.

First, we used previously determined DNA microarray data of the ileum [28] to establish how HGB flour intake affects gene expression related to the immune system in the high-fat diets. KEGG enrichment analysis of DEGs suggested that HGB flour altered gene expression involved in several metabolic pathways of the ileum. Specifically, the B cell receptor signaling pathway (mmu04662) was up-regulated in mice fed the HGB diet. The B cell receptor is connected with the function and regulation of B cells differentiated into IgA plasma cells. The IgA antigen contributes an important role in the immune response in mucous membranes, such as the digestive tract, and is the most abundant antigen on the mucosal surface [32]. In a recent study, IgA was shown to mediate the modulation of microbiota in the digestive tract [33]. A previous study revealed that the microbiota composition differed significantly between immunodeficient mice and wild-type mice [34]. Thus, our findings suggest that barley affects the immune system, such as IgA secretion, via regulation of gut microbiota.

Next, we investigated changes to the immune system and microbiota using high-fat model mice. A previous study indicated that C57BL/6J mice fed a high-fat diet (60 kcal%fat) for 14 weeks had reduced IgA-producing cells in the gut-associated immune system and sIgA in the colon compared to mice fed a normal diet (16 kcal%fat) [35]. Nonetheless, we found an increase in the concentration of sIgA in the cecum and serum of the HGB group compared to the C group. Additionally, these results were also supported by the increase in the mRNA expression of *pIgR*, which binds to IgA and transports sIgA to the intestinal lumen side. Moreover, a positive correlation was found between the concentration of sIgA in the serum or cecum with the increased number of gut microbiota in mice from the HGB group. Thus, this observation may be the result of a prebiotic effect. Several studies have reported the effects of cereal dietary fiber on gut microbiota and the immune response. Yuri et al. showed that C57BL/6J mice fed oat-derived β-glucan formulations had higher levels of total serum immunoglobulins, which increased their resistance to the murine pathogen *Eimeria vermiformis* [36]. Another study indicated that a long-term intake of wheat bran significantly increased fecal sIgA and IgA bacteria in male BALB/c mice via modulation of the gut microbiota [37]. These reports back up our present findings. Moreover, our results show that ingestion of HGB flour up-regulates immune function, especially the secretion of IgA under a high-fat diet.

The concentration of total SCFAs and propionate in the cecum digesta were significantly higher in the HBG group compared to the C group. Smith et al. reported that acetate and propionate activated the migratory properties of Tregs through mediated G protein-coupled receptor 43 (*GPR43*) [38]. Several studies have addressed how Tregs regulate the IgA secretion of germinal center B cells [39,40]. Moreover, IL-10 secreted by Tregs suppresses the production of inflammatory cytokines by acting on macrophages and dendritic cells. Our results showed that HGB flour increased mRNA expression of *IL-10* in the ileum, and a positive correlation was confirmed between propionate and *IL-10*. Thus, intake of barley flour may affect the migration of Tregs and secretion of IL-10 via propionate. Indeed, a previous study showed that gut microbiota-derived SCFAs promote IL-10 production [41]. However, no significant difference was found in other inflammatory cytokines in the ileum and IL-6 and IL-10 level in serum. Although other studies reported that a long-term intake of oat β-glucan lowered the levels of inflammatory cytokines in the colon [42], we found no convincing evidence of an anti-inflammatory effect. Further studies are needed to investigate the anti-inflammatory effect of HGB flour.

A previous study indicated that butyrate strongly induces the differentiation of T regs in the colon through activation of the forkhead box protein P3 (*Foxp3*) gene [43]. Foxp3 plays an essential role in Tregs differentiation, functional expression, and maintenance of differentiation state [44]. However, our results showed no significant difference between the experimental diets in terms of the concentration of butyrate or butyrate-producing bacteria, such as *Clostridium leptum group* and the *Clostridium coccoides subgroup*. Moreover, there was no significant difference in the gene expression level of *Foxp3* (data not shown). Therefore, we concluded that the observed increase in sIgA is due to HGB intake promoting Tregs migration rather than the differentiation of Tregs.

In this study, a positive correlation was observed between the concentration of sIgA in the serum or cecum and *Lactobacillus* or lactate. Moreover, the bacterial count of the *Atopobium* cluster was significantly higher in the HGB group. Since the *Atopobium* cluster produces lactate as major metabolite, we proposed that these bacteria contributed to the observed increase in the concentration of lactate [45]. There are numerous reports on the antigenic effect of some *Lactobacillus* species on the gut immune system [46,47]. In particular, in vitro studies have shown that *Lactobacillus reuteri* and *Lactobacillus casei* increase the expression of *IL-10* [48]. Therefore, the correlation coefficient observed in this study suggests that increased levels of lactic acid-producing bacteria brought about by an intake of HGB flour may have enhanced the concentration of sIgA through the intestinal immune response. Our previous investigations have shown that the ingestion of HGB flour increases the levels of lactic acid-producing bacteria [49], but this is the first such study to report an effect on immune function. Further research is needed to clarify the species of lactic acid bacteria increased by the intake of HGB flour and to investigate their functions.

The β-(1-3), (1-6) glucan derived from yeast and mushrooms bind to the dectin-1 and Toll-like receptors present in immune cells, such as monocytes and dendritic cells, and influence the adaptive immune response, such as the production of cytokines and chemokines [50,51]. These effects had been attributed to the β-(1-3) linkage and structure of the β-(1-6) branching at a certain site that enhances interaction with specific receptors [52]. Results from a previous in vitro study suggested that barley-derived β-glucan may, at least in part, also affect the immune system via dectin-1 mediated changes [53]. However, it was shown that the affinity for barley β-glucan was weaker than for the continuous β-(1-3) glucan (sizofiran) [54]. Furthermore, because β-(1-3), (1-4) glucans of cereals do not contain β-(1-6) branches, β-glucans of cereals and yeast/mushrooms differ considerably in terms of solubility, molecular weight, and branching structure. Therefore, we propose that the increased level of sIgA observed in the HGB group was influenced by SCFAs and gut microbiota rather than dectin-1 signaling. Further studies are required to clarify the affinity and function for dectin-1 binding of β-(1-3), (1-4) glucan.

5. Conclusions

Our results indicated that an intake of HGB flour increases the concentration of sIgA in the serum and cecum under high-fat diet conditions. These findings suggest that this effect is mediated by an alternation in the gut microbiota and a subsequent increase in the levels of organic acids, including SCFAs, generated by intestinal fermentation of barley β-glucan. Since the mRNA expression level of *IL-10* was elevated in the ileum, it may be affected by Tregs, which are IL-10-producing cells. These findings will help to explain how the prebiotic effects of barley flour can improve the immune system or alleviate a weakened immune system deteriorated by a poor diet.

Supplementary Materials: The following are available online at https://www.mdpi.com/2072-6643/13/3/907/s1, Table S1: composition of the control and HGB diets, Table S2: List of standard bacterial strain and primer sequence for real-time PCR, Table S3: List of primers used for real-time PCR, Table S4: KEGG enrichment analysis with up-regulated DEGs in the ileum of barley group by using DAVID database, Table S5: Final weight, body weight gain, food intake, and organ weight in mice fed the control and HGB diets, Figure S1: Flow chart showing the study design of the animal studies, Figure S2: DEGs up-regulated by barley group involved in B cell receptor signaling pathway (extracted from Kyoto Encyclopedia of Genes and Genomes pathway), Figure S3: The concentrations of *IL-10* and *IL-6* in serum, Figure S4: The concentrations of SCFAs and organic acids in feces.

Author Contributions: Conceptualization, K.M., S.N., T.M. and S.A.; methodology, K.M. and S.A.; formal analysis, K.M. and N.O.; writing—original draft preparation, K.M.; writing—review and editing, K.M., N.O., S.N., T.M. and S.A.; supervision, S.A. All authors have read and agreed to the published version of the manuscript.

Funding: This research received no external funding.

Institutional Review Board Statement: The study was conducted according to the Declaration of Helsinki, and approved by the Otsuma Women's University Animal Research Committee (Tokyo, Japan, No.19013, 13 December 2019) and implemented in accordance with animal experimentation according to their regulations.

Conflicts of Interest: The authors declare no conflict of interest.

References

1. Allaire, J.M.; Crowley, S.M.; Law, H.T.; Chang, S.-Y.; Ko, H.-J.; Vallance, B.A. The Intestinal Epithelium: Central Coordinator of Mucosal Immunity. *Trends Immunol.* **2018**, *39*, 677–696. [CrossRef] [PubMed]
2. Gerriets, V.A.; MacIver, N.J. Role of T Cells in Malnutrition and Obesity. *Front. Immunol.* **2014**, *5*, 379. [CrossRef] [PubMed]
3. Rühl, R. Effects of dietary retinoids and carotenoids on immune development. *Proc. Nutr. Soc.* **2007**, *66*, 458–469. [CrossRef] [PubMed]
4. Sherman, P.M.; Ossa, J.C.; Johnson-Henry, K. Unraveling Mechanisms of Action of Probiotics. *Nutr. Clin. Pract.* **2009**, *24*, 10–14. [CrossRef]
5. Antoine, J.M. Probiotics: Beneficial factors of the defence system. *Proc. Nutr. Soc.* **2010**, *69*, 429–433. [CrossRef]
6. Mantis, N.J.; Rol, N.; Corthésy, B. Secretory IgA's complex roles in immunity and mucosal homeostasis in the gut. *Mucosal Immunol.* **2011**, *4*, 603–611. [CrossRef]
7. Nagai, M.; Noguchi, R.; Takahashi, D.; Morikawa, T.; Koshida, K.; Komiyama, S.; Ishihara, N.; Yamada, T.; Kawamura, Y.I.; Muroi, K.; et al. Fasting-Refeeding Impacts Immune Cell Dynamics and Mucosal Immune Responses. *Cell* **2019**, *178*, 1072–1087.e14. [CrossRef]
8. Kawamoto, S.; Maruya, M.; Kato, L.M.; Suda, W.; Atarashi, K.; Doi, Y.; Tsutsui, Y.; Qin, H.; Honda, K.; Okada, T.; et al. Foxp3+ T Cells Regulate Immunoglobulin A Selection and Facilitate Diversification of Bacterial Species Responsible for Immune Homeostasis. *Immunity* **2014**, *41*, 152–165. [CrossRef] [PubMed]
9. Fadlallah, J.; El Kafsi, H.; Sterlin, D.; Juste, C.; Parizot, C.; Dorgham, K.; Autaa, G.; Gouas, D.; Almeida, M.; Lepage, P.; et al. Microbial ecology perturbation in human IgA deficiency. *Sci. Transl. Med.* **2018**, *10*, eaan1217. [CrossRef]
10. Slavin, J. Fiber and prebiotics: Mechanisms and health benefits. *Nutrients* **2013**, *5*, 1417–1435. [CrossRef]
11. Jang, E.Y.; Hong, K.-B.; Chang, Y.B.; Shin, J.; Jung, E.Y.; Jo, K.; Suh, H.J. In Vitro Prebiotic Effects of Malto-Oligosaccharides Containing Water-Soluble Dietary Fiber. *Molecules* **2020**, *25*, 5201. [CrossRef] [PubMed]
12. Arena, M.; Caggianiello, G.; Fiocco, D.; Russo, P.; Torelli, M.; Spano, G.; Capozzi, V. Barley β-Glucans-Containing Food Enhances Probiotic Performances of Beneficial Bacteria. *Int. J. Mol. Sci.* **2014**, *15*, 3025–3039. [CrossRef] [PubMed]

13. Russo, P.; López, P.; Capozzi, V.; de Palencia, P.F.; Dueñas, M.T.; Spano, G.; Fiocco, D. Beta-Glucans Improve Growth, Viability and Colonization of Probiotic Microorganisms. *Int. J. Mol. Sci.* **2012**, *13*, 6026–6039. [CrossRef] [PubMed]

14. Arena, M.P.; Russo, P.; Capozzi, V.; Rascón, A.; Felis, G.E.; Spano, G.; Fiocco, D. Combinations of cereal β-glucans and probiotics can enhance the anti-inflammatory activity on host cells by a synergistic effect. *J. Funct. Foods* **2016**, *23*, 12–23. [CrossRef]

15. De Angelis, M.; Montemurno, E.; Vannini, L.; Cosola, C.; Cavallo, N.; Gozzi, G.; Maranzano, V.; Di Cagno, R.; Gobbetti, M.; Gesualdo, L. Effect of Whole-Grain Barley on the Human Fecal Microbiota and Metabolome. *Appl. Environ. Microbiol.* **2015**, *81*, 7945–7956. [CrossRef]

16. Martínez, I.; Lattimer, J.M.; Hubach, K.L.; Case, J.A.; Yang, J.; Weber, C.G.; Louk, J.A.; Rose, D.J.; Kyureghian, G.; Peterson, D.A.; et al. Gut microbiome composition is linked to whole grain-induced immunological improvements. *ISME J.* **2013**, *7*, 269–280. [CrossRef]

17. Hughes, S.A.; Shewry, P.R.; Gibson, G.R.; McCleary, B.V.; Rastall, R.A. In vitro fermentation of oat and barley derived β-glucans by human faecal microbiota. *FEMS Microbiol. Ecol.* **2008**, *64*, 482–493. [CrossRef] [PubMed]

18. Corrêa-Oliveira, R.; Fachi, J.L.; Vieira, A.; Sato, F.T.; Vinolo, M.A.R. Regulation of immune cell function by short-chain fatty acids. *Clin. Transl. Immunol.* **2016**, *5*, e73. [CrossRef]

19. Chen, K.; Magri, G.; Grasset, E.K.; Cerutti, A. Rethinking mucosal antibody responses: IgM, IgG and IgD join IgA. *Nat. Rev. Immunol.* **2020**, *20*, 427–441. [CrossRef] [PubMed]

20. Park, J.; Kim, M.; Kang, S.G.; Jannasch, A.H.; Cooper, B.; Patterson, J.; Kim, C.H. Short-chain fatty acids induce both effector and regulatory T cells by suppression of histone deacetylases and regulation of the mTOR–S6K pathway. *Mucosal Immunol.* **2015**, *8*, 80–93. [CrossRef] [PubMed]

21. Zimmerman, M.A.; Singh, N.; Martin, P.M.; Thangaraju, M.; Ganapathy, V.; Waller, J.L.; Shi, H.; Robertson, K.D.; Munn, D.H.; Liu, K. Butyrate suppresses colonic inflammation through HDAC1-dependent Fas upregulation and Fas-mediated apoptosis of T cells. *Am. J. Physiol. Liver Physiol.* **2012**, *302*, G1405–G1415. [CrossRef] [PubMed]

22. Ramakers, J.D.; Volman, J.J.; Biörklund, M.; Önning, G.; Mensink, R.P.; Plat, J. Fecal water from ileostomic patients consuming oat β-glucan enhances immune responses in enterocytes. *Mol. Nutr. Food Res.* **2007**, *51*, 211–220. [CrossRef]

23. Volman, J.J.; Mensink, R.P.; Ramakers, J.D.; de Winther, M.P.; Carlsen, H.; Blomhoff, R.; Buurman, W.A.; Plat, J. Dietary (1→3), (1→4)-β-d-glucans from oat activate nuclear factor-κB in intestinal leukocytes and enterocytes from mice. *Nutr. Res.* **2010**, *30*, 40–48. [CrossRef] [PubMed]

24. Cox, A.J.; West, N.P.; Cripps, A.W. Obesity, inflammation, and the gut microbiota. *Lancet Diabetes Endocrinol.* **2015**, *3*, 207–215. [CrossRef]

25. de Heredia, F.P.; Gómez-Martínez, S.; Marcos, A. Obesity, inflammation and the immune system. *Proc. Nutr. Soc.* **2012**, *71*, 332–338. [CrossRef]

26. Patterson, E.; Ryan, P.M.; Cryan, J.F.; Dinan, T.G.; Ross, R.P.; Fitzgerald, G.F.; Stanton, C. Gut microbiota, obesity and diabetes. *Postgrad. Med. J.* **2016**, *92*, 286–300. [CrossRef]

27. MuhomahH, T.A.; Nishino, N.; Katsumata, E.; Haoming, W.; Tsuruta, T. High-fat diet reduces the level of secretory immunoglobulin A coating of commensal gut microbiota. *Biosci. Microbiota Food Health* **2019**, *38*, 55–64. [CrossRef] [PubMed]

28. Mio, K.; Yamanaka, C.; Matsuoka, T.; Kobayashi, T.; Aoe, S. Effects of β-glucan Rich Barley Flour on Glucose and Lipid Metabolism in the Ileum, Liver, and Adipose Tissues of High-Fat Diet Induced-Obesity Model Male Mice Analyzed by DNA Microarray. *Nutrients* **2020**, *12*, 3546. [CrossRef] [PubMed]

29. Atarashi, K.; Tanoue, T.; Oshima, K.; Suda, W.; Nagano, Y.; Nishikawa, H.; Fukuda, S.; Saito, T.; Narushima, S.; Hase, K.; et al. Treg induction by a rationally selected mixture of Clostridia strains from the human microbiota. *Nature* **2013**, *500*, 232–236. [CrossRef] [PubMed]

30. Matsuda, K.; Tsuji, H.; Asahara, T.; Kado, Y.; Nomoto, K. Sensitive Quantitative Detection of Commensal Bacteria by rRNA-Targeted Reverse Transcription-PCR. *Appl. Environ. Microbiol.* **2007**, *73*, 6695. [CrossRef]

31. Matsuki, T. Development of quantitative PCR detection method with 16S rRNA gene-targeted genus- and species-specific primers for the analysis of human intestinal microflora and its application. *Nihon Saikingaku Zasshi* **2007**, *62*, 255–261. [CrossRef] [PubMed]

32. Lycke, N.Y.; Bemark, M. The regulation of gut mucosal IgA B-cell responses: Recent developments. *Mucosal Immunol.* **2017**, *10*, 1361–1374. [CrossRef] [PubMed]

33. Dollé, L.; Tran, H.Q.; Etienne-Mesmin, L.; Chassaing, B. Policing of gut microbiota by the adaptive immune system. *BMC Med.* **2016**, *14*, 27. [CrossRef] [PubMed]

34. Zhang, H.; Sparks, J.B.; Karyala, S.V.; Settlage, R.; Luo, X.M. Host adaptive immunity alters gut microbiota. *ISME J.* **2015**, *9*, 770–781. [CrossRef] [PubMed]

35. Luck, H.; Khan, S.; Kim, J.H.; Copeland, J.K.; Revelo, X.S.; Tsai, S.; Chakraborty, M.; Cheng, K.; Tao Chan, Y.; Nøhr, M.K.; et al. Gut-associated IgA+ immune cells regulate obesity-related insulin resistance. *Nat. Commun.* **2019**, *10*, 3650. [CrossRef] [PubMed]

36. Yuri, C.; Estrada, A.; Van Kessel, A.; Gajadhar, A.; Redmond, M.; Laarveld, B. Immunomodulatory Effects of Oat β-Glucan Administered Intragastrically or Parenterally on Mice Infected with Eimeria vermiformis. *Microbiol. Immunol.* **1998**, *42*, 457–465. [CrossRef] [PubMed]

37. Matsuzaki, K.; Iwai, K.; Yoshikawa, Y.; Shimamura, Y.; Miyoshi, N.; Hiramoto, S.; Asada, K.; Fukutomi, R.; Su, H.; Ohashi, N. Wheat Bran Intake Enhances the Secretion of Bacteria-Binding IgA in a Lumen of the Intestinal Tract by Incrementing Short Chain Fatty Acid Production Through Modulation of Gut Microbiota. *Nat. Prod. Commun.* **2020**, *15*. [CrossRef]

38. Smith, P.M.; Howitt, M.R.; Panikov, N.; Michaud, M.; Gallini, C.A.; Bohlooly, Y.M.; Glickman, J.N.; Garrett, W.S. The Microbial Metabolites, Short-Chain Fatty Acids, Regulate Colonic Treg Cell Homeostasis. *Science (80-)* **2013**, *341*, 569–573. [CrossRef] [PubMed]

39. Tsuji, M.; Komatsu, N.; Kawamoto, S.; Suzuki, K.; Kanagawa, O.; Honjo, T.; Hori, S.; Fagarasan, S. Preferential Generation of Follicular B Helper T Cells from Foxp3+ T Cells in Gut Peyer's Patches. *Science (80-)* **2009**, *323*, 1488–1492. [CrossRef]

40. Linterman, M.A.; Pierson, W.; Lee, S.K.; Kallies, A.; Kawamoto, S.; Rayner, T.F.; Srivastava, M.; Divekar, D.P.; Beaton, L.; Hogan, J.J.; et al. Foxp3+ follicular regulatory T cells control the germinal center response. *Nat. Med.* **2011**, *17*, 975–982. [CrossRef]

41. Sun, M.; Wu, W.; Chen, L.; Yang, W.; Huang, X.; Ma, C.; Chen, F.; Xiao, Y.; Zhao, Y.; Ma, C.; et al. Microbiota-derived short-chain fatty acids promote Th1 cell IL-10 production to maintain intestinal homeostasis. *Nat. Commun.* **2018**, *9*, 3555. [CrossRef]

42. Shi, H.; Yu, Y.; Lin, D.; Zheng, P.; Zhang, P.; Hu, M.; Wang, Q.; Pan, W.; Yang, X.; Hu, T.; et al. β-glucan attenuates cognitive impairment via the gut-brain axis in diet-induced obese mice. *Microbiome* **2020**, *8*, 143. [CrossRef] [PubMed]

43. Furusawa, Y.; Obata, Y.; Fukuda, S.; Endo, T.A.; Nakato, G.; Takahashi, D.; Nakanishi, Y.; Uetake, C.; Kato, K.; Kato, T.; et al. Commensal microbe-derived butyrate induces the differentiation of colonic regulatory T cells. *Nature* **2013**, *504*, 446–450. [CrossRef] [PubMed]

44. Hori, S.; Nomura, M.; Sakaguchi, S. Control of Regulatory T Cell Development by the Transcription Factor Foxp3. *Science (80-)* **2003**, *299*, 1057–1061. [CrossRef] [PubMed]

45. Harmsen, H.J.M.; Wildeboer-Veloo, A.C.M.; Grijpstra, J.; Knol, J.; Degener, J.E.; Welling, G.W. Development of 16S rRNA-Based Probes for theCoriobacterium Group and the Atopobium Cluster and Their Application for Enumeration of Coriobacteriaceaein Human Feces from Volunteers of Different Age Groups. *Appl. Environ. Microbiol.* **2000**, *66*, 4523–4527. [CrossRef] [PubMed]

46. Perdigón, G.; Maldonado Galdeano, C.; Valdez, J.; Medici, M. Interaction of lactic acid bacteria with the gut immune system. *Eur. J. Clin. Nutr.* **2002**, *56*, S21–S26. [CrossRef] [PubMed]

47. Sashihara, T.; Sueki, N.; Ikegami, S. An Analysis of the Effectiveness of Heat-Killed Lactic Acid Bacteria in Alleviating Allergic Diseases. *J. Dairy Sci.* **2006**, *89*, 2846–2855. [CrossRef]

48. Smits, H.H.; Engering, A.; van der Kleij, D.; de Jong, E.C.; Schipper, K.; van Capel, T.M.M.; Zaat, B.A.J.; Yazdanbakhsh, M.; Wierenga, E.A.; van Kooyk, Y. Selective probiotic bacteria induce IL-10–producing regulatory T cells in vitro by modulating dendritic cell function through dendritic cell–specific intercellular adhesion molecule 3–grabbing nonintegrin. *J. Allergy Clin. Immunol.* **2005**, *115*, 1260–1267. [CrossRef] [PubMed]

49. Mio, K.; Yamanaka, C.; Ichinose, Y.; Kohyama, N.; Yanagisawa, T.; Aoe, S. Effects of barley β-glucan with various molecular weights partially hydrolyzed by endogenous β-glucanase on glucose tolerance and lipid metabolism in mice. *Cereal Chem.* **2020**, *97*, 1056–1065. [CrossRef]

50. Stier, H.; Ebbeskotte, V.; Gruenwald, J. Immune-modulatory effects of dietary Yeast Beta-1,3/1,6-D-glucan. *Nutr. J.* **2014**, *13*, 38. [CrossRef]

51. Brown, G.D.; Herre, J.; Williams, D.L.; Willment, J.A.; Marshall, A.S.J.; Gordon, S. Dectin-1 Mediates the Biological Effects of β-Glucans. *J. Exp. Med.* **2003**, *197*, 1119–1124. [CrossRef] [PubMed]

52. Legentil, L.; Paris, F.; Ballet, C.; Trouvelot, S.; Daire, X.; Vetvicka, V.; Ferrières, V. Molecular Interactions of β-(1→3)-Glucans with Their Receptors. *Molecules* **2015**, *20*, 9745–9766. [CrossRef] [PubMed]

53. Tada, R.; Ikeda, F.; Aoki, K.; Yoshikawa, M.; Kato, Y.; Adachi, Y.; Tanioka, A.; Ishibashi, K.; Tsubaki, K.; Ohno, N. Barley-derived β-d-glucan induces immunostimulation via a dectin-1-mediated pathway. *Immunol. Lett.* **2009**, *123*, 144–148. [CrossRef] [PubMed]

54. Tada, R.; Adachi, Y.; Ishibashi, K.; Tsubaki, K.; Ohno, N. Binding Capacity of a Barley β-D-Glucan to the β-Glucan Recognition Molecule Dectin-1. *J. Agric. Food Chem.* **2008**, *56*, 1442–1450. [CrossRef] [PubMed]

Article

Low Molecular Weight Barley β-Glucan Affects Glucose and Lipid Metabolism by Prebiotic Effects

Seiichiro Aoe [1,2,*], Kento Mio [1], Chiemi Yamanaka [2] and Takao Kuge [3]

[1] Studies in Human Life Sciences, Graduate School of Studies in Human Culture, Otsuma Women's University, Chiyoda-ku, Tokyo 102-8357, Japan; mio.kento@hakubaku.co.jp
[2] The Institute of Human Culture Studies, Otsuma Women's University Chiyoda-ku, Tokyo 102-8357, Japan; chiemiy@gmail.com
[3] ADEKA Corporation, Tokyo 116-8553, Japan; kuge@adeka.co.jp
* Correspondence: s-aoe@otsuma.ac.jp; Tel.: +81-3-5275-6048

Abstract: We investigated the effect of low molecular weight barley β-glucan (LMW-BG) on cecal fermentation, glucose, and lipid metabolism through comparisons to high molecular weight β-glucan (HMW-BG). C57BL/6J male mice were fed a moderate-fat diet for 61 days. LMW-BG or HMW-BG was added to the diet corresponding to 4% β-glucan. We measured the apparent absorption of fat, serum biomarkers, the expression levels of genes involved in glucose and lipid metabolism in the liver and ileum, and bacterial counts of the major microbiota groups using real time PCR. The concentration of short-chain fatty acids (SCFAs) in the cecum was analyzed by GC/MS. Significant reductions in serum leptin, total- and LDL-cholesterol concentrations, and mRNA expression levels of sterol regulatory element-binding protein-1c (SREBP-1c) were observed in both BG groups. HMW-BG specific effects were observed in inhibiting fat absorption and reducing abdominal deposit fat, whereas LMW-BG specific effects were observed in increasing bacterial counts of *Bifidobacterium* and *Bacteroides* and cecal total SCFAs, acetate, and propionate. mRNA expression of neurogenin 3 was increased in the LMW-BG group. We report that LMW-BG affects glucose and lipid metabolism via a prebiotic effect, whereas the high viscosity of HMW-BG in the digestive tract is responsible for its specific effects.

Keywords: barley; β-glucan; low molecular weight; fermentation; prebiotics

Citation: Aoe, S.; Mio, K.; Yamanaka, C.; Kuge, T. Low Molecular Weight Barley β-Glucan Affects Glucose and Lipid Metabolism by Prebiotic Effects. *Nutrients* **2021**, *13*, 130. https://doi.org/10.3390/nu13010130

Received: 30 November 2020
Accepted: 29 December 2020
Published: 31 December 2020

Publisher's Note: MDPI stays neutral with regard to jurisdictional claims in published maps and institutional affiliations.

1. Introduction

Barley and oats are rich in β-glucan, which has several positive effects on inhibiting a postprandial glucose rise [1–3] and improvement in serum cholesterol concentrations [4–8]. Barley and oat β-glucans have β-1,3 and β-1,4 glycosidic polysaccharides with high molecular weight [9,10]. It is generally recognized that β-glucans in barley and oats with higher molecular weights are essential for a reduction in postprandial blood glucose rise and incremental area under the blood concentration curve (IAUC) [11–13]. It was also reported that an increase in excretion of neutral and acidic sterols reduced serum LDL cholesterol concentrations by promoting catabolism and reduced cholesterol absorption [14]. This mechanism is positively related to the high viscosity of β-glucan in the small intestine and consequent increase in the excretion of neutral and acidic sterols [15]. However, a previous study reported that the serum lipid profile in mice fed a high-fat diet with added β-glucan of three different molecular weights (1450, 730, and 370 kDa) did not differ among the groups [16]. Another report showed that all of the molecular weight forms of β-glucan (2348, 1311, 241, 56, 21 or <10 kDa) significantly reduced plasma cholesterol concentrations when compared with the control diet [17]. This discrepancy was explained by the experimental condition, such as excessive doses of β-glucan (6.8–8.4%). Our previous study showed that partially hydrolyzed barley β-glucan (50 and 110 kDa, 2.5% β-glucan in the diet) demonstrated the physiological functions similar to intact barley β-glucan which improved glucose and lipid metabolism [18].

Propionate, one of the major short-chain fatty acids (SCFAs), plays a significant role in the cholesterol-lowering effect [19]. A lot of evidence has accumulated to support the effect of β-glucan intake on cholesterol metabolism and gut microbiota metabolism, and it is clear that intake of β-glucans modifies the balance of gut microbiota [20]. Studies have shown that both cholesterol and bile acid metabolism is regulated by the metabolism of gut microbiota. A prebiotic effect was previously reported in humans: β-glucan altered the microbiota and led to an improvement in bile acid metabolism by the gut microbial community [21]. Generally, low molecular weight dietary fiber is more fermentable compared to high molecular weight fiber. However, an increase in SCFAs in the feces of subjects supplemented with high molecular weight barley β-glucan (HMW-BG) increased fecal bile acids [22]. The same effects were not observed in subjects supplemented with low molecular weight barley β-glucan (LMW-BG). In contrast to this result, a positive effect of β-glucans, such as promotion of fecal SCFAs, especially the low molecular weight form, was also reported in the colon tissue of both healthy and LPS-induced enteritis rats [23]. These results suggest that effects of low and high molecular weight β-glucans on their physiological functions were still controversial because of different experimental conditions.

We hypothesized that HMW-BG, with high viscosity, attenuates the glycemic response and lipid absorption, whereas LMW-BG, with high fermentability affects the glycemic response and lipid metabolism by prebiotic actions, such as SCFA production. The main objective of this study was to investigate the effect of LMW-BG on cecal fermentation, and glucose and lipid metabolism in mice fed a moderate-fat diet, compared with the effects of HMW-BG.

2. Materials and Methods

2.1. Chemical Analysis of β-Glucan

LMW-BG, partially hydrolyzed by cellulase, was obtained from ADEKA Corp. (30SP, Tokyo, Japan). HMW-BG was purchased from Megazyme Ltd. (Bray, Ireland). The average molecular weights of LMW-BG and HMW-BG were approximately 12 and 500 kDa, respectively. The total dietary fiber (TDF) content of LMW-BG and HMW-BG was determined to be 45% and 94%, respectively, using the method of Prosky et al. [24]. The β-glucan content of LMW-BG and HMW-BG was determined to be 33% and 94%, respectively, using the method of McCleary et al. (AOAC 995.16) [25].

2.2. Animals and Study Design

Male C57BL/6J 4-week-old mice were purchased from Charles River Laboratories Japan, Inc. (Yokohama, Japan). Each mouse was housed individually in polycarbonate cages. Mice were maintained on a 12 h light/dark cycle (lights on at 08:00 h). The studies were approved by the Otsuma Women's University Animal Research Committee (Tokyo, Japan) and were performed in accordance with the Regulation on Animal Experimentation at Otsuma Women's University (No.19007). After acclimatization for 7 days, the mice were randomized into 3 groups (n = 8 per group) and shifted to a 25% fat energy diet supplemented with LMW-BG or HMW-BG powder. LMW-BG and HMW-BG were added to the diets at concentrations corresponding to 4% β-glucan. The total dietary fiber content of the LMW-BG test diet was 5.48%; therefore, the other diets were supplemented with the amount of cellulose necessary to adjust the total dietary fiber content to 5.48%. The compositions of the experimental diets are shown in Table 1. Purified HMW-BG has poor solubility at body temperature; therefore, it was precipitated in ethanol and then dissolved in hot water, according to the manufacturer's instruction. Dissolved HMW-BG (31.6%) was mixed with corn starch (68.4%), freeze-dried, and then the freeze-dried mixture was supplemented into the experimental diet. The protein and available carbohydrate contents in the LMW-BG diet were adjusted with casein and dextrinized corn starch, because the

LMW-BG contained 4.2% protein and 33.6% available carbohydrate. Mice were fed the experimental diets ad libitum for 61 days. Food intake and body weights were monitored 3 times per week throughout the study period. Feces were collected for the final 5 days of the study period. The feces were freeze dried after washing the surface with distilled water to remove attached diet powder, milled, and kept at −20 °C until measurement. After fasting for 8 h, mice were sacrificed by isoflurane/CO_2 anesthesia, then the cecum with digesta, adipose tissues (retroperitoneal, mesenteric, epididymal depot fats), and liver were dissected and weighed. Small samples of liver (200 mg) were suspended in RNA Stabilization Reagent (RNAlater, Qiagen, Hilden, Germany), and the remainder was freeze-dried, milled, and stored at −20 °C until required for cholesterol and triglyceride analysis. Small samples of ileum (40 mm portion from the cecum) were suspended in RNA Stabilization Reagent (RNAlater, Qiagen, Hilden, Germany). Cecum with digesta was stored at −20 °C until required for major microbiota and short chain fatty acid analysis. Blood samples, collected from the heart under isoflurane/CO_2 anesthesia at sacrifice, were centrifuged, and serum was collected for biochemical analysis.

Table 1. Composition of the experimental diets.

			(g/kg Diet)
	Control	**LMW-BG**	**HMW-BG**
Casein	200	183	200
L-cystin	3	3	3
Corn starch	350.7	300.9	350.7
Dextrinized corn starch	132	132	45.4
Sucrose	100	100	100
Soybean oil	70	70	70
Lard	42	42	42
Cellulose	54.8	-	14.8
LMW-BG	-	121.6	-
HMW-BG mixed with corn starch	-	-	126.6
AIN-93G mineral mixture	35	35	35
AIN-93 vitamin mixture	10	10	10
Choline bitartrate	2.5	2.5	2.5
t-Butylhydroquinone	0.014	0.014	0.014

LMW-BG: Low molecular weight β-glucan; 32.9%, total dietary fiber (TDF); 45.1%. HMW-BG mixed with corn starch was prepared from a freeze-dried mixture of previously resolved high molecular weight β-glucan (31.6%) in hot water and corn starch (68.4%).

2.3. Biochemical Analysis of the Serum and Concentration of Liver Lipids

Total, LDL, and HDL cholesterol, triglycerides and non-esterified fatty acids (NEFA) were measured in mouse sera using Hitachi 71,870 auto-analyzers at the Nagahama Research Institute (Oriental Yeast Co., Ltd., Shiga, Japan). Serum leptin and insulin concentrations were measured using enzyme-linked immunosorbent assay (ELISA) kits (Mouse Leptin Immunoassay Kit, R&D Systems, Inc., MN, USA; Mouse Insulin ELISA Kit (TMB), Shibayagi Co., Ltd., Gunma, Japan). A 2:1 (*v/v*) chloroform–methanol solution was used to extract lipids from the liver [26], which were then dissolved in isopropanol containing 10% Triton X-100 (FUJIFILM Wako Pure Chemical Corporation, Osaka, Japan). Hepatic cholesterol and triglyceride levels were measured enzymatically with commercial kits (Cholesterol E-test and Triglyceride E-test; FUJIFILM Wako Pure Chemical Corporation, Osaka, Japan).

2.4. Total Fecal Lipid and β-Glucan Analysis

Fecal lipids were extracted using a 2:1 (*v/v*) chloroform–methanol mixture under acidic conditions (acetic acid was added to a final concentration of 4%) [26,27] and determined gravimetrically. Fecal β-glucans were analyzed to assess the fermentation rate according to the method of McCleary et al. (AOAC 995.16) [25].

2.5. Analysis of Short Chain Fatty Acids in Cecal Digesta

The concentration of cecal SCFAs was determined by the method previously described, using a gas chromatography–mass spectrometry (GC/MS) system [28]. GC/MS used a 7890B GC system (Agilent, Tokyo, Japan) equipped with a 5977A MSD (Agilent). A DB-5MS capillary column (30 m × 0.53 mm) (Agilent) was used to separate the SCFAs. SCFA concentrations were calculated by comparing their peak areas with the internal standard (crotonic acid) and were expressed as μmol/cecum.

2.6. Analysis of Major Microbiota in Cecal Digesta

Cecal bacterial counts were analyzed by real-time PCR according to the methods previously described [29,30]. DNA extraction from cecal digesta was performed by using a QIAamp Fast DNA Stool Mini Kit (QIAGEN GmbH). Two μl of DNA solution from 200 mg cecum digesta was used for the real-time PCR. Each real-time PCR analysis was performed in a 20 μL reaction mixture containing DNA and PowerUp SYBR Green Master Mix (Thermo Fisher Scientific, Waltham, MA, USA). Real-time PCR amplification and detection were performed in 96-well optical plates with a QuantStudio 3 real-time PCR system (Thermo Fisher Scientific K.K., MA, USA). A standard curve was generated with the real-time PCR data and the corresponding cell count, for dilution series of the following standard strains: *Bacteroides fragilis* JCM11019 (for *Bacteroides fragilis* group), *Prevotella melaninogenica* JCM6325 (for *Prevotella*), *Bifidobacterium longum* JCM1217 (for *Bifidobacterium*), *Lactobacillus rhamnosus* ATCC8530 (for *Lactobacillus*), *Clostridium coccoides* JCM1395T (for *Clostridium coccoides* group), *Ruminococcus albus* JCM14654 (for *Clostridium leptum subgroup*), *Collinsella aerofaciens* JCM10188 (for *Atopobium cluster*). Group-specific primers for the real-time PCR are shown in Supplementary Table S1. Colony formation units (CFU) were calculated using the standard curve for the representative strain of each group obtained as described above.

2.7. Expression Analysis of mRNAs in Liver and Ileum

Primer sequences are presented in Supplementary Table S2. Total RNAs in the liver and ileum were prepared using RNeasy mini kits (Qiagen, Hilden, Germany) according to the manufacturer's instructions. mRNA expression was measured with a Quant3 Real-Time PCR System and PowerUp SYBR Green Master Mix (Thermo Fisher Scientific, Waltham, MA, USA) using cDNA prepared by RT-PCR. The reaction mixture for RT-PCR was prepared with 10 mM dNTPs (TOYOBO Co., Ltd., Osaka, Japan), 0.3 μg/mL Random primer (Life Technologies Japan, Tokyo, Japan), ReverTra ace Buffer (ReverTra ace, TOYOBO Co., Ltd., Osaka, Japan) and mixed into 5μg RNA/11.5μL RNAase-free water. The reaction was carried out by RT-PCR at 30 °C for 10 min, 42 °C for 60 min, and 99 °C for 5 min. The sample was diluted 20-fold with RNAase-free water to form cDNA template, which was used to measure mRNA expression. The $2^{-\Delta\Delta CT}$ method was utilized for data analysis. The reference gene was 36B4. The $\Delta\Delta CT$ is the difference between the ΔCT for the BG diets and control diet. Relative expression levels are presented as fold changes to the control group (arbitrary unit).

2.8. Statistical Analysis

Sample sizes were determined from our previous study [18]. Twenty-four mice (8 mice per group) were used. Data are presented as mean ± standard error of the mean. Significant difference ($p < 0.05$) between group means was determined by Tukey–Kramer's test or the Student's *t*-test. The relationships among the SCFAs and parameters related to the prebiotic effect were assessed using Spearman's rank correlation coefficient. JMP (Version 14.1, SAS Institute Inc., Cary, NC, USA) was used to perform the statistical analyses.

3. Results

3.1. Food Intake, Body Weight and Organ Weight

Body weight gain, food intake, and food efficiency ratio in mice fed LMW-BG or HMW-BG are shown in Table 2. Body weight gain was significantly lower in the HMW-BG group than the control group ($p < 0.05$); however, food intake was significantly lower in both LMW-BG and HMW-BG groups than the control group. A significant difference in the food efficiency ratio between the LMW-BG and HMW-BG group was observed ($p < 0.05$). The organ weights in mice fed LMW-BG and HMW-BG are shown in Table 3. The weights of the cecum with digesta were significantly higher in both LMW-BG and HMW-BG groups compared with the control group ($p < 0.05$). Liver and total abdominal, retroperitoneal, epididymal and mesenteric fat weights were significantly lower in the HMW-BG group compared with the control group ($p < 0.05$).

Table 2. Body weight gain, food intake, and food efficiency ratio.

	Control	LMW-BG	HMW-BG
Initial weight (g)	20.6 ± 0.5	20.6 ± 0.4	20.6 ± 0.4
Final weight (g)	31.3 ± 0.9	31.0 ± 0.8	28.9 ± 0.7
Body weight gain (g/d)	0.19 ± 0.01 [a]	0.17 ± 0.01 [ab]	0.14 ± 0.01 [b]
Food intake (g/d)	4.1 ± 0.2 [a]	3.2 ± 0.1 [b]	3.4 ± 0.2 [ab]
Food efficiency ratio (%)	4.56 ± 0.23 [ab]	5.30 ± 0.25 [a]	4.01 ± 0.22 [b]

Values are means $\pm$ SE, $n = 8$. Means with suffixed superscript letters differ significantly (Tukey–Kramer's test, $p < 0.05$). LMW-BG; low molecular weight β-glucan, HMW-BG; high molecular weight β-glucan.

Table 3. Organ weights.

	Control	LMW-BG	HMW-BG
Liver (g)	1.11 ± 0.03 [a]	1.04 ± 0.05 [ab]	0.92 ± 0.02 [b]
Cecum with digesta (g)	0.30 ± 0.03 [a]	0.51 ± 0.03 [b]	0.44 ± 0.03 [b]
Total abdominal fat	2.09 ± 0.11 [a]	1.54 ± 0.23 [ab]	1.25 ± 0.19 [b]
Retroperitoneal fat (g)	0.46 ± 0.03 [a]	0.32 ± 0.06 [ab]	0.25 ± 0.05 [b]
Epididymal fat (g)	1.22 ± 0.06 [a]	0.89 ± 0.08 [ab]	0.75 ± 0.10 [b]
Mesenteric fat (g)	0.42 ± 0.03 [a]	0.33 ± 0.05 [ab]	0.26 ± 0.05 [b]

Values are means $\pm$ SE, $n = 8$. Means with suffixed superscript letters differ significantly (Tukey–Kramer's test, $p < 0.05$). LMW-BG; low molecular weight β-glucan, HMW-BG; high molecular weight β-glucan.

3.2. Fecal Total Fat Excretion and Apparent Absorption of Fat

Fecal total fat excretion and apparent digestibility of fat in the final five days are shown in Table 4. The apparent digestibility of fat in the HMW-BG group was significantly lower than the control group. The same tendency was observed in the LMW-BG group, but the difference was not significant.

Table 4. Fecal fat excretion and apparent digestibility of fat in the final 5 days.

	Control	LMW-BG	HMW-BG
Fat intake (mg/d)	404 ± 1 [a]	339 ± 1 [b]	344 ± 1 [b]
Fecal fat excretion (mg/d)	11 ± 1 [a]	21 ± 3 [ab]	39 ± 10 [b]
Apparent digestibility of fat (%)	97.2 ± 0.2 [a]	93.8 ± 0.7 [ab]	88.2 ± 3.2 [b]

Values are means $\pm$ SE, $n = 8$. Means with suffixed superscript letters differ significantly (Tukey–Kramer's test, $p < 0.05$). LMW-BG; low molecular weight β-glucan, HMW-BG; high molecular weight β-glucan.

3.3. Fecal β-Glucan Excretion and Fermentability of β-Glucan

Fecal β-glucan excretion and fermentability of β-glucan are shown in Figure 1. Fecal β-glucan was significantly higher in the HMW-BG group than the LMW-BG group. Fecal β-glucan was not detected in the control group. The fermentability of β-glucan in the LMW-

BG group was almost 100%, and a significant difference was observed when compared with the HMW-BG group.

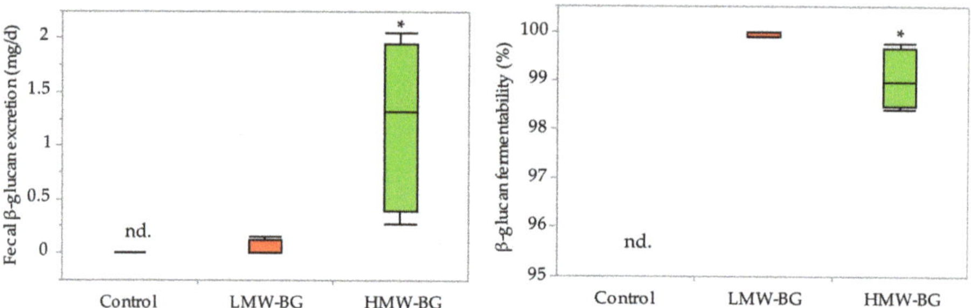

Figure 1. Box-and-whisker plots of fecal β-glucan excretion and the fermentability of β-glucan. Box marked by an asterisk differs significantly compared to the LMW-BG group (Student's *t*-test, *$p < 0.05$). nd., not detected; LMW-BG, low molecular weight β-glucan; HMW-BG, high molecular weight β-glucan.

3.4. Short-Chain Fatty Acid (SCFA) Concentrations in Cecal Digesta

The concentrations of short-chain fatty acids in cecal digesta are shown in Figure 2. Total SCFA concentrations, as well as acetate and propionate concentrations, were significantly higher in the LMW-BG group than the control group. The same tendency was observed in the HMW-BG group, but the difference was not significant.

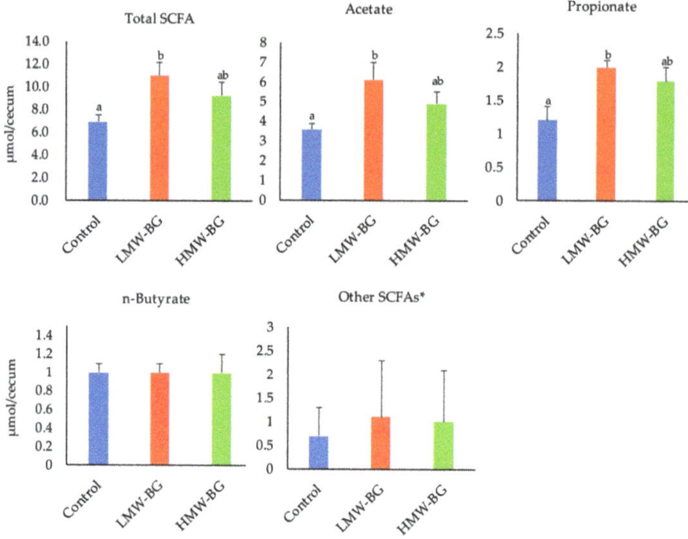

Figure 2. Short-chain fatty acid (SCFA) concentrations in the cecal digesta of mice fed the test diets. Bars represent means and SE, $n = 8$. * Other SCFAs, the sum of the concentrations of formate, iso-butyrate, iso-valerate, and valerate is shown. Means with suffixed superscript letters differ significantly (Tukey–Kramer's test, $p < 0.05$). LMW-BG; low molecular weight β-glucan, HMW-BG; high molecular weight β-glucan.

3.5. Bacterial Counts of the Major Microbiota Groups in the Cecal Digesta of Mice Fed the Test Diets

Bacterial counts of the major microbiota groups in the cecal digesta are shown in Figure 3 and Supplementary Table S3. Bacterial counts of *Bifidobacterium* were significantly higher in the LMW-BG group than the control group. Bacterial counts of *Bacteroides fragilis group* were significantly higher in the LMW-BG group than the control and HMW-BG groups. No significant differences in the other bacterial counts were observed.

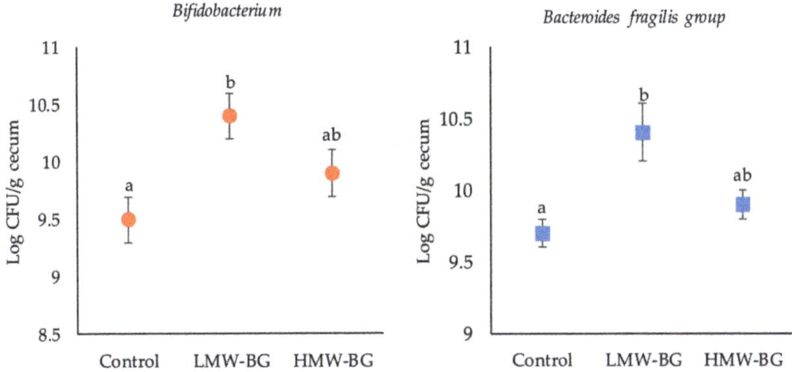

Figure 3. Bacterial counts of Bifidobacterium and Bacteroides fragilis group in the cecal digesta of mice fed the test diets. Dots and error bars represent means and SE, *n* = 8. Means with suffixed superscript letters differ significantly (Tukey–Kramer's test, $p < 0.05$). LMW-BG; low molecular weight β-glucan, HMW-BG; high molecular weight β-glucan.

3.6. Biochemical Analysis of the Serum and Liver Lipids

Serum biochemical concentrations are shown in Table 5. Serum total- and LDL-cholesterol and leptin concentrations were significantly reduced in both BG groups compared with the control ($p < 0.05$). Serum HDL-cholesterol concentration was also significantly lower in the LMW-BG group compared with the control group ($p < 0.05$). Serum glucose concentration was significantly lower in the LMW-BG group than the control group, whereas serum insulin concentration was significantly lower in the HMW-BG group than the control group ($p < 0.05$). Serum NEFA concentration was significantly lower in the HMW-BG group than the LMW-BG group ($p < 0.05$), whereas no significant difference was observed between the control and BG groups. There was no significant difference in serum triglyceride concentration among the groups. Liver lipid levels are shown in Supplementary Table S4: cholesterol and triglyceride accumulation (mmol/liver) and triglyceride concentration (mmol/g liver) were not statistically different among the groups.

Table 5. Serum biochemical concentrations.

	Control	**LMW-BG**	**HMW-BG**
Total cholesterol (mmol/L)	3.42 ± 0.09 [a]	2.57 ± 0.23 [b]	2.66 ± 0.12 [b]
LDL-cholesterol (mmol/L)	0.15 ± 0.01 [a]	0.09 ± 0.01 [b]	0.09 ± 0.01 [b]
HDL-cholesterol (mmol/L)	1.88 ± 0.05 [a]	1.52 ± 0.15 [b]	1.54 ± 0.07 [ab]
Triglyceride (mmol/L)	0.52 ± 0.05	0.45 ± 0.07	0.31 ± 0.04
NEFA (μmol/L)	635.9 ± 23.1 [ab]	654.9 ± 36.6 [a]	533.5 ± 37.1 [b]
Glucose (mmol/L)	21.1 ± 0.69 [a]	17.21 ± 1.17 [b]	18.64 ± 0.78 [ab]
Insulin (ng/mL)	1.20 ± 0.17 [a]	1.05 ± 0.24 [ab]	0.39 ± 0.13 [b]
Leptin (ng/mL)	15.30 ± 1.29 [a]	9.58 ± 2.05 [b]	5.74 ± 1.19 [b]

Values are means ± SE, *n* = 8. Means with suffixed superscript letters differ significantly (Tukey–Kramer's test, $p < 0.05$). LMW-BG; low molecular weight β-glucan, HMW-BG; high molecular weight β-glucan, NEFA; non-esterified fatty acid.

3.7. Expression of mRNAs Related to Liver Lipid Metabolism

Hepatic mRNA expression levels are shown in Figure 4. The mRNA expression level of sterol regulatory element-binding protein-1c (SREBP-1c) was significantly lower in both BG groups when compared with the control group ($p < 0.05$). No significant differences in the mRNA expression levels of peroxisome proliferator-activated receptorα (PPARα) and liver X receptor (LXR) were observed. The mRNA expression level of diacyl glycerol acyl-transferase 1 (DGAT1) was significantly lower in the HMW-BG group than the control group, whereas the mRNA expression level of carnitine palmitoyl transferase 1 (CPT1) was significantly higher in the HMW-BG group when compared with the control group ($p < 0.05$). No significant differences were observed in the mRNA expression levels of glucose 6-phosphate dehydrogenase (G6PD), pyruvate kinase (PK), fatty acid synthase (FAS), acetyl-CoA carboxylase (ACC), acyl-coenzyme A oxidase (ACOX), stearoyl coenzyme A desaturase 1 (SCD1), fatty acid translocase (CD36), and 3-hydroxy-3-methyl-glutaryl-CoA reductase, (HMG-CoA reductase).

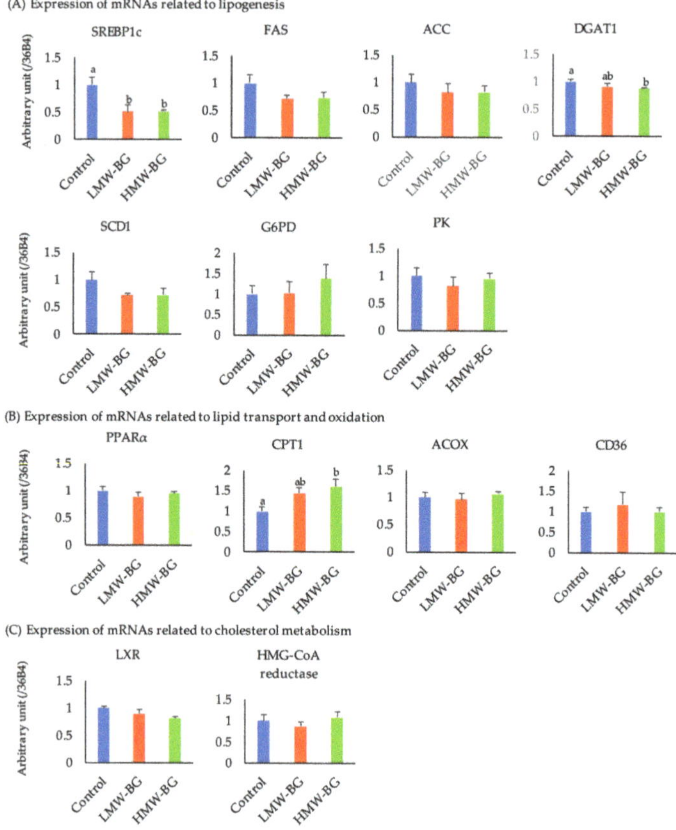

Figure 4. The expression of mRNAs related to lipogenesis (**A**), lipid transport and oxidation (**B**), and cholesterol metabolism (**C**) in liver. Bars represent means and SE, n = 8. Means with suffixed superscript letters differ significantly (Tukey–Kramer's test, $p < 0.05$). LMW-BG; low molecular weight β-glucan, HMW-BG; high molecular weight β-glucan. SREBP-1c, sterol regulatory element-binding protein-1c; FAS, fatty acid synthase; ACC, acetyl-CoA carboxylase; DGAT1, diacyl glycerol acyl-transferase 1; SCD1, stearoyl coenzyme A desaturase 1; G6PD, glucose 6-phosphate dehydrogenase; PK, pyruvate kinase; PPPARα, peroxisome proliferator-activated receptorα; CPT1, carnitine palmitoyl transferase 1; ACOX, acyl-coenzyme A oxidase; CD36 (FAT), fatty acid translocase; LXR, liver X receptor; 3-hydroxy-3-methyl-glutaryl-CoA reductase, (HMG-CoA reductase).

3.8. Expression of mRNAs Related to Ileal L Cell Function

Ileal mRNA expression levels related to L cell function are shown in Figure 5. The mRNA expression level of neurogenic differentiation factor (NGN3) was significantly higher in the LMW-BG group when compared with the control group ($p < 0.05$). No significant differences were observed in the mRNA expression levels of neurogenic differentiation factor (Neuro D), prohormone convertases1/3 (PC1/3), proglucagon (PGCG), G-protein-coupled bile acid receptor 1 (GPBAR1), and G-protein-coupled receptor 43 (GPR43).

(A) Expression of mRNAs related to transcription factor for L cell differentiation

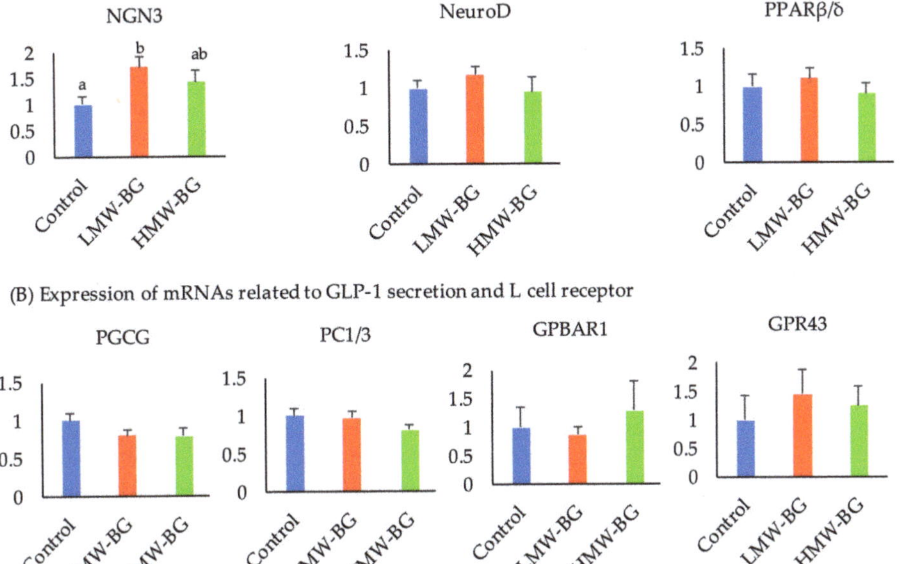

(B) Expression of mRNAs related to GLP-1 secretion and L cell receptor

Figure 5. Expression of mRNAs related to transcription factor for L cell differentiation (**A**) and GLP-1 secretion and L cell receptor (**B**) in ileum. Bars represent means and SE, $n = 8$. Means with suffixed superscript letters differ significantly (Tukey–Kramer's test, $p < 0.05$). LMW-BG; low molecular weight β-glucan, HMW-BG; high molecular weight β-glucan. NGN3, neurogenin 3; Neuro D, neurogenic differentiation factor; PPPARβ/δ, peroxisome proliferator-activated receptorβ/δ; PGCG, proglucagon; PC1/3, prohormone convertases1/3; GPBAR1, G-protein-coupled bile acid receptor 1; GPR43, G-protein-coupled receptor 43.

4. Discussion

The effects of LMW-BG and HMW-BG on cecal fermentation, glucose and lipid metabolism in mice fed a moderate-fat diet were compared. HMW-BG specific effects were observed in inhibiting dietary fat absorption and reducing abdominal depot fat. LMW-BG specific effects were observed in increasing bacterial counts of *Bifidobacterium* and *Bacteroides*, and consequently increasing cecal total SCFAs, acetate, and propionate. mRNA expression of NGN3, an L cell marker, was increased in the LMW-BG group compared with the control group. These results indicate that LMW-BG has the potential to improve glucose and lipid metabolism by a different mechanism from HMW-BG. Many studies have reported that HMW-BG has a greater impact than LMW-BG on the improvement of glucose and lipid metabolism; however, our results suggested that LMW-BG improves glucose and lipid metabolism through prebiotic effects. The prebiotic effects are expected to protect from pro-inflammation of organs and promote the gut immune system. Further studies are needed to clarify the prebiotic effects of LMW-BG.

Similar significant reductions in the mRNA expression of SREBP-1c and serum total- and LDL-cholesterol and leptin concentrations were observed in both BG groups compared with the control group. LMW-BG was almost 100% fermented and cecal bacterial counts of *Bifidobacterium* and *Bacteroides* were significantly increased in the LMW-BG group, resulting in an increase in the cecal contents of acetic acid and propionic acid. The results indicated that decreases in abdominal depot fats, serum cholesterol and leptin concentrations in the HMW-BG group were caused by the inhibition of nutrient absorption due to high viscosity in the digestive tract, whereas the changes in the LMW-BG group were due to prebiotic effects. It is reported that an increase in SCFAs decreases serum cholesterol, fasting blood glucose, and leptin concentrations [19,21]. The blood was collected under fasting conditions in this study; therefore, the serum GLP-1 concentration was not measured. NGN3 is a key factor which initiates endocrine differentiation [31]. NGN3 and BETA2/Neuro D (Neuro D) specifically induce certain types of enteroendocrine cells, such as L cells [32]. It is therefore possible that the serum glucose concentration in the LMW-BG group might be lowered through the action of GLP-1. Spearman's rank correlation coefficient analysis related to the prebiotic effect is shown in Supplementary Table S5. Significant positive correlation coefficients between cecal total SCFAs, especially acetate and propionate, and serum biomarkers were observed. It is suggested that serum glucose and lipid concentrations were improved by the SCFAs, directly or indirectly. Significant positive correlation coefficients between cecal SCFAs and the mRNA expression of ileal NGN3 were also observed. Furthermore, significant positive correlation coefficients between ileal mRNA expressions of NGN3, NeuroD, and GPBAR1 were observed (Supplementary Table S5); the increases in mRNA levels might decrease serum glucose and lipid concentrations through an L cell function. It was reported that GPBAR1 is a selective regulator of intestinal L cell differentiation [33]. Further studies are needed to elucidate the mechanism of LMW-BG on glucose metabolism through L cell function. HMW-BG had a high fermentation rate but the microbiota influence and the amount of SCFAs were relatively small when compared with LMW-BG, suggesting that the high viscosity of HMW-BG contributes more than the prebiotic effect.

Significant reductions in food intake were observed in both BG groups. HMW-BG increases viscosity, thereby delaying gastric emptying time [34], suggesting that the reduction in food intake in the HMW-BG group in this study might be due to delayed gastric emptying. The reduction in food intake in the LMW-BG group might be caused by a different mechanism, such as intestinal hormone secretion which is promoted by SCFAs. It was reported that propionate in the colon of rats and mice stimulated the release of both glucagon-like peptide 1 (GLP-1) and peptide PYY, which are anorexigenic gut hormones [35]. Anorexigenic gut hormones related to food intake were not analyzed in our study. Further studies are needed to elucidate the mechanism of reduction in food intake in the LMW-BG and HMW-BG groups.

Serum insulin concentrations were significantly reduced in the HMW-BG group only; however, hepatic mRNA expression of SREBP-1c controlled by insulin was significantly reduced in both groups. Hepatic SREBP-1c has been reported as an insulin-mediated transcriptional activator of genes involved in carbohydrate and lipid metabolism [36,37]. The reduction in SREBP-1c mRNA expression in the HMW-BG group was caused by decreasing insulin secretion through the suppression of the digestion and absorption of carbohydrates. Fasting serum glucose concentrations were decreased in the LMW-BG group; therefore, the mRNA expression of SREBP-1c was decreased through the modification of glucose metabolism by the secretion of incretins such as GLP-1, whose secretion was enhanced by SCFAs. In this study, there were no significant differences in mRNA expression levels of lipogenic enzymes, such as FAS and ACC, among the groups; however, we suggest that there was an effect on lipid metabolism through suppression of SREBP-1c mRNA expression. In our previous study using diet-induced obesity mice, we confirmed a decrease in the mRNA expression of fatty acid synthase and glucose-6-phosphate dehydrogenase due to a reduction in SREBP-1c mRNA expression [38]. Significant negative correlation between cecal SCFAs and mRNA expression of hepatic SREBP-1c, and significant positive correlation between mRNA expression of hepatic SREBP-1c and FAS, CD36, LXR, PPARα was observed (Supplementary Table S6). It is concluded that the regulation of lipid metabolism mediated by SREBP-1c is exerted in HMW-BG and LMW-BG by different mechanisms.

A previous study using barley flour with high β-glucan reported different results [39]: the abundance of *Bacteroides* was significantly higher in the β-glucan rich barley flour group compared with the β-glucan free barley flour group, whereas the abundance of *Clostridium* clusters was significantly lower in the β-glucan rich barley flour group compared with the β-glucan free barley flour group. We suggest that the food matrix is responsible for the differences between β-glucan rich barley flour and isolated β-glucan because it affects the release of carbohydrates and β-glucan, thereby affecting the digestibility and absorption rate. Consumption of barley flour leads to the slow release of carbohydrates and β-glucan, which would promote reductions in a postprandial glucose rise and serum cholesterol concentration. Barley flour also contains arabinoxylan which has a high prebiotic effect [40]. It is reported that arabinoxylan from wheat modulates both the gut microbiota and lipid metabolism in high-fat diet-induced obese mice [41]. When barley flour is ingested, it is expected that the combination of β-glucan and arabinoxylan in barley would have a synergistic affect.

5. Conclusions

Our results indicated that LMW-BG and HMW-BG affect glucose and lipid metabolism by different mechanisms. The results from this study and previous reports also suggested that the physiological responses to ingested β-glucan rich barley flour or isolated HMW-BG differed. The prebiotic effect of LMW-BG is expected to be applied to several foods and beverages. It was revealed that the expected function of barley β-glucan differs according to the molecular weight; therefore, it is suggested that different nutritional interventions may be possible, depending on the purpose of the treatment.

Supplementary Materials: The following are available online at https://www.mdpi.com/2072-6643/13/1/130/s1. Table S1: Group-specific primers for real-time reverse transcription polymerase chain reaction (PCR), Table S2: Primers used in the real-time reverse transcription polymerase chain reaction (PCR), Table S3: Bacterial counts of major genus microbiota in the cecal digesta of mice fed the test diets, Table S4: Liver lipid accumulation, Table S5: Spearman's rank correlation coefficient related to the prebiotic effect, Table S6: Spearman's rank correlation coefficients for the relationship between parameters related to liver lipid metabolism (A) and ileal L cell function (B).

Author Contributions: Conceptualization, S.A.; data curation, S.A. and C.Y.; formal analysis, K.M.; investigation, S.A.; methodology, K.M. and C.Y.; project administration, S.A.; resources, T.K.; supervision, S.A.; writing—original draft, S.A., K.M. and T.K.; writing—review and editing, S.A., K.M. and T.K. All authors have read and agreed to the published version of the manuscript.

Funding: This research received no external funding.

Institutional Review Board Statement: The studies were approved by the Otsuma Women's University Animal Research Committee (Tokyo, Japan) and were performed in accordance with the Regulation on Animal Experimentation at Otsuma Women's University (No. 19007, 2 July 2019).

Data Availability Statement: Data are contained within the article or Supplementary Materials.

Acknowledgments: This study was financially supported by the ADEKA corporation. (Tokyo, Japan).

Conflicts of Interest: One of the authors (T.K.) is a salaried employee of the ADEKA corporation. The remaining authors (S.A., K.M. and C.Y.) have no conflicts of interest to disclose. The LMW-BG (30SP) used in this study was provided by the ADEKA Corporation.

Abbreviations

LMW-BG, low molecular weight β-glucan; HMW-BG, high molecular weight β-glucan; NEFA, non-esterified fatty acid; SCFA, short-chain fatty acid; SREBP-1c, sterol regulatory element-binding protein-1c; FAS, fatty acid synthase; ACC, acetyl-CoA carboxylase; DGAT1, diacyl glycerol acyltransferase 1; SCD1, stearoyl coenzyme A desaturase 1; G6PD, glucose 6-phosphate dehydrogenase; PK, pyruvate kinase; PPPARα, peroxisome proliferator-activated receptorα; CPT1, carnitine palmitoyl transferase 1; ACOX, acyl-coenzyme A oxidase; CD36 (FAT), fatty acid translocase; LXR, liver X receptor; 3-hydroxy-3-methyl-glutaryl-CoA reductase, (HMG-CoA reductase); NGN3, neurogenin 3; Neuro D, neurogenic differentiation factor; PPPARβ/δ, peroxisome proliferator-activated receptorβ/δ; PGCG, proglucagon; PC1/3, prohormone convertases1/3; GPBAR1, G-protein-coupled bile acid receptor 1; GPR43, G-protein-coupled receptor 43.

References

1. Juntunen, K.S.; Niskanen, L.K.; Liukkonen, K.H.; Poutanen, K.S.; Holst, J.J.; Mykkänen, H.M. Postprandial glucose, insulin, and incretin responses to grain products in healthy subjects. *Am. J. Clin. Nutr.* **2002**, *75*, 254–262. [CrossRef] [PubMed]
2. Soong, Y.Y.; Quek, R.Y.; Henry, C.J. Glycemic potency of muffins made with wheat, rice, corn, oat and barley flours: A comparative study between in vivo and in vitro. *Eur. J. Nutr.* **2015**, *54*, 1281–1285. [CrossRef] [PubMed]
3. Behall, K.M.; Scholfield, D.J.; Hallfrisch, J. Comparison of hormone and glucose responses of overweight women to barley and oats. *J. Am. Coll. Nutr.* **2005**, *2*, 182–188. [CrossRef] [PubMed]
4. Brown, L.; Rosner, B.; Willett, W.W.; Sacks, F.M. Cholesterol-lowering effects of dietary fiber: A meta-analysis. *Am. J. Clin. Nutr.* **1999**, *69*, 30–42. [CrossRef] [PubMed]
5. Berg, A.; Konig, D.; Deibert, P.; Grathwohl, D.; Baumstark, M.W.; Franz, I.W. Effect of an oat bran enriched diet on the atherogenic lipid profile in patients with an increased coronary heart disease risk. A controlled randomized lifestyle intervention study. *Ann. Nutr. Metab.* **2003**, *47*, 306–311. [CrossRef] [PubMed]
6. Othman, R.A.; Moghadasian, M.H.; Jones, P.J. Cholesterol-lowering effects of oat beta-glucan. *Nutr. Rev.* **2011**, *69*, 299–309. [CrossRef]
7. Ho, H.V.; Sievenpiper, J.L.; Zurbau, A.; Blanco Mejia, S.; Jovanovski, E.; Au-Yeung, F.; Jenkins, A.L.; Vuksan, V. The effect of oat beta-glucan on LDL-cholesterol, non-HDL-cholesterol and apoB for CVD risk reduction: A systematic review and meta-analysis of randomised-controlled trials. *Br. J. Nutr.* **2016**, *116*, 1369–1382. [CrossRef]
8. Ho, H.V.; Sievenpiper, J.L.; Zurbau, A.; Blanco Mejia, S.; Jovanovski, E.; Au-Yeung, F.; Jenkins, A.L.; Vuksan, V. A systematic review and meta-analysis of randomized controlled trials of the effect of barley beta-glucan on LDL-C, non-HDL-C and apoB for cardiovascular disease risk reduction(i–iv). *Eur. J. Clin. Nutr.* **2016**, *70*, 1239–1245. [CrossRef]
9. Izydorczyk, M.S.; Macri, L.J.; MacGregor, L.W. Structure and physicochemical properties of barley non-starch polysaccharides—I. Water extractable β-Glucans and arabinoxylans. *Carbohydr. Polym.* **1998**, *35*, 249–258. [CrossRef]
10. Henrion, M.; Francey, C.; Lê, K.-A.; Lamothe, L. Cereal B-Glucans: The Impact of Processing and How it affects physiological responses. *Nutrients* **2019**, *11*, 1729. [CrossRef]
11. Kwong, M.G.; Wolever, T.M.; Brummer, Y.; Tosh, S.M. Increasing the viscosity of oat beta-glucan beverages by reducing solution volume does not reduce glycaemic responses. *Br. J. Nutr.* **2013**, *110*, 1465–1471. [CrossRef] [PubMed]
12. Ostman, E.; Rossi, E.; Larsson, H.; Brighenti, F.; Björck, I. Glucose and insulin responses in healthy men to barley bread with different levels of (1→3;1→4)-β-glucans; predictions using fluidity measurements of in vitro enzyme digests. *J. Cereal Sci.* **2006**, *43*, 230–235. [CrossRef]
13. Regand, A.; Tosh, S.M.; Wolever, T.M.; Wood, P.J. Physicochemical properties of beta-glucan in differently processed oat foods influence glycemic response. *J. Agric. Food. Chem.* **2009**, *57*, 8831–8838. [CrossRef] [PubMed]
14. Lia, A.; Hallmans, G.; Sandberg, A.S.; Sundberg, B.; Aman, P.; Andersson, H. Oat beta-glucan increases bile acid excretion and a fiber-rich barley fraction increases cholesterol excretion in ileostomy subjects. *Am. J. Clin. Nutr.* **1995**, *62*, 1245–1251. [CrossRef] [PubMed]
15. El Khoury, D.; Cuda, C.; Luhovy, B.L.; Anderson, G.H. Beta glucan: Health benefits in obesity and metabolic syndrome. *J. Nutr. Metab.* **2012**, *2012*, 851362. [CrossRef] [PubMed]
16. Baea, I.N.; Lee, N.; Kim, S.M.; Lee, H.G. Effect of partially hydrolyzed oat β-glucan on the weight gain and lipid profile of mice. *Food Hydrocoll.* **2009**, *23*, 2016–2021. [CrossRef]
17. Immerstrand, T.; Andersson, K.E.; Wange, C.; Rascon, A.; Hellstrand, P.; Nyman, M.; Cui, K.E.; Bergenstahl, B.; Tragardh, C.; Oste, R. Effects of oat bran, processed to different molecular weights of beta-glucan, on plasma lipids and caecal formation of SCFA in mice. *Br. J. Nutr.* **2010**, *104*, 364–373. [CrossRef]
18. Mio, K.; Yamanaka, C.; Ichinose, Y.; Kohyama, N.; Yanagisawa, T.; Aoe, S. Effects of barley β-glucan with various molecular weights partially hydrolyzed by endogenous β-glucanase on glucose tolerance and lipid metabolism in mice. *Cereal Chem.* **2020**, *97*, 1056–1065. [CrossRef]
19. Hosseini, E.; Grootaert, C.; Verstraete, W.; Van de Wiele, T. Propionate as a health-promoting microbial metabolite in the human gut. *Nutr. Rev.* **2011**, *69*, 245–258. [CrossRef]
20. Joyce, S.A.; Kamil, A.; Fleige, L.; Gahan, C.G.M. The Cholesterol-lowering effect of oats and oat beta glucan: Modes of action and potential role of bile acids and the microbiome. *Front. Nutr.* **2019**, *6*, 171. [CrossRef]

21. Connolly, M.L.; Tzounis, X.; Tuohy, K.M.; Lovegrove, J.A. Hypocholesterolemic and prebiotic effects of a whole-grain oat-based granola breakfast cereal in a cardio-metabolic "At Risk" population. *Front Microbiol.* **2016**, *7*, 1675. [CrossRef] [PubMed]

22. Thandapilly, S.J.; Ndou, S.P.; Wang, Y.; Nyachoti, C.M.; Ames, N.P. Barley beta-glucan increases fecal bile acid excretion and short chain fatty acid levels in mildly hypercholesterolemic individuals. *Food Funct.* **2018**, *9*, 3092–3096. [CrossRef] [PubMed]

23. Wilczak, J.; Błaszczyk, K.; Kamola, D.; Gajewska, M.; Harasym, J.P.; Jałosińska, M.; Gudej, S.; Suchecka, D.; Oczkowski, M.; Gromadzka-Ostrowska, J. The effect of low or high molecular weight oat beta-glucans on the inflammatory and oxidative stress status in the colon of rats with LPS-induced enteritis. *Food Funct.* **2015**, *6*, 590–603. [CrossRef] [PubMed]

24. Lee, S.C.; Rodriguez, F.; Storey, M.; Farmakalidis, E.; Prosky, L. Determination of soluble and insoluble dietary fiber in psyllium containing cereal products. *J. AOAC Int.* **1995**, *78*, 724–729. [CrossRef] [PubMed]

25. McCleary, B.V.; Codd, R. Measurement of (1→3), (1→4)-β-D-glucan in barley and oats: A streamlined enzymic procedure. *J. Sci. Food Agric.* **1991**, *55*, 303–312. [CrossRef]

26. Folch, J.; Lees, M.; Sloane-Stanley, G.H. A simple method for the isolation and purification of total lipids from animal tissues. *J. Biol. Chem.* **1962**, *226*, 497–509.

27. Kojima, M.; Arishima, T.; Shimizu, R.; Kohno, M.; Kida, H.; Hirotsuka, M.; Ikeda, I. Consumption of a structured triacylglycerol containing behenic and oleic acids increases fecal fat excretion in humans. *J. Oleo Sci.* **2013**, *62*, 997–1001. [CrossRef]

28. Atarashi, K.; Tanoue, T.; Oshima, K.; Suda, W.; Nagano, Y.; Nishikawa, H.; Fukuda, S.; Saito, T.; Narushima, T.; Hase, K.; et al. Treg induction by a rationally selected mixture of *Clostridia* strains from the human microbiota. *Nature* **2013**, *500*, 232–236. [CrossRef]

29. Matsuda, K.; Tsuji, H.; Asahara, T.; Kado, Y.; Nomoto, K. Sensitive quantitative detection of commensal bacteria by rRNA-targeted reverse transcription-PCR. *Appl. Environ. Microbiol.* **2007**, *73*, 32–39. [CrossRef]

30. Matsuki, T. Development of quantitative PCR detection method with 16S rRNA gene-targeted genus- and species-specific primers for the analysis of human intestinal microflora and its application. *Jpn. J. Bacteriol.* **2007**, *62*, 255–261. [CrossRef]

31. Fujita, Y.; Cheung, A.T.; Kieffer, T.J. Harnessing the gut to treat diabetes. *Pediatr. Diabetes* **2004**, *5*, 57–69. [CrossRef] [PubMed]

32. Cani, P.D.; Hoste, S.; Guiot, Y.; Delzenne, N.M. Dietary non-digestible carbohydrates promote L-cell differentiation in the proximal colon of rats. *Br. J. Nutr.* **2007**, *98*, 32–37. [CrossRef] [PubMed]

33. Lund, M.L.; Sorrentino, G.; Egerod, K.L.; Kroone, C.; Mortensen, B.; Knop, F.K.; Reimann, F.; Gribble, F.M.; Drucker, D.J.; de Koning, E.J.P.; et al. L-cell differentiation is induced by bile acids through GPBAR1 and paracrine GLP-1 and serotonin signaling. *Diabetes* **2020**, *69*, 614–623. [CrossRef] [PubMed]

34. Thondre, P.S.; Shafat, A.; Clegg, M.E. Molecular weight of barley β-glucan influences energy expenditure, gastric emptying and glycaemic response in human subjects. *Br. J. Nutr.* **2013**, *110*, 2173–2179. [CrossRef]

35. Psichas, A.; Sleeth, M.L.; Murphy, K.G.; Brooks, L.; Bewick, G.A.; Hanyaloglu, A.C.; Ghatei, M.A.; Bloom, S.R.; Frost, G. The short chain fatty acid propionate stimulates GLP-1 and PYY secretion via free fatty acid receptor 2 in rodents. *Int. J. Obes.* **2015**, *39*, 424–429. [CrossRef]

36. Horton, J.D. Sterol regulatory element-binding proteins: Transcriptional activators of lipid synthesis. *Biochem. Soc. Trans.* **2002**, *30*, 1091–1095. [CrossRef]

37. Kim, S.Y.; Kim, H.J.I.; Kim, T.H.; Im, S.S.; Park, S.K.; Lee, I.K.; Kim, K.S.; Ahn, Y.H. SREBP-1c mediates the insulin-dependent hepatic glucokinase expression. *J. Biol. Chem.* **2004**, *279*, 30823–30829. [CrossRef]

38. Aoe, S.; Watanabe, N.; Yamanaka, C.; Ikegami, S. Effects of barley on carbohydrate metabolism and abdominal fat accumulation in diet-induced obese mice. *J. Jpn. Assoc. Diet. Fiber Res.* **2010**, *14*, 55–65.

39. Aoe, S.; Ichinose, Y.; Kohyama, N.; Komae, K.; Takahashi, A.; Yoshioka, T.; Yanagisawa, T. Effects of β-glucan content and pearling of barley in diet-induced obese mice. *Cereal Chem.* **2017**, *94*, 956–962. [CrossRef]

40. Guo, R.; Xu, Z.; Wu, S.; Li, X.; Li, J.; Hu, H.; Wu, Y.; Ai, L. Molecular properties and structural characterization of an alkaline extractable arabinoxylan from hull-less barley bran. *Carbohydr. Polym.* **2019**, *15*, 250–260. [CrossRef]

41. Neyrinck, A.M.; Possemiers, S.; Druart, C.; Van de Wiele, T.; De Backer, F.; Cani, P.D.; Larondelle, Y.; Delzenne, M.N. Prebiotic effects of wheat arabinoxylan related to the increase in bifidobacteria, Roseburia and Bacteroides/Prevotella in diet-induced obese mice. *PLoS ONE* **2011**, *6*, e20944. [CrossRef] [PubMed]

nutrients

Article

Oat and Barley in the Food Supply and Use of Beta Glucan Health Claims

Jaimee Hughes [1] and Sara Grafenauer [1,2,*]

1   Grains & Legumes Nutrition Council, 1 Rivett Road, North Ryde, NSW 2113, Australia; j.hughes@glnc.org.au
2   School of Medicine, University of Wollongong, Northfields Avenue, Wollongong, NSW 2522, Australia
*   Correspondence: sarag@glnc.org.au; Tel.: +61-401-265-142

Abstract: Beta glucan is a type of soluble dietary fibre found in oats and barley with known cholesterol-lowering benefits. Many countries globally have an approved beta glucan health claim related to lowering blood cholesterol, an important biomarker for cardiovascular disease. However, the use of these claims has not been examined. The aim of this study was to explore the range and variety of oat and barley products in the Australian and global market within a defined range of grain food and beverage categories and examine the frequency of beta glucan health claims. Australian data were collected via a recognised nutrition audit process from the four major Australian supermarkets in metropolitan Sydney (January 2018 and September 2020) and Mintel Global New Product Database was used for global markets where a claim is permitted. Categories included breakfast cereals, bread, savoury biscuits, grain-based muesli bars, flour, noodles/pasta and plant-based milk alternatives and information collected included ingredients lists and nutrition and health claims. Products from Australia (*n* = 2462) and globally (*n* = 44,894) were examined. In Australia, 37 products (1.5%) made use of the beta glucan claim (84% related to oat beta glucan and 16% related to barley beta glucan, specifically BARLEYmax®). Of products launched globally, 0.9% (*n* = 403) displayed beta glucan cholesterol-lowering claims. Despite the number of products potentially eligible to make beta glucan claims, their use in Australia and globally is limited. The value of dietary modification in cardiovascular disease treatment and disease progression deserves greater focus, and health claims are an opportunity to assist in communicating the role of food in the management of health and disease. Further assessment of consumer understanding of the available claims would be of value.

Keywords: beta glucan; oats; barley; health claim; regulation; food-health relationship

Citation: Hughes, J.; Grafenauer, S. Oat and Barley in the Food Supply and Use of Beta Glucan Health Claims. *Nutrients* **2021**, *13*, 2556. https://doi.org/10.3390/nu13082556

Academic Editors: Maria Luz Fernandez and Iain A. Brownlee

Received: 4 June 2021
Accepted: 24 July 2021
Published: 26 July 2021

Publisher's Note: MDPI stays neutral with regard to jurisdictional claims in published maps and institutional affiliations.

1. Introduction

Cardiovascular disease (CVD) is a major health problem in Australia, affecting 1.2 million adults [1], yet is largely preventable through the modification of risk factors such as physical inactivity, poor dietary habits and smoking, which together account for up to 90% of the risk of myocardial infarction [2]. Hypercholesterolemia or elevated blood cholesterol is a key risk factor for the development of CVD, and in 2015 accounted for 37% of the burden of coronary heart disease [3]. It is reported that 1.5 million Australian adults had high cholesterol in 2017/18, a condition which often presents with no signs or symptoms. There is a significant opportunity for prevention and treatment of elevated cholesterol and reduction in CVD risk through diet therapy, and one such strategy is focused on adequate consumption of whole grains [4,5]. According to Food Standards Australia New Zealand (FSANZ), whole grain is defined as the intact grain or the dehulled, ground, milled, cracked or flaked grain where the constituents—endosperm, germ and bran—are present in such proportions that represent the typical ratio of those fractions occurring in the whole cereal, and includes wholemeal [6].

Whole grain oats and barley are within the top four largest grain crops in Australia, with 1.6 million tonnes (up 89%) of oats and 12 million tonnes (up 33%) of barley produced in 2020–2021 as a result of a very good growing season [7]. Oats and barley share a favourable, chemically similar polysaccharide beta glucan, (1→3), (1→4)–β–D–glucan and depending on the specific variety, oats contain 6–8% and barley 4–10% w/w β-glucan [8], which has known cholesterol-lowering properties. The daily consumption of 3 g of beta glucan has been shown to significantly lower blood cholesterol concentrations and reduce circulating low-density lipoprotein cholesterol [9–11], key risk factors for CVD. Whole grain oats and barley are the richest sources of beta glucan [8], although specialised plant breeding programs have also produced cultivars with naturally higher levels of beta glucan, such as BARLEYmax®, developed by the CSIRO (Commonwealth Scientific and Industrial Research Organisation) [12]. BARLEYmax® is a whole grain barley line that is high in resistant starch and beta glucan (6.4% w/w beta glucan) and is used as an ingredient in a range of food products in Australia, New Zealand, Japan, USA, Singapore and Malaysia [13].

In Australia and New Zealand, nutrition and health claims are legislated under the Food Standards Code (Standard 1.2.7 Nutrition, Health and Related Claims [14]), which is developed and administered by FSANZ. Currently, there are two existing pre-approved food–health relationships related to beta glucan outlined in Schedule 4 of the Food Standards Code (FSC) [15]. Manufacturers can utilise 'beta glucan reduces blood cholesterol', which can be used as the basis for making a high level health claim, or 'beta glucan reduces dietary and biliary cholesterol absorption', which can be used as the basis for making a general level health claim [14]. To be eligible, the food product must (1) contain either oat bran, whole grain oats or whole grain barley, (2) provide at least 1 g of beta glucan per manufacturer serving, (3) be accompanied by a dietary context statement (e.g., as part of a balanced diet low in saturated fatty acids and containing 3 g of beta glucan per day) and (4) pass the Nutrient Profiling Scoring Criteria (NPSC), which prevents health claims being carried on products high in energy, saturated fat, sugars and sodium [14,16].

Similar beta glucan claims are approved for use in the United States, Europe, Canada, Brazil, Malaysia, Singapore, Indonesia and South Korea [17]. However, there is no consistency with respect to the prescribed wording of claims or the eligibility criteria between the regulatory bodies [17], except for the universally adopted efficacious daily dose of 3 g beta glucan. The general level and high level health claims on beta glucan and blood lipids, as outlined in the FSC, were based on similar claims that were approved by the Food and Drug Administration and the European Food Safety Authority, respectively.

Although pre-approved beta glucan cholesterol-lowering claims have been permitted in Australia since the introduction of Standard 1.2.7 in 2013, research to date has not examined the frequency of such claims on pack labels. Given the prevalence of hypercholesterolemia globally and in Australia, beta glucan cholesterol-lowering health claims are an opportunity to assist in communicating the role of food in the management of health and disease. Whole grain oats, barley and oat bran have known cholesterol-lowering capabilities, yet the use of these grains in the local and global food supply is unknown. The aim of this study was to explore the range and variety of oat and barley products in the Australian and global market within a defined range of grain food and beverage categories and examine the frequency of beta glucan health claims.

2. Materials and Methods

Separate nutrition audits of grain food and beverage categories were conducted from the four major Australian supermarkets (Woolworths, Coles, Aldi, IGA) in metropolitan Sydney and undertaken between January 2018 and September 2020. Approximately 79% of the Australian market share is represented by these supermarkets [18] and were therefore chosen to reflect food choices that the majority of Australians are faced with during food shopping. This recognised process [19] was replicated for each audit category. The food categories included were breakfast cereals (including ready-to-eat cereals, hot cereals, breakfast biscuits, muesli, granola, clusters), bread, savoury biscuits (crackers, crispbread, rice crackers, rice/other grain cakes), grain-based muesli bars, flour, noodles/pasta and plant-based milk alternatives. With permission from store managers, smartphones were used to capture all information on food packaging, including ingredient lists, nutrition panel information and nutrition and health claims. A supplementary internet search was conducted through retailer and identified manufacturer websites to ensure all available products were collected. As a number of products are sold in multiple supermarkets, each product was only recorded once. For products that had multiple pack sizes, the largest pack size was used for data analysis as it was more likely to display a greater number of claims.

As beta glucan health claims are only eligible to be displayed on products made with oat bran, whole grain oats and barley [15], data analysis specifically focused on products containing whole grain oats (e.g., rolled oats, flakes, flour), oat bran, whole grain barley (e.g., barley flour, kibble, puffs) and BARLEYmax®. According to FSANZ, a claim is defined as an express or implied statement, representation, design or information in relation to a food or a property of food [20]. For this reason, claims conveyed as text and/or images were included in the analysis. Beta glucan health claims were categorised as either general level or high level as outlined in Standard 1.2.7 of the FSC [14]. General level health claims refer to a nutrient or food component and its influence on general health (e.g., beta glucan reduces dietary and biliary cholesterol absorption), whereas high level health claims are statements that refer to a serious disease or a biomarker of serious disease (e.g., beta glucan reduces blood cholesterol). Compliance of claims according to Standard 1.2.7 was not assessed.

The prevalence of beta glucan heart health/cholesterol-lowering health claims in countries with approved food health relationships was assessed using the Mintel Global New Product Database (Mintel GNPD) [21], an extensive database of food and beverage products launched globally. The number of food and beverage products on the market between January 2018 and September 2020 made with whole grain oats, oat bran, whole grain barley and/or BARLEYmax® were recorded. The time frame and product categories for the Mintel GNPD search were selected to align with the data collection process used for the Australian market. The Mintel GNPD was also used to record the number of products that displayed at least one beta glucan heart health/cholesterol-lowering claim in selected markets conveyed in text or image form. A recent global review of heart health beta glucan claims was used to determine the eligibility for inclusion in the present analysis [17]. According to Mathews et al. (2020), oat beta glucan heart health/cholesterol-lowering claims are permitted in the United States, Europe, Canada, New Zealand, Malaysia, Brazil, Singapore, Indonesia and South Korea [17]. Barley beta glucan claims are permitted in the United States, Europe, Canada and New Zealand. South Korea was excluded from this analysis as the permitted beta glucan claim referred to extracted oat fibre in supplement form, rather than from whole foods [17]. The Mintel GNPD search was conducted on 15 April 2021 using the input parameters outlined in Table 1. Following the search, all products were manually reviewed to ensure that the on-pack beta glucan claims were eligible for inclusion. For non-English food labels, the translated Mintel product description was used to assess on pack claims. Claims not specifically related to oat and/or barley beta glucan and blood cholesterol/heart health were excluded (e.g., whole grain heart health claims permitted in the United States).

Statistics

Data from photographs taken at each nutrition audit and the Mintel GNPD search output were transcribed into a Microsoft® Excel® spreadsheet (Microsoft 365 MSO Version 16.0.13426.20306, Redmond, WA, USA) for analysis, and checked for errors by an independent reviewer prior to analysis. Descriptive statistics were used to determine the number and proportion of products that contained whole grain oats and/oat bran, whole grain barley and BARLEYmax® and the number and proportion of products that displayed beta glucan heart health/cholesterol-lowering claims within each category.

Table 1. Mintel GNPD search strategy.

Search Variables	Parameters
Market	United States, Europe, Canada, New Zealand, Malaysia, Brazil, Singapore, Indonesia
Date published	Between January 2018 and September 2020
Mintel GNPD categories included	Breakfast cereals; savoury biscuits/crackers; bread and bread products; plant-based drinks; snack/cereal/energy bars; noodles; wheat and other grain-based snacks; pasta
Ingredients included—Oats	Oats and all child ingredients; oat fibre and all child ingredients; oat bran; oat bran flour; oat flour; whole oat flour; oat milk
Ingredients included—Barley	Barley flakes; barley grit; barley groats; barley meal; barley semolina; barley flour; BARLEYmax
Claims	Functional cardiovascular claim

3. Results

Between January 2018 and September 2020, 2462 products from seven food categories were collected from Australian nutrition audits, including 769 breads, 543 breakfast cereals, 363 savoury crackers, 337 noodles/pastas, 173 flours, 165 grain-based muesli bars and 112 plant-based milk alternatives.

3.1. Australian Market

A quarter of all food and beverage products were made with whole grain oats and/or oat bran (Table 2), including 78.8% of grain-based muesli bars (n = 130), 68% of breakfast cereals (n = 369), 9.6% of breads (n = 74) and 6.9% of savoury biscuits (n = 25). Grain-based muesli bars, breakfast cereals and savoury crackers were made predominantly with rolled oats, while the noodles/pasta products contained added oat fibre.

Table 2. Number and proportion of food and beverage products made with whole grain oats and/or oat bran, whole grain barely and BARLEYmax®, per category.

Food Category	Contains Whole Grain Oats and/or Oat Bran. n (% of Total Market)	Contains Whole Grain Barley * n (% of Total Market)	Contains BARLEYmax® n (% of Total Market)
Breakfast cereal (n = 543)	369 (68.0)	73 (13.4)	7 (1.3)
Bread (n = 769)	74 (9.6)	110 (14.3)	3 (0.4)
Savoury biscuits (n = 363)	25 (6.9)	3 (0.8)	0
Noodles/pasta (n = 337)	9 (2.7)	0	0
Flour (n = 173)	3 (1.7)	1 (0.6)	0
Grain-based bars (n = 165)	130 (78.8)	5 (3.0)	4 (2.4)
Plant-based milk alternatives (n = 112)	3 (2.7)	3 (2.7)	0
Total (n = 2462)	613 (24.9)	195 (7.9)	14 (0.6)

* Excluding BARLEYmax®.

As shown in Table 2, fewer products contained barley and BARLEYmax® compared to oats (n = 195, 14 and 613, respectively). Barley was most commonly found in breads (n = 110, 14.3%) as whole grain barley flour or kibbled barley. Seventy-three breakfast cereals (13.4% of subcategory) and a smaller proportion of savoury biscuits, flours, grain-based muesli bars and plant-based milk alternatives contained barley as an ingredient. There were no noodle/pasta products made with barley. BARLEYmax® was found in 14 Australian products overall, most frequently in breakfast cereals, followed by grain-based muesli bars and breads (Table 2).

Table 3 outlines the frequency of beta glucan general and high level health claims in Australia. Across all food and beverage categories examined, only 37 products made a beta glucan health claim (1.5%), and high level health claims were more common than general level health claims. Claims were most often displayed on breakfast cereals (n = 32, 5.9% of subcategory). Savoury biscuits, noodles/pasta products and grain-based muesli bars did not display beta glucan health claims. Overall, 84% of all claims were related to oat beta glucan, while the remaining were BARLEYmax® beta glucan claims (16%, n = 6). There were no beta glucan claims related to whole grain barley.

Table 3. Frequency of beta glucan general and high level health claims on products from seven food and beverage categories in Australia.

Food Category	Beta Glucan General Level Health Claim n (% of Total Market)	Beta Glucan High Level Health Claim n (% of Total Market)
Breakfast cereal (n = 543)	0	32 (5.9)
Bread (n = 769)	2 (0.3)	0
Savoury crackers (n = 363)	0	0
Noodles/pasta (n = 337)	0	0
Flour (n = 173)	1 (0.6)	0
Grain-based bars (n = 165)	0	0
Plant-based milk alternatives (n = 112)	0	2 (1.8)
Total (n = 2462)	3 (0.1)	34 (1.4)

3.2. Global Market

Of all products launched in the selected markets, 24% contained whole grain oats and/or oat bran (n = 10,763), 2.6% contained whole grain barley (n = 1188) and nine products contained BARLEYmax® (Table 4). The United States had the highest proportion of oat-containing products relative to all products launched (31.3%, n = 1926), followed by New Zealand (24%, n = 541), Canada (23.9%, n = 679) and Europe (23.7%, n = 7068). A small proportion of food and beverage products in Europe were made with whole grain barley (3%, n = 940), while less of 2% of products launched in Singapore, Malaysia, New Zealand and Indonesia contained whole grain barley. BARLEYmax® was found in a limited number of products from Malaysia and Singapore only (n = 6 and 3 products, respectively).

Table 4. Food and beverage products launched between January 2018 and September 2020, containing whole grain oats and/or oat bran, whole grain barley and BARLEYmax® and use of beta glucan heart health/cholesterol-lowering health claims in selected markets.

Market	Contains Whole Grain Oats and/or Oat Bran n (% of Total Market)	Contains Whole Grain Barley n (% of Total Market)	Contains BARLEYmax® n (% of Total Market)	Beta Glucan Health Claim n (% of Total Market)
Europe (n = 29,780)	7068 (23.7)	940 (3.2)	0	254 (0.8)
United States (n = 6145)	1926 (31.3)	78 (1.3)	0	52 (0.8)
Brazil (n = 2991)	493 (16.5)	63 (2.1)	0	17 (0.6)
Canada (n = 2838)	679 (23.9)	74 (2.6)	0	31 (1.1)
Indonesia (n = 1604)	246 (15.3)	8 (0.5)	0	23 (1.4)
Malaysia (n = 552)	116 (21.0)	9 (1.6)	6 (1.1)	10 (1.8)
New Zealand (n = 541)	130 (24.0)	8 (1.5)	0	10 (1.8)
Singapore (n = 443)	105 (23.7)	8 (1.8)	3 (0.7)	6 (1.4)
Total (n = 44,894)	10,763 (24.0)	1188 (2.6)	9 (0.0)	403 (0.9)

The initial Mintel GNPD search returned a total of 651 products that made at least one cardiovascular functional claim, including 181 products from the United States, 347 from Europe, 24 from Brazil, 38 from Canada, 26 from Indonesia, 17 from Malaysia, 10 from New Zealand and 8 from Singapore. Following a visual examination of all product labels, 248 products were excluded as the claim did not specifically relate to beta glucan and blood cholesterol. Of note, 52.7% (n = 68/129) of the excluded products in the United States made a whole grain and heart health claim ('Diets rich in whole grain foods and other plant foods and low in total fat, saturated fat and cholesterol may help reduce the risk of heart disease and certain cancers'), which does not relate to beta glucan specifically.

As shown in Table 4, a small proportion of products displayed beta glucan cholesterol-lowering claims on pack labels, indicating that the low usage is not limited to Australia. Europe had by far the greatest number of products that displayed beta glucan claims ($n = 254$), although this only accounted for less than 1% of the total market. In the United States, 52 products made a beta glucan cholesterol-lowering claim, accounting for 0.8% of the US market. Interestingly, the number of products making a whole grain heart health claim, exceeded those making beta glucan claims ($n = 68$ and 52, respectively).

Overall, the majority of claims were made on breakfast cereals (91%, $n = 365$), while a smaller proportion of claims were made on pasta products ($n = 8$), breads ($n = 7$), crackers/biscuits ($n = 6$), wheat- and grain-based snacks ($n = 5$), plant-based milk alternatives ($n = 3$) and cereal bars ($n = 3$), all of which were from the European market.

4. Discussion

A range of whole grain oat and barley products are available in the Australian and global market, from intact grains, and as ingredients in a range of food products including breakfast cereals, grain-based muesli bars and breads. However, relative to the number of products that contain whole grain oats, oat bran and barley, the number of claims related to beta glucan appear to be in limited use with <2% of products carrying a claim. The efficacious daily dose of 3 g beta glucan is provided in 75 g of whole grain oats (minimum 5.5% beta glucan) or 55 g of oat bran (4% beta glucan) [22]. For an average adult, this quantity is likely to be difficult to achieve on a regular basis. Whole grain oats, barley and oat bran were the richest sources of beta glucan [8], until the more recent introduction of BARLEYmax®, a cultivar with naturally higher levels of beta glucan (6.4% w/w). BARLEYmax® is 40% higher in beta glucan than rolled oats (personal communications with The Healthy Grain), with 3 g of beta glucan provided within a 45 g serving. Given the smaller volume required to achieve the beta glucan daily target, there is an opportunity for greater use of BARLEYmax® in new product development. It should be noted that work is also underway to develop new wheat lines with higher levels of soluble beta glucan [23].

The data suggests there is limited variety and use of whole grain oats and barley in products other than breakfast cereals and grain-based muesli bars. Previous studies have explored the feasibility of incorporating oats and/or barley into products that require further processing such as bread-making [24] and noodles/pasta [25,26], proving that these whole grains can be utilised as a wheat alternative in a range of food products [27]. This innovation would provide an opportunity to extend the use of oats and barley in products for other meal occasions to assist in achieving the efficacious dose of beta glucan more regularly. Innovative products such as oat rice and oat noodles/pasta that are eligible to carry beta glucan cholesterol-lowering health claims are available in parts of Europe and the Asian market [21], although to our knowledge, such products are not yet commercially available in Australia, presenting an opportunity to expand the use of oats and/or barley in other market sectors. However, it is known that the level of processing of beta glucan is critical for its ability to reduce serum cholesterol levels [28–30]. Product innovation should therefore consider use of minimally processed whole grain oats and barley to maintain the food matrix and maximise the beneficial hypocholesteric effects.

Nutrition and health claims have been shown to be helpful in assisting consumers make informed choices and identify healthier food products [31]. Claims also appear to be highly desirable for food industry as evidenced by the increase in product sales following the introduction of a beta glucan heart health claim by Quaker Oats in 1997 [32]. Despite this, the use of beta glucan cholesterol-lowering health claims now appears to be limited, as only 1.5% ($n = 37$) of all Australian products examined carried a claim and just 1–2% of products in other global markets. A possible limitation to the use of beta glucan claims is the serving size required to achieve the beta glucan 1 g dose which may apply to breakfast cereals and the smaller portion grain products (e.g., muesli bars). Furthermore, the additional requirement to pass the NPSC via FSANZ in Australia may place limits on the use of such claims within the snack food category. Greater use of beta glucan health claims may assist in providing consumer awareness of the hypocholesteric benefits of consuming oats and barley, as suggested by Smulders et al. [33] and Ames et al. [27]. Investigation into Australian consumer perceptions of beta glucan cholesterol-lowering claims is an area for future research, using methods utilised in the literature [34,35].

Interestingly, in the United States, the general whole grain heart health claim appeared to be more common than beta glucan related claims although this claim is only available in the US and has been rejected by several other regulatory bodies, FSANZ included. The use of this more general health claim may be more appealing than those relating to beta glucan which is a single food component. The greater use may also relate to perceptions regarding consumer understanding or may simply be aligned with research outcomes for whole grains and CVD. The benefits of regular consumption of whole grain foods are well documented [4,36–39], with every 16 g increase (one serving) associated with a 9% reduction in cardiovascular disease risk [40,41]. In addition to exploring the Australian consumer understanding of beta glucan claims, a comparison to more general whole grain claims would also be of value especially as whole grain claims are growing rapidly, with a 39% increase in the number of products making whole grain claims in Australia in the past 5 years [21].

Furthermore, a recent cost of illness analysis based on the Australian healthcare system reported the government could save AUD 717.4 million annually in direct and indirect cost for prevention of CVD if all Australian adults (>20 years) consumed the recommended 48 g whole grain daily target intake (DTI) [42]. Despite the profound benefits for individual health and the Australian healthcare system, current consumption of whole grains is poor, at 21 g/day for adults (19–85 years), less than half of the 48 g DTI [43]. As previously suggested, whole grain oats are well placed to form part of nutritious products for the future [44], and the same applies to barley. This presents a significant opportunity for innovation using Australian whole grains to increase intakes in line with dietary recommendations which may have powerful impacts on individual and population health.

To our knowledge, the present study is the first to comprehensively review the use of whole grain oats/oat bran, whole grain barley and BARLEYmax® in the Australian and global food supply, within a defined range of grain food and beverage categories and examine the frequency of beta glucan cholesterol-lowering claims on pack labels. However, some limitations should be noted. While all efforts were made to capture the Australian market in its entirety, differences may exist between geographic areas. Although the MINTEL GNPD reportedly captures 70–80% of all product launches from over 86 economies globally, products may be missed due to the nature of the data collection processes utilised by this organisation. As the beta glucan content of a product is not included in the nutrition information panel unless a nutrition claim is made on the label, eligibility to make beta glucan cholesterol-lowering claims could not assessed.

5. Conclusions

Australia is thought to be a world leader in the production of high-quality milling oats and barley for the international market. Given the favourability of growing conditions, there is value in expanding the future application and opportunities for whole grain oats and barley. Despite the sheer number of oat and barley products in the global food market, beta glucan health claims were in limited use and were mostly found on breakfast cereals. Manufacturers could consider selection of oat varieties with higher beta glucan content or enriching products with BARLEYmax® so that products are more likely to be eligible to make beta glucan health claims. Furthermore, the success of beta glucan cholesterol-lowering claims in guiding healthy dietary decisions is dependent on their perception by consumers. Exploring perceptions of beta glucan health claims, and in comparison to whole grain claims, is warranted to gain insight into consumer understanding and value of claims in an Australian context.

Author Contributions: Conceptualization, S.G. and J.H.; methodology, S.G. and J.H.; formal analysis, J.H.; writing—original draft preparation, J.H. and S.G.; writing—review and editing, S.G. All authors have read and agreed to the published version of the manuscript.

Funding: This research received no external funding but was supported by the Grains & Legumes Nutrition Council, a not-for-profit charity.

Institutional Review Board Statement: Not applicable.

Informed Consent Statement: Not applicable.

Data Availability Statement: All data for this study are contained within the article.

Acknowledgments: The authors wish to thank the student dietitians from the University of Wollongong, NSW, and University of Newcastle, NSW, who were involved in the Australian data collection as part of their university studies.

Conflicts of Interest: J.H. and S.G. are employed by the Grains & Legumes Nutrition Council, a not-for-profit charity.

References

1. Australian Institute of Health and Welfare. Cardiovascular Disease. Available online: https://www.aihw.gov.au/reports/heart-stroke-vascular-diseases/cardiovascular-health-compendium (accessed on 30 April 2021).
2. Yusuf, S.; Hawken, S.; Ounpuu, S.; Dans, T.; Avezum, A.; Lanas, F.; McQueen, M.; Budaj, A.; Pais, P.; Varigos, J.; et al. Effect of potentially modifiable risk factors associated with myocardial infarction in 52 countries (the INTERHEART study): Case-control study. *Lancet* **2004**, *364*, 937–952. [CrossRef]
3. Australian Institute of Health and Welfare. Coronary Heart Disease. Available online: https://www.aihw.gov.au/reports/australias-health/coronary-heart-disease (accessed on 7 May 2021).
4. Reynolds, A.; Mann, J.; Cummings, J.; Winter, N.; Mete, E.; Te Morenga, L. Carbohydrate quality and human health: A series of systematic reviews and meta-analyses. *Lancet* **2019**, *393*, 434–445. [CrossRef]
5. Theuwissen, E.; Mensink, R.P. Water-soluble dietary fibers and cardiovascular disease. *Physiol. Behav.* **2008**, *94*, 285–292. [CrossRef]
6. Australia New Zealand Food Standards Code. Standard 2.1.1. Cereal and Cereal Products. Available online: https://www.legislation.gov.au/Details/F2015L00420 (accessed on 25 May 2021).
7. Australian Bureau of Agricultural and Resource Economics and Sciences: Department of Agriculture, Water and the Environment. Australian Crop Report. Available online: https://daff.ent.sirsidynix.net.au/client/en_AU/search/asset/1031421/0 (accessed on 19 May 2021).
8. Burton, R.; Fincher, G. Current challenges in cell wall biology in the cereals and grasses. *Front. Plant Sci.* **2012**, *3*. [CrossRef]
9. Whitehead, A.; Beck, E.J.; Tosh, S.; Wolever, T.M. Cholesterol-lowering effects of oat β-glucan: A meta-analysis of randomized controlled trials. *Am. J. Clin. Nutr.* **2014**, *100*, 1413–1421. [CrossRef]
10. Wolever, T.M.S.; Tosh, S.M.; Gibbs, A.L.; Brand-Miller, J.; Duncan, A.M.; Hart, V.; Lamarche, B.; Thomson, B.A.; Duss, R.; Wood, P.J. Physicochemical properties of oat β-glucan influence its ability to reduce serum LDL cholesterol in humans: A randomized clinical trial. *Am. J. Clin. Nutr.* **2010**, *92*, 723–732. [CrossRef]
11. Tiwari, U.; Cummins, E. Meta-analysis of the effect of β-glucan intake on blood cholesterol and glucose levels. *Nutrition* **2011**, *27*, 1008–1016. [CrossRef] [PubMed]
12. The Healthy Grain. BARLEYmax®. Available online: https://thehealthygrain.com/barleymax/ (accessed on 24 May 2021).
13. Commonwealth Scientific and Industrial Research Organisation (CSIRO). BARLEYmax. Available online: https://www.csiro.au/en/research/production/food/BARLEYmax-BUcase-study (accessed on 24 May 2021).
14. Food Standards Australia New Zealand. Australia New Zealand Food Standards Code—Standard 1.2.7—Nutrition, Health and Related Claims. Available online: www.comlaw.gov.au/Series/F2013L00054 (accessed on 7 May 2021).
15. Food Standards Australia New Zealand. Schedule 4 Nutrition, Health and Related Claims. Available online: https://www.legislation.gov.au/Series/F2015L00474 (accessed on 24 May 2021).
16. Australian New Zealand Food Standards Code. Overview of the Nutrient Profiling Scoring Criterion. Available online: http://www.foodstandards.gov.au/industry/labelling/Pages/Consumer-guide-to-NPSC.aspx (accessed on 24 May 2021).
17. Mathews, R.; Kamil, A.; Chu, Y. Global review of heart health claims for oat beta-glucan products. *Nutr. Rev.* **2020**, *78*, 78–97. [CrossRef]
18. Roy Morgan. Looking Beyond the Panic-Buying, Australia's Big Supermarket Story Is Aldi's Growing Market Share. Available online: http://www.roymorgan.com/findings/8336-fresh-food-and-grocery-report-december-2019-202003230634 (accessed on 25 May 2021).

19. Hughes, J.; Vaiciurgis, V.; Grafenauer, S. Flour for Home Baking: A Cross-Sectional Analysis of Supermarket Products Emphasising the Whole Grain Opportunity. *Nutrients* **2020**, *12*, 2058. [CrossRef] [PubMed]
20. Food Standards Australia New Zealand. Standard 1.1.2 Definitions Used throughout the Code. Available online: http://www.foodstandards.gov.au/code/Pages/default.aspx (accessed on 5 July 2021).
21. GNPD—Global New Products Database: Mintel Group Ltd. Available online: https://www.mintel.com/global-new-products-database (accessed on 19 April 2021).
22. Charlton, K.E.; Tapsell, L.C.; Batterham, M.J.; O'Shea, J.; Thorne, R.; Beck, E.; Tosh, S.M. Effect of 6 weeks' consumption of β-glucan-rich oat products on cholesterol levels in mildly hypercholesterolaemic overweight adults. *Br. J. Nutr.* **2012**, *107*, 1037–1047. [CrossRef]
23. Commonwealth Scientific and Industrial Research Organisation (CSIRO). Developing Wheat with Cholesterol Lowering Properties. Available online: https://www.csiro.au/en/research/plants/crops/Grains/b-glucan-wheat (accessed on 25 May 2021).
24. Londono, D.M.; Gilissen, L.J.W.J.; Visser, R.G.F.; Smulders, M.J.M.; Hamer, R.J. Understanding the role of oat β-glucan in oat-based dough systems. *J. Cereal Sci.* **2015**, *62*, 1–7. [CrossRef]
25. Liao, M.-Y.; Shen, Y.-C.; Chiu, H.-F.; Ten, S.-M.; Lu, Y.-Y.; Han, Y.-C.; Venkatakrishnan, K.; Yang, S.-F.; Wang, C.-K. Down-regulation of partial substitution for staple food by oat noodles on blood lipid levels: A randomized, double-blind, clinical trial. *J. Food Drug Anal.* **2019**, *27*, 93–100. [CrossRef]
26. Nguyen, T.T.L.; Gilbert, R.G.; Gidley, M.J.; Fox, G.P. The contribution of β-glucan and starch fine structure to texture of oat-fortified wheat noodles. *Food Chem.* **2020**, *324*, 126858. [CrossRef] [PubMed]
27. Ames, N.P.; Rhymer, C.R. Issues Surrounding Health Claims for Barley. *J. Nutr.* **2008**, *138*, 1237S–1243S. [CrossRef] [PubMed]
28. Grundy, M.M.L.; Fardet, A.; Tosh, S.M.; Rich, G.T.; Wilde, P.J. Processing of oat: The impact on oat's cholesterol lowering effect. *Food Funct.* **2018**, *9*, 1328–1343. [CrossRef] [PubMed]
29. Henrion, M.; Francey, C.; Lê, K.A.; Lamothe, L. Cereal B-Glucans: The Impact of Processing and How It Affects Physiological Responses. *Nutrients* **2019**, *11*, 1729. [CrossRef]
30. Goudar, G.; Sharma, P.; Janghu, S.; Longvah, T. Effect of processing on barley β-glucan content, its molecular weight and extractability. *Int. J. Biol. Macromol.* **2020**, *162*, 1204–1216. [CrossRef]
31. Williams, P. Consumer Understanding and Use of Health Claims for Foods. *Nutr. Rev.* **2005**, *63*, 256–264. [CrossRef]
32. Paul, G.L.; Ink, S.L.; Geiger, C.J. The Quaker Oats Health Claim. *J. Nutraceuticals Funct. Med. Foods* **1999**, *1*, 5–32. [CrossRef]
33. Smulders, M.J.M.; van de Wiel, C.C.M.; van den Broeck, H.C.; van der Meer, I.M.; Israel-Hoevelaken, T.P.M.; Timmer, R.D.; van Dinter, B.-J.; Braun, S.; Gilissen, L.J.W.J. Oats in healthy gluten-free and regular diets: A perspective. *Food Res. Int.* **2018**, *110*, 3–10. [CrossRef]
34. Wong, C.L.; Mendoza, J.; Henson, S.J.; Qi, Y.; Lou, W.; L'Abbé, M.R. Consumer attitudes and understanding of cholesterol-lowering claims on food: Randomize mock-package experiments with plant sterol and oat fibre claims. *Eur. J. Clin. Nutr.* **2014**, *68*, 946–952. [CrossRef]
35. Stancu, V.; Grunert, K.G.; Lähteenmäki, L. Consumer inferences from different versions of a beta-glucans health claim. *Food Qual. Prefer.* **2017**, *60*, 81–95. [CrossRef]
36. Aune, D.; Keum, N.; Giovannucci, E.; Fadnes, L.T.; Boffetta, P.; Greenwood, D.C.; Tonstad, S.; Vatten, L.J.; Riboli, E.; Norat, T. Whole grain consumption and risk of cardiovascular disease, cancer, and all cause and cause specific mortality: Systematic review and dose-response meta-analysis of prospective studies. *BMJ* **2016**, *353*, i2716. [CrossRef]
37. Aune, D.; Norat, T.; Romundstad, P.; Vatten, L.J. Whole grain and refined grain consumption and the risk of type 2 diabetes: A systematic review and dose-response meta-analysis of cohort studies. *Eur. J. Epidemiol.* **2013**, *28*, 845–858. [CrossRef]
38. Schwingshackl, L.; Schwedhelm, C.; Hoffmann, G.; Lampousi, A.-M.; Knüppel, S.; Iqbal, K.; Bechthold, A.; Schlesinger, S.; Boeing, H. Food groups and risk of all-cause mortality: A systematic review and meta-analysis of prospective studies. *Am. J. Clin. Nutr.* **2017**. [CrossRef]
39. Marshall, S.; Petocz, P.; Duve, E.; Abbott, K.; Cassettari, T.; Blumfield, M.; Fayet-Moore, F. The Effect of Replacing Refined Grains with Whole Grains on Cardiovascular Risk Factors: A Systematic Review and Meta-Analysis of Randomized Controlled Trials with GRADE Clinical Recommendation. *J. Acad. Nutr. Diet.* **2020**. [CrossRef] [PubMed]
40. Chen, G.-C.; Tong, X.; Xu, J.-Y.; Han, S.-F.; Wan, Z.-X.; Qin, J.-B.; Qin, L.-Q. Whole-grain intake and total, cardiovascular, and cancer mortality: A systematic review and meta-analysis of prospective studies. *Am. J. Clin. Nutr.* **2016**. [CrossRef] [PubMed]
41. Zong, G.; Gao, A.; Hu, F.B.; Sun, Q. Whole Grain Intake and Mortality from All Causes, Cardiovascular Disease, and Cancer: A Meta-Analysis of Prospective Cohort Studies. *Circulation* **2016**, *133*, 2370–2380. [CrossRef]
42. Abdullah, M.M.H.; Hughes, J.; Grafenauer, S. Healthcare Cost Savings Associated with Increased Whole Grain Consumption among Australian Adults. *Nutrients* **2021**, *13*, 1855. [CrossRef]
43. Galea, L.; Beck, E.; Probst, Y.; Cashman, C. Whole grain intake of Australians estimates from a cross-sectional analysis of dietary intake data from the 2011-13 Australian Health Survey. *Public Health Nutr.* **2017**, *20*, 2166–2172. [CrossRef]
44. Clemens, R.; van Klinken, B.J.-W. The future of oats in the food and health continuum. *Br. J. Nutr.* **2014**, *112*, S75–S79. [CrossRef] [PubMed]

Article

Euglena Gracilis and β-Glucan Paramylon Induce Ca²⁺ Signaling in Intestinal Tract Epithelial, Immune, and Neural Cells

Kosuke Yasuda [1], Ayaka Nakashima [1,*], Ako Murata [1], Kengo Suzuki [1] and Takahiro Adachi [2,*]

[1]   Euglena Co., Ltd., Tokyo 108-0014, Japan; kosuke.yasuda@euglena.jp (K.Y.); ako.murata@euglena.jp (A.M.); suzuki@euglena.jp (K.S.)

[2]   Department of Immunology, Medical Research Institute, Tokyo Medical and Dental University, Tokyo 113-8510, Japan

*   Correspondence: nakashima@euglena.jp (A.N.); tadachi.imm@mri.tmd.ac.jp (T.A.); Tel.: +81-3-5442-4907 (A.N.); +81-3-5803-4591 (T.A.)

Received: 12 June 2020; Accepted: 27 July 2020; Published: 30 July 2020

Abstract: The intestinal tract contains over half of all immune cells and peripheral nerves and manages the beneficial interactions between food compounds and the host. Paramylon is a β-1,3-glucan storage polysaccharide from *Euglena gracilis* (*Euglena*) that exerts immunostimulatory activities by affecting cytokine production. This study investigated the signaling mechanisms that regulate the beneficial interactions between food compounds and the intestinal tract using cell type-specific calcium (Ca²⁺) imaging in vivo and in vitro. We successfully visualized *Euglena*- and paramylon-mediated Ca²⁺ signaling in vivo in intestinal epithelial cells from mice ubiquitously expressing the Yellow Cameleon 3.60 (YC3.60) Ca²⁺ biosensor. Moreover, in vivo Ca²⁺ imaging demonstrated that the intraperitoneal injection of both *Euglena* and paramylon stimulated dendritic cells (DCs) in Peyer's patches, indicating that paramylon is an active component of *Euglena* that affects the immune system. In addition, in vitro Ca²⁺ imaging in dorsal root ganglia indicated that *Euglena*, but not paramylon, triggers Ca²⁺ signaling in the sensory nervous system innervating the intestine. Thus, this study is the first to successfully visualize the direct effect of β-1,3-glucan on DCs in vivo and will help elucidate the mechanisms via which *Euglena* and paramylon exert various effects in the intestinal tract.

Keywords: β-1,3-glucan; *Euglena gracilis*; Ca²⁺ signaling; intestinal epithelial cell; intravital imaging; small intestine; immune system

---

1. Introduction

The intestinal tract is the first line of defense against pathogenic microorganisms and manages beneficial interactions between food compounds and the host [1]. These interactions are mediated by the neural, endocrine, and immune systems to maintain intestinal homeostasis [2]; however, homeostatic dysfunction can significantly affect host immunity and the course of chronic inflammation [3]. Some probiotics and polysaccharides have been found to regulate intestinal and immune homeostasis; for instance, probiotic *Bifidobacterium bifidum* alleviates dysbiosis and constipation in mice induced by a low-fiber diet [4], while grains fermented with *Aspergillus oryzae* can protect against chronic constipation and gastrointestinal damage [5,6]. In addition, the soluble dietary fiber β-1,3-glucan from seaweed has been shown to suppress intestinal inflammation in mouse models of human inflammatory bowel disease by increasing the *Lactobacillus* population and the number of regulatory T cells in the colon [7]. However, different polysaccharides can exert varying effects on immune homeostasis [8–10].

Euglena gracilis (*Euglena*) is a microalga that contains a wide range of nutrients, including vitamins, minerals, amino acids, and fatty acids. Since it combines the properties of both plants and animals, it is often used as a food or dietary supplement [11]. The storage polysaccharide paramylon is an insoluble dietary fiber unique to *Euglena* that has a triple-helical polymer structure composed of straight-chain β-1,3-glucans. Like other β-1,3-glucans, paramylon has various beneficial effects on health, such as modulating immune function [12–15] and suppressing visceral fat accumulation, likely by improving the intestinal environment [16–19]. In addition, paramylon can bind to and stimulate Dectin-1, the primary receptor on epithelial cells, macrophages, and dendritic cells (DCs) to exert immunomodulatory effects [20–22]. Following their ingestion, polysaccharides and probiotics are thought to be taken up into M cells, a type of intestinal epithelial cell (IEC) that lies over Peyer's patches (PPs), allowing them to access gut immune cells such as macrophages and DCs [23,24]; however, it is difficult to monitor these biological events in real time.

Calcium imaging using genetically encoded calcium biosensors has enabled nutrition-sensing mechanisms in the intestinal tract to be visualized. For example, Calcium ion (Ca^{2+}) signaling in specific cell populations has been visualized in conditional transgenic mice expressing Yellow Cameleon 3.60 (YC3.60) [25,26]. Ca^{2+} is a universal secondary messenger that performs multiple functions in most cells, including lymphocytes, epithelial cells, and neurons [27–30]. Intravital Ca^{2+} imaging in transgenic mice expressing YC3.60 under the control of the CD11c gene promoter has demonstrated that oral propolis administration stimulates DCs in lymphoid tissues [31], while in vivo Ca^{2+} imaging has revealed that probiotics induce Ca^{2+} signaling in IECs under physiological conditions [32]. Enteroendocrine cells are a form of IEC that constitute the largest endocrine organ in the human body, while the autonomic nervous system that innervates the intestine exerts strong modulatory effects on the motor, secretory, and immunologic functions of the intestinal tract [33,34]. To elucidate the complex sensing mechanisms that recognize *Euglena* and paramylon in the intestinal tract, we performed cell type-specific Ca^{2+} imaging in three conditional transgenic mouse lines with specific or ubiquitous YC3.60 expression [26,31] to visualize *Euglena*- and paramylon-induced Ca^{2+} signaling in IECs, DCs in intestinal PPs, and nerve cells innervating the intestine.

2. Materials and Methods

2.1. Animals

Conditional YC3.60 expressing transgenic mice were as described previously [26]. The floxed YC3.60 reporter (YC3.60^{flox}) mouse line was crossed with cell type-specific Cre mouse lines (CD11c-Cre, Nestin-Cre, and CAG-Cre) to produce YC3.60^{flox}/CD11c-Cre, YC3.60^{flox}/Nestin-Cre, and YC3.60^{flox}/CAG-Cre mice, respectively [26,31]. These mice were maintained in our animal facility under specific pathogen-free conditions in accordance with the animal care guidelines of the Tokyo Medical and Dental University, while animal procedures were approved by its Animal Care Committee (approval number A2019-207C4, date of approval 3 December 2019).

2.2. Dorsal Root Ganglia (DRG) Cells

Dorsal root ganglia (DRG) cells were prepared from YC3.60^{flox}/Nestin-Cre mice as described previously [35]. Briefly, mice were euthanized by cervical dislocation and their DRG excised before being treated with collagenase (1 mg/mL) and trypsin (0.25 mg/mL), as described previously [36]. The cells were then washed twice with Dulbecco's modified Eagle's medium (DMEM) and cultured at 37 °C on a gelatin-coated plate with DMEM containing 10% fetal calf serum and 100 ng/mL of nerve growth factor.

2.3. Test Substances

Euglena powder and paramylon were obtained from Euglena Co., Ltd. (Tokyo, Japan). The nutritional composition of the *Euglena* powder was as follows: carbohydrates 55.0%, protein 29.9%, and lipid 9.0%. Approximately 70–80% of the carbohydrate content was paramylon. Paramylon was prepared according to the standard method, as follows: cultured *Euglena gracilis* Z cells were collected by continuous centrifugation and washed with water. After being suspending in water, the cells were disintegrated using ultrasonic waves and the cell contents (containing paramylon) were collected. The crude paramylon preparation was treated with 1% sodium dodecyl sulfate (SDS) solution for 1 h at 95 °C followed by 0.1% SDS for 30 min at 50 °C to remove lipids and proteins. After centrifugation, paramylon was refined by repeated washing with water, acetone, and ether.

2.4. Intravital and In Vitro Imaging

IECs and PPs from anesthetized mice were imaged as described previously [32]. Small intestinal tracts were surgically opened lengthwise, placed on a glass cover slip, and immobilized on a microscope stage. To observe IECs, 0.1 mL of *Euglena* or paramylon in phosphate buffered saline (PBS; 1 mg/mL) was added to the intestinal tract, with PBS as a control. Images were acquired using a Nikon A1 laser-scanning confocal microscope with a 20× objective lens, dichronic mirrors (DM457/514), and two bandpass emission filters (482/35 for cyan fluorescent protein, CFP, 540/30 for yellow fluorescent protein, YFP), as described previously [26]. The YFP/CFP ratio was obtained by excitation at 458 nm. Images of purified spleen cells in PBS were obtained using the same method. Acquired images were analyzed using NIS-Elements software (Nikon, Tokyo, Japan).

2.5. In Vivo Stimulation Assay

To observe PPs, the peritoneal cavity of each mouse was injected with 200 μg of *Euglena* or paramylon in PBS (1 mg/mL), with PBS as a control. After 2 hours, the mice were subjected to intravital imaging analysis, as described previously [31].

2.6. Statistical Analysis

Statistical analysis was performed with Pearson's chi-square test to compare the proportions of cells. R version 3.4.1 were used to conduct the statistical analyses. $p < 0.05$ was considered significant.

3. Results

3.1. Euglena and Paramylon Induce Ca^{2+} Signaling in the IECs of Mice With Ubiquitous Yc3.60 Expression

To examine whether *Euglena* and paramylon directly stimulate IECs, we carried out intravital Ca^{2+} imaging. *Euglena* induced transient Ca^{2+} signaling in most IECs (Figure 1a) and intracellular Ca^{2+} levels increased following stimulation (Figure 1b). Conversely, paramylon induced robust but sparse Ca^{2+} signaling limited to minor IEC subpopulations (Figure 1). Together, these findings suggest that *Euglena* and paramylon directly stimulate IECs.

3.2. Euglena and Paramylon Induce Ca^{2+} Signaling in DCs

To assess the potential immune-stimulatory effects of *Euglena* and paramylon on immune cells in vivo, we performed intravital Ca^{2+} imaging on PPs in YC3.60^{flox}/CD11c-Cre mice [31] injected intraperitoneally (IP) with *Euglena* or paramylon (1 mg/mL). After 2 hours of intraperitoneal administration, intravital imaging analysis showed that *Euglena* increased intracellular Ca^{2+} levels in DCs, with 31.2% of DCs exhibiting higher intracellular Ca^{2+} concentrations (Figure 2). However, paramylon induced robust Ca^{2+} signaling in 78.2% of DCs, thus exerted stronger effects than *Euglena* (Figure 2b). These results indicate that *Euglena* and paramylon possess immune-stimulatory functions.

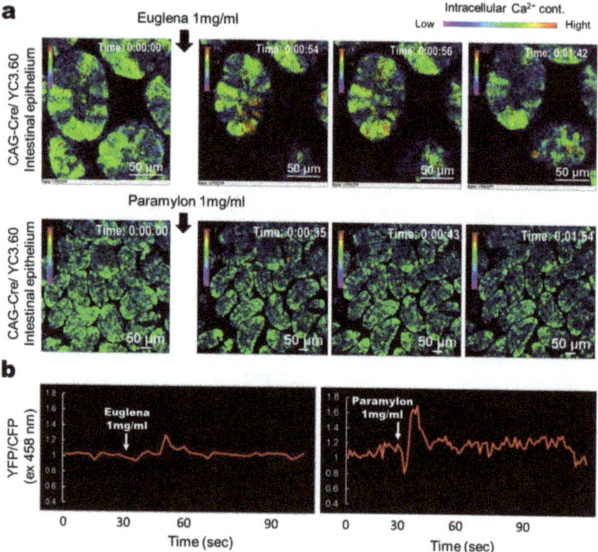

Figure 1. Intravital calcium (Ca^{2+}) signaling mediated by *Euglena gracilis* (*Euglena*) or paramylon in the intestinal tract. (**a**) Representative ratiometric Ca^{2+} signaling images from the intestinal tract of a YC3.60[flox]/CAG-Cre mouse with ubiquitous YC3.60 expression showing yellow fluorescent protein/cyan fluorescent protein (YFP/CFP) intensity at 458 nm excitation. *Euglena* or paramylon (0.1 mL) in phosphate buffered saline (PBS) (1 mg/mL) was added at the indicated time point. The color scale indicates relative Ca^{2+} concentration. (**b**) Time course of YFP/CFP fluorescence intensity at 458 nm excitation. *Euglena* or paramylon (0.1 mL) in PBS (1 mg/mL) was added at the indicated time point. Results are representative of at least three independent experiments (*n* = 3 mice; scale bars, 50 µm).

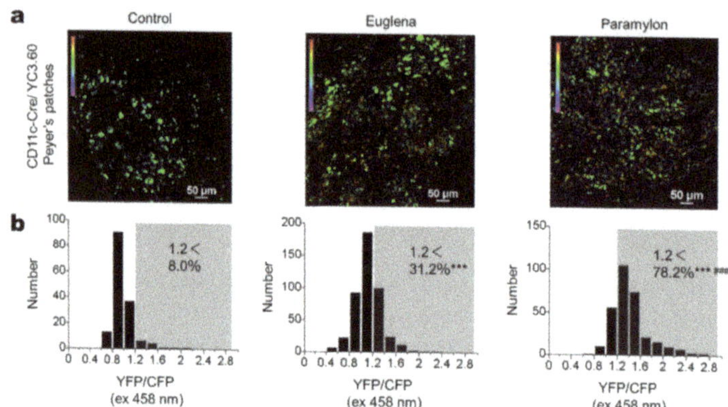

Figure 2. Intravital Ca^{2+} signaling in Peyer's patches (PPs). (**a**) Representative Ca^{2+} signaling images of PPs in YC3.60[flox]/CD11c-Cre mice intraperitoneally injected with PBS (control, left) *Euglena*/PBS (center), or paramylon/PBS (right). Intravital ratiometric imaging was carried out 2 hours after injection and shows YFP/CFP excitation at 458 nm. The results are representative of at least three independent experiments (*n* = 3 mice; scale bars, 50 µm). (**b**) Distribution of intracellular Ca^{2+} levels in randomly selected cells (control, *n* = 152; *Euglena*, *n* = 455; paramylon, *n* = 303). YFP/CFP > 1.2 was defined as cells of relatively high Ca^{2+} concentration. Pearson's chi-square test, *** $p < 0.001$ vs. Control, ### $p < 0.001$ *Euglena* vs. Paramylon.

3.3. Euglena Elicits In Vitro Ca²⁺ Signaling in DRG-Derived Neurons From YC3.60^{flox}/Nestin-Cre Mice

To test whether *Euglena* or paramylon have the potential to stimulate sensory neurons in the intestine, we performed Ca^{2+} imaging on primary neurons dissected from the DRG of YC3.60^{flox}/Nestin-Cre mice [32] and cultured on a dish for several days [35]. *Euglena* induced robust Ca^{2+} signaling in the DRG neurons (Figure 3a), whereas no Ca^{2+} signaling was induced by paramylon (Figure 3b), indicating that *Euglena* stimulates sensory neurons but its component paramylon does not.

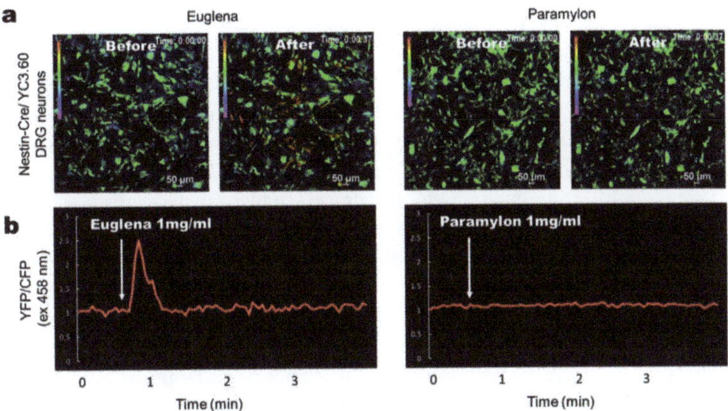

Figure 3. Ca^{2+} signaling images using *Euglena* and Paramylon in dorsal root ganglia (DRG) neurons in vitro. (**a**) Representative ratiometric Ca^{2+} signaling images in DRG cells from YC3.60^{flox}/Nestin-Cre mice showing YFP/CFP excitation at 458 nm. *Euglena* or paramylon (0.1 mL) in PBS (1 mg/mL) was added to the cell culture at the indicated time point. The color scale indicates relative Ca^{2+} concentration. (**b**) Time course of ratiometric YFP/CFP fluorescence intensity at 458 nm excitation. Results are representative of at least three independent experiments (n = 9 mice; scale bars, 50 µm).

4. Discussion

This study successfully visualized *Euglena*- and paramylon-mediated Ca^{2+} signaling in IECs, DCs, and DRG-derived neurons using Ca^{2+} imaging in YC3.60 mice. *Euglena* and paramylon both exhibited stimulatory activities in IECs in vivo and possessed immune-stimulating properties against DCs in PPs in vivo. Furthermore, *Euglena* directly induced Ca^{2+} signaling in DRG-derived neurons.

Understanding the mechanisms underlying the interaction between food compounds and IECs is important for evaluating the physiological benefits of food compounds, since stimulated IECs can produce cytokines and/or peptide hormones [30]. Probiotic microbes bind with a series of pattern recognition receptors, including toll-like receptors, nucleotide-binding sites, leucine-rich repeat-containing receptors, and retinoic acid-inducible gene-I-like receptors [37]. On IECs, the β-glucan receptor Dectin-1 triggers the secretion of pro-inflammatory cytokines, such as Interleukin-8 (IL-8) and monocyte chemoattractant protein 1 (CCL2) [22]. Although enteroendocrine cells make up less than 1% of the IEC population, they form the largest endocrine organ in the body and secrete multiple peptide hormones such as ghrelin, serotonin (5-hydroxytryptamine), Cholecystokinin (CCK), peptide tyrosine-tyrosine (PYY), glucagon-like peptide-1 (GLP-1) and glucose-dependent insulinotropic peptide (GIP) [38]. In response to food intake, enteroendocrine cells produce GLP-1, a 30-amino acid peptide hormone that exerts various metabolic actions, such as reducing appetite and food intake [39]. In addition, GLP-1 promotes insulin secretion and may contribute toward improved insulin sensitivity [40]. Thus, food components and dietary supplements that modulate nutrient-sensing pathways may have therapeutic potential for treating obesity and metabolic diseases [38,41]. A previous study showed that *Bacillus subtilis* var. *natto*, which has been shown to modulate immune responses, triggers gradual and sustained Ca^{2+} signaling in IECs [32]. Conversely, the probiotic *Lactococcus lactis*, which is similar to commensal bacteria, does not induce Ca^{2+} signaling in IECs, likely due to hyporesponsiveness following chronic exposure to bacteria [32]. In this study, we found that both *Euglena* and paramylon evoked Ca^{2+} signaling in IECs, similar to other prebiotics, suggesting that they can directly access IECs. It should be noted that *Euglena* and paramylon showed different bioactivity for IECs. *Euglena* stimulation evoked transient Ca^{2+} signaling throughout IECs with distinct Ca^{2+} signaling kinetics to those evoked by paramylon. Previous studies have shown that oral paramylon and *Euglena* intake exert preventive effects against obesity, likely due to the presence of paramylon [17,18]. Here, paramylon induced sparse transient Ca^{2+} signaling in minor IEC subpopulations, with a similar spatial pattern of Ca^{2+} signaling to the cellular distribution of enteroendocrine cells [42]. Thus, nutrient sensing in the intestinal tract may transmit neuronal signals to the brain by secreting multiple gastrointestinal hormones to modulate the physiological response to food components. However, further studies of specific cell types are required to clarify the biological properties of paramylon.

Biologically-active polysaccharides may be a potential method of preventing dysfunctional immune homeostasis [9,43]. For instance, β-glucan extracted from the maitake mushroom (*Grifola frondosa*) has been shown to control the cytokine balance between T lymphocyte Th-1 and Th-2 subsets, resulting in enhanced cellular immunity [8]. In addition, the subcutaneous application of β-1,3-glucan extracted from *Saccharomyces cerevisiae* in 20 children with asthma increased serum levels of the anti-inflammatory cytokine IL-10 and simultaneously reduced symptom scores [10]. The oral intake of paramylon has been found to reduce cytokine secretion and relieve arthritis symptoms in mice by modulating Th17 immunity [44], while studies have suggested that *Euglena* and paramylon could reduce upper respiratory tract infection symptoms and protect against influenza virus infection [13,45]. Here, we found that IP *Euglena* and paramylon injection stimulated DCs in PPs in vivo, with stronger Ca^{2+} signaling observed with paramylon treatment; thus, paramylon may be the active component of *Euglena* that affects the immunological system. This finding is consistent with previous studies indicating that β-1,3-glucans stimulate macrophages and DCs via a Dectin-1-dependent pathway [20,22]. Orally administered β-glucans are thought to access gut immune cells, such as macrophages and DCs, by being taken up into M cells, a type of IEC that lies over PPs [23,24]. On macrophages, the β-glucan receptor Dectin-1 specifically binds to paramylon and exerts immune stimulatory effects [20,46]. Alongside phagocytosis in macrophages, Dectin-1 also induces proinflammatory cytokine secretion [47]. Due to its crystalline structure, paramylon typically exists in the form of insoluble granules that are 2–3 μm in size, suggesting that paramylon may be transported across IECs via the same mechanism as pathogenic bacteria. Our findings, together with those of previous studies, suggest that oral paramylon intake could stimulate intestinal DCs in PPs to enhance host immune function. Consistently, some studies in human subjects suggest that *Euglena* and paramylon may support a healthy immune system and protect overall health [45,48]. Intravital imaging using several transgenic mice with biosensors expressed in the intestinal epithelium, nervous system, and immune cells will be a powerful tool to dissect the mechanism of a beneficial effect of the active ingredient in food products on human immunity [32].

In this study, we demonstrated that *Euglena*, but not paramylon, directly triggers Ca^{2+} signaling in DRG neurons, suggesting that *Euglena* can excite visceral afferents. Although paramylon has various bioactive functionalities [21], other bioactive components of water-soluble fraction seem to activate the neurons. Indeed, the water extract partially purified from *Euglena* also contains bioactive materials, as this is crucial for preventing lung carcinoma growth and intracellular lipid accumulation [19,49]. The sensory innervation of the small intestine is due to spinal and vagus nerves, which have cell bodies in the DRG and nodose ganglion, respectively [50]. DRG afferents are largely peptidergic and express the calcitonin gene-related peptide Substance P and/or transient receptor potential vanilloid 1 [51,52]. The calcitonin gene-related peptide (CGRP) neuropeptide modifies macrophages, DCs, and other immune cells, suggesting that it plays a key role in neuro-immune cross-talk and allows sensory fibers to mediate immune function [52]. The release of neuropeptides from sensory afferents has been associated to nociceptive transmission, energy homeostasis, and longevity [53,54], whereas the vagal afferent is required for the activity of sympathetic nerves innervating brown adipose tissue triggered by a capsaicin analog, indicating that sensory afferents play important roles in inducing autonomous nerve activity [55]. Some prebiotic bacteria have been shown to alter emotional behavior by regulating the vagal nerve [56], whereas capsaicin in hot peppers can trigger Ca^{2+} signaling in the intestinal tract, which is transmitted to the nervous system and results in a transient shift toward higher arousal levels in the brain [35]. Further studies will be required to better understand the physiological role of food compound-mediated visceral afferents in homeostasis, behavior, emotion or cognitive function. Although the mechanisms underlying our observations remain unclear, the ability of *Euglena* to excite neurons may underlie its beneficial effects on health-related Quality of Life observed in human subjects [45]. In the near future, after the components involved in *Euglena* responsible for these signaling responses have been identified and a method for extracting the active ingredients has been established, *Euglena* can be utilized to produce useful ingredients on an industrial scale [57].

5. Conclusions

Interactions between food and the intestinal tract transmit signals via the gut–immune–brain axis by secreting multiple cytokines and hormones that modulate physiological homeostasis [38,58]. Thus, our findings help to elucidate the mechanisms via which *Euglena* and paramylon exert various effects from the intestinal tract.

Author Contributions: Conceptualization, K.Y., A.N., A.M., K.S. and T.A.; methodology, T.A.; formal analysis, T.A.; resources, K.Y., A.N., A.M. and K.S.; original draft preparation and writing, K.Y.; review and editing, K.Y., A.N., A.M., K.S. and T.A.; supervision, K.S. and T.A. All authors have read and agreed to the published version of the manuscript.

Funding: This research received no external funding.

Conflicts of Interest: K.Y., A.N., A.M. and K.S. are salaried employees of Euglena Co., Ltd. which produced some of the *Euglena* used in this study. All research funding for this study was provided by Euglena Co., Ltd.

References

1. Mitsuoka, T. Development of Functional Foods. *Biosci. Microbiota Food Health* **2014**, *33*, 117–128. [CrossRef] [PubMed]
2. Furness, J.B.; Rivera, L.R.; Cho, H.J.; Bravo, D.M.; Callaghan, B. The gut as a sensory organ. *Nat. Rev. Gastroenterol. Hepatol.* **2013**, *10*, 729–740. [CrossRef]
3. Ott, S.J.; Musfeldt, M.; Wenderoth, D.F.; Hampe, J.; Brant, O.; Fölsch, U.R.; Timmis, K.N.; Schreiber, S. Reduction in diversity of the colonic mucosa associated bacterial microflora in patients with active inflammatory bowel disease. *Gut* **2004**, *53*, 685–693. [CrossRef] [PubMed]
4. Makizaki, Y.; Maeda, A.; Oikawa, Y.; Tamura, S.; Tanaka, Y.; Nakajima, S.; Yamamura, H. Alleviation of low-fiber diet-induced constipation by probiotic *Bifidobacterium bifidum* G9-1 is based on correction of gut microbiota dysbiosis. *Biosci. Microbiota Food Health* **2019**, *38*, 49–53. [CrossRef] [PubMed]
5. Ochiai, M.; Shiomi, S.; Ushikubo, S.; Inai, R.; Matsuo, T. Effect of a fermented brown rice extract on the gastrointestinal function in methotrexate-treated rats. *Biosci. Biotechnol. Biochem.* **2013**, *77*, 243–248. [CrossRef] [PubMed]
6. Kataoka, K.; Ogasa, S.; Kuwahara, T.; Bando, Y.; Hagiwara, M.; Arimochi, H.; Nakanishi, S.; Iwasaki, T.; Ohnishi, Y. Inhibitory effects of fermented brown rice on induction of acute colitis by dextran sulfate sodium in rats. *Dig. Dis. Sci.* **2008**, *53*, 1601–1608. [CrossRef] [PubMed]
7. Tang, C.; Kamiya, T.; Liu, Y.; Kadoki, M.; Kakuta, S.; Oshima, K.; Hattori, M.; Takeshita, K.; Kanai, T.; Saijo, S.; et al. Inhibition of dectin-1 signaling ameliorates colitis by inducing lactobacillus-mediated regulatory T cell expansion in the intestine. *Cell Host Microbe* **2015**, *18*, 183–197. [CrossRef]
8. Inoue, A.; Kodama, N.; Nanba, H. Effect of maitake (*Grifola frondosa*) D-Fraction on the control of the T lymph node Th-1/Th-2 proportion. *Biol. Pharm. Bull.* **2002**, *25*, 536–540. [CrossRef]
9. Jesenak, M.; Banovcin, P.; Rennerova, Z.; Majtan, J. β-Glucans in the treatment and prevention of allergic diseases. *Allergol. Immunopathol. (Madr.)* **2014**, *42*, 149–156. [CrossRef]

10. Sarinho, E.; Medeiros, D.; Schor, D.; Rego Silva, A.; Sales, V.; Motta, M.E.; Costa, A.; Azoubel, A.; Rizzo, J.A. Production of interleukin-10 in asthmatic children after Beta-1-3-glucan. *Allergol. Immunopathol.* **2009**, *37*, 188–192. [CrossRef]

11. Schwartzbach, S.D.; Shigeoka, S. *Euglena: Biochemistry, Cell and Molecular Biology*; Springer International Publishing: Cham, Switzerland, 2017.

12. Kondo, Y.; Kato, A.; Hojo, H.; Nozoe, S.; Takeuchi, M.; Ochi, K. Cytokine-Related Immunopotentiating Activities of Paramylon, a (3-1-≫3)-d-Glucan from *Euglena gracilis*. *J. Pharmacobiodyn.* **1992**, *15*, 617–621. [CrossRef] [PubMed]

13. Nakashima, A.; Suzuki, K.; Asayama, Y.; Konno, M.; Saito, K.; Yamazaki, N.; Takimoto, H. Oral administration of *Euglena gracilis* Z and its carbohydrate storage substance provides survival protection against influenza virus infection in mice. *Biochem. Biophys. Res. Commun.* **2017**, *494*, 379–383. [CrossRef] [PubMed]

14. Russo, R.; Barsanti, L.; Evangelista, V.; Frassanito, A.M.; Longo, V.; Pucci, L.; Penno, G.; Gualtieri, P. *Euglena gracilis* paramylon activates human lymphocytes by upregulating pro-inflammatory factors. *Food Sci. Nutr.* **2017**, *5*, 205–214. [CrossRef] [PubMed]

15. Yasuda, K.; Ogushi, M.; Nakashima, A.; Nakano, Y.; Suzuki, K. Accelerated wound healing on the skin using a film dressing with β-glucan paramylon. *In Vivo* **2018**, *32*, 799–805. [CrossRef]

16. Aoe, S.; Yamanaka, C.; Nishioka, M.; Onaka, N.; Nishida, N.; Takahashi, M. Effects of paramylon extracted from *Euglena gracilis* EOD-1 on parameters related to metabolic syndrome in diet-induced obese mice. *Nutrients* **2019**, *11*, 1674. [CrossRef] [PubMed]

17. Sakanoi, Y.; Shuang, E.; Yamamoto, K.; Ota, T.; Seki, K.; Imai, M.; Ota, R.; Asayama, Y.; Nakashima, A.; Suzuki, K.; et al. Simultaneous intake of *Euglena gracilis* and vegetables synergistically exerts an anti-inflammatory effect and attenuates visceral fat accumulation by affecting gut microbiota in mice. *Nutrients* **2018**, *10*, 1417. [CrossRef]

18. Okouchi, R.; Shuang, E.; Yamamoto, K.; Ota, T.; Seki, K.; Imai, M.; Ota, R.; Asayama, Y.; Nakashima, A.; Suzuki, K.; et al. Simultaneous intake of *Euglena gracilis* and vegetables exerts synergistic anti-obesity and anti-inflammatory effects by modulating the gut microbiota in diet-induced obese mice. *Nutrients* **2019**, *11*, 204. [CrossRef]

19. Sugimoto, R.; Ishibashi-Ohgo, N.; Atsuji, K.; Miwa, Y.; Iwata, O.; Nakashima, A.; Suzuki, K. *Euglena* extract suppresses adipocyte-differentiation in human adipose-derived stem cells. *PLoS ONE* **2018**, *13*, 1–14. [CrossRef]

20. Brown, G.D.; Gordon, S. Immune recognition: A new receptor for β-glucans. *Nature* **2001**, *413*, 36–37. [CrossRef]

21. Nakashima, A.; Yamada, K.; Iwata, O.; Sugimoto, R.; Atsuji, K.; Ogawa, T.; Ishibashi-Ohgo, N.; Suzuki, K. β-Glucan in Foods and Its Physiological Functions. *J. Nutr. Sci. Vitaminol. (Tokyo)* **2018**, *64*, 8–17. [CrossRef]

22. Cohen-Kedar, S.; Baram, L.; Elad, H.; Brazowski, E.; Guzner-Gur, H.; Dotan, I. Human intestinal epithelial cells respond to β-glucans via Dectin-1 and Syk. *Eur. J. Immunol.* **2014**, *44*, 3729–3740. [CrossRef] [PubMed]

23. Macpherson, A.J.; Harris, N.L. Interactions between commensal intestinal bacteria and the immune system. *Nat. Rev. Immunol.* **2004**, *4*, 478–485. [CrossRef] [PubMed]

24. Yanagihara, S.; Kanaya, T.; Fukuda, S.; Nakato, G.; Hanazato, M.; Wu, X.R.; Yamamoto, N.; Ohno, H. Uromodulin-SlpA binding dictates *Lactobacillus acidophilus* uptake by intestinal epithelial M cells. *Int. Immunol.* **2017**, *29*, 357–363. [CrossRef] [PubMed]

25. Nagai, T.; Yamada, S.; Tominaga, T.; Ichikawa, M.; Miyawaki, A. Expanded dynamic range of fluorescent indicators for Ca^{2+} by circularly permuted yellow fluorescent proteins. *Proc. Natl. Acad. Sci. USA* **2004**, *101*, 10554–10559. [CrossRef]

26. Yoshikawa, S.; Usami, T.; Kikuta, J.; Ishii, M.; Sasano, T.; Sugiyama, K.; Furukawa, T.; Nakasho, E.; Takayanagi, H.; Tedder, T.F.; et al. Intravital imaging of Ca^{2+} signals in lymphocytes of Ca^{2+} biosensor transgenic mice: Indication of autoimmune diseases before the pathological onset. *Sci. Rep.* **2016**, *6*, 1–13. [CrossRef]

27. Kurosaki, T.; Shinohara, H.; Baba, Y.B. Cell Signaling and Fate Decision. *Annu. Rev. Immunol.* **2010**, *28*, 21–55. [CrossRef]

28. Nathanson, M.H. Cellular and subcellular calcium signaling in gastrointestinal epithelium. *Gastroenterology* **1994**, *106*, 1349–1364. [CrossRef]

29. Silva, A.J.; Paylor, R.; Wehner, J.M.; Tonegawa, S. Impaired spatial learning in α-calcium-calmodulin kinase II mutant mice. *Science* **1992**, *257*, 206–211. [CrossRef]

30. Feske, S. Calcium signalling in lymphocyte activation and disease. *Nat. Rev. Immunol.* **2007**, *7*, 690–702. [CrossRef]

31. Adachi, T.; Yoshikawa, S.; Tezuka, H.; Tsuji, N.M.; Ohteki, T.; Karasuyama, H.; Kumazawa, T. Propolis induces Ca^{2+} signaling in immune cells. *Biosci. Microbiota Food Health* **2019**, *38*, 141–149. [CrossRef]

32. Adachi, T.; Kakuta, S.; Aihara, Y.; Kamiya, T.; Watanabe, Y.; Osakabe, N.; Hazato, N.; Miyawaki, A.; Yoshikawa, S.; Usami, T.; et al. Visualization of probiotic-mediated Ca^{2+} signaling in intestinal epithelial cells *in vivo*. *Front. Immunol.* **2016**, *7*, 601. [CrossRef] [PubMed]

33. Elenkov, I.J.; Wilder, R.L.; Chrousos, G.P.; Vizi, E.S. The sympathetic nerve—An integrative interface between two supersystems: The brain and the immune system. *Pharmacol. Rev.* **2000**, *52*, 595–638. [PubMed]

34. Horii, Y.; Nakakita, Y.; Misonou, Y.; Nakamura, T.; Nagai, K. The serotonin receptor mediates changes in autonomic neurotransmission and gastrointestinal transit induced by heat-killed *Lactobacillus brevis* SBC8803. *Benef. Microbes* **2015**, *6*, 817–822. [CrossRef]

35. Nishimura, Y.; Fukuda, Y.; Okonogi, T.; Yoshikawa, S.; Karasuyama, H.; Osakabe, N.; Ikegaya, Y.; Sasaki, T.; Adachi, T. Dual real-time in vivo monitoring system of the brain-gut axis. *Biochem. Biophys. Res. Commun.* **2020**, *524*, 340–345. [CrossRef] [PubMed]

36. Okazawa, M.; Takao, K.; Hori, A.; Shiraki, T.; Matsumura, K.; Kobayashi, S. Ionic Basis of Cold Receptors Acting as Thermostats. *J. Neurosci.* **2002**, *22*, 3994–4001. [CrossRef]

37. Lavelle, E.C.; Murphy, C.; O'Neill, L.A.J.; Creagh, E.M. The role of TLRs, NLRs, and RLRs in mucosal innate immunity and homeostasis. *Mucosal Immunol.* **2010**, *3*, 17–28. [CrossRef]

38. Raka, F.; Farr, S.; Kelly, J.; Stoianov, A.; Adeli, K. Metabolic control via nutrient-sensing mechanisms: Role of taste receptors and the gut-brain neuroendocrine axis. *Am. J. Physiol. Endocrinol. Metab.* **2019**, *317*, E559–E572. [CrossRef]

39. Holst, J.J. The physiology of glucagon-like peptide 1. *Physiol. Rev.* **2007**, *87*, 1409–1439. [CrossRef]

40. Cani, P.; Delzenne, N. The Role of the Gut Microbiota in Energy Metabolism and Metabolic Disease. *Curr. Pharm. Des.* **2009**, *15*, 1546–1558. [CrossRef]

41. Shibakami, M.; Shibata, K.; Akashi, A.; Onaka, N.; Takezaki, J.; Tsubouchi, G.; Yoshikawa, H. Correction to: Creation of Straight-Chain Cationic Polysaccharide-Based Bile Salt Sequestrants Made from Euglenoid β-1,3-Glucan as Potential Antidiabetic Agents. *Pharm. Res.* **2019**, *36*, 31. [CrossRef]

42. Gerbe, F.; Legraverend, C.; Jay, P. The intestinal epithelium tuft cells: Specification and function. *Cell. Mol. Life Sci.* **2012**, *69*, 2907–2917. [CrossRef] [PubMed]

43. Vetvicka, V.; Vannucci, L.; Sima, P.; Richter, J. Beta glucan: Supplement or drug? From laboratory to clinical trials. *Molecules* **2019**, *24*, 1251. [CrossRef] [PubMed]

44. Suzuki, K.; Nakashima, A.; Igarashi, M.; Saito, K.; Konno, M.; Yamazaki, N.; Takimoto, H. *Euglena gracilis* Z and its carbohydrate storage substance relieve arthritis symptoms by modulating Th17 immunity. *PLoS ONE* **2018**, *13*, e0191462. [CrossRef] [PubMed]

45. Ishibashi, K.; Nishioka, M.; Onaka, N.; Takahashi, M.; Yamanaka, D.; Adachi, Y.; Ohno, N. Effects of *Euglena gracilis* EOD-1 Ingestion on Salivary IgA Reactivity and Health-Related Quality of Life in Humans. *Nutrients* **2019**, *11*, 1144. [CrossRef]

46. Ujita, M.; Nagayama, H.; Kanie, S.; Koike, S.; Ikeyama, Y.; Ozaki, T.; Okumura, H. Carbohydrate binding specificity of recombinant human macrophage β-glucan receptor dectin-1. *Biosci. Biotechnol. Biochem.* **2009**, *73*, 237–240. [CrossRef]

47. Kankkunen, P.; Teirilä, L.; Rintahaka, J.; Alenius, H.; Wolff, H.; Matikainen, S. (1,3)-β-Glucans Activate Both Dectin-1 and NLRP3 Inflammasome in Human Macrophages. *J. Immunol.* **2010**, *184*, 6335–6342. [CrossRef]

48. Evans, M.; Falcone, P.H.; Crowley, D.C.; Sulley, A.M.; Campbell, M.; Zakaria, N.; Lasrado, J.A.; Fritz, E.P.; Herrlinger, K.A. Effect of a *Euglena gracilis* Fermentate on Immune Function in Healthy, Active Adults: A Randomized, Double-Blind, Placebo-Controlled Trial. *Nutrients* **2019**, *11*, 2926. [CrossRef]

49. Ishiguro, S.; Upreti, D.; Robben, N.; Burghart, R.; Loyd, M.; Ogun, D.; Le, T.; Delzeit, J.; Nakashima, A.; Thakkar, R.; et al. Water extract from *Euglena gracilis* prevents lung carcinoma growth in mice by attenuation of the myeloid-derived cell population. *Biomed. Pharmacother.* **2020**, *127*, 110166. [CrossRef]

50. Riera, C.E.; Dillin, A. Emerging Role of Sensory Perception in Aging and Metabolism. *Trends Endocrinol. Metab.* **2016**, *27*, 294–303. [CrossRef]

51. Christianson, J.A.; Davis, B.M. The role of visceral afferents in disease. In *Translational Pain Research: From Mouse to Man*; CRC Press: Bosca Raton, FL, USA, 2009; pp. 51–76.

52. Assas, B.M.; Pennock, J.I.; Miyan, J.A. Calcitonin gene-related peptide is a key neurotransmitter in the neuro-immune axis. *Front. Neurosci.* **2014**, *8*, 23. [CrossRef]

53. Riera, C.E.; Huising, M.O.; Follett, P.; Leblanc, M.; Halloran, J.; Van Andel, R.; De Magalhaes Filho, C.D.; Merkwirth, C.; Dillin, A. TRPV1 pain receptors regulate longevity and metabolism by neuropeptide signaling. *Cell* **2014**, *157*, 1023–1036. [CrossRef] [PubMed]

54. Benemei, S.; Nicoletti, P.; Capone, J.G.; Geppetti, P. CGRP receptors in the control of pain and inflammation. *Curr. Opin. Pharmacol.* **2009**, *9*, 9–14. [CrossRef] [PubMed]

55. Ono, K.; Tsukamoto-Yasui, M.; Hara-Kimura, Y.; Inoue, N.; Nogusa, Y.; Okabe, Y.; Nagashima, K.; Kato, F. Intragastric administration of capsiate, a transient receptor potential channel agonist, triggers thermogenic sympathetic responses. *J. Appl. Physiol.* **2011**, *110*, 789–798. [CrossRef] [PubMed]

56. Bravo, J.A.; Forsythe, P.; Chew, M.V.; Escaravage, E.; Savignac, H.M.; Dinan, T.G.; Bienenstock, J.; Cryan, J.F. Ingestion of *Lactobacillus* strain regulates emotional behavior and central GABA receptor expression in a mouse via the vagus nerve. *Proc. Natl. Acad. Sci. USA* **2011**, *108*, 16050–16055. [CrossRef] [PubMed]

57. Harada, R.; Nomura, T.; Yamada, K.; Mochida, K. Genetic engineering strategies for *Euglena gracilis* and its industrial contribution to sustainable development goals: A review. *Front. Bioeng. Biotechnol.* **2020**, *8*, 1–10. [CrossRef]

58. Van Sadelhoff, J.H.J.; Pardo, P.P.; Wu, J.; Garssen, J.; Van Bergenhenegouwen, J.; Hogenkamp, A.; Hartog, A.; Kraneveld, A.D. The gut-immune-brain axis in autism spectrum disorders; a focus on amino acids. *Front. Endocrinol. (Lausanne)* **2019**, *10*, 247. [CrossRef]

Article

Microarray Analysis of Paramylon, Isolated from Euglena Gracilis EOD-1, and Its Effects on Lipid Metabolism in the Ileum and Liver in Diet-Induced Obese Mice

Seiichiro Aoe [1,2,*], Chiemi Yamanaka [1] and Kento Mio [2]

[1]   The Institute of Human Culture Studies, Otsuma Women's University, Chiyoda-ku, Tokyo 102-8357, Japan; chiemiy@gmail.com
[2]   Studies in Human Life Sciences, Graduate School of Studies in Human Culture, Otsuma Women's University, Chiyoda-ku, Tokyo 102-8357, Japan; d5120401@cst.otsuma.ac.jp
*   Correspondence: s-aoe@otsuma.ac.jp; Tel.: +81-3-5275-6048

Abstract: We previously showed that supplementation of a high fat diet with paramylon (PM) reduces the postprandial glucose rise, serum total and LDL cholesterol levels, and abdominal fat accumulation in mice. The purpose of this study was to explore the underlying mechanism of PM using microarray analysis. Male mice (C57BL/BL strain) were fed an experimental diet (50% fat energy) containing 5% PM isolated from *Euglena gracilis* EOD-1 for 12 weeks. After confirming that PM had an improving effect on lipid metabolism, we assessed ileal and hepatic mRNA expression using DNA microarray and subsequent analysis by gene ontology (GO) classification and Kyoto Encyclopedia of Genes and Genomes (KEGG) pathway analysis. The results suggested that dietary supplementation with PM resulted in decreased abdominal fat accumulation and serum LDL cholesterol concentrations via suppression of the digestion and absorption pathway in the ileum and activation of the hepatic PPAR signaling pathway. Postprandial glucose rise was reduced in mice fed PM, whereas changes in the glucose metabolism pathway were not detected in GO classification and KEGG pathway analysis. PM intake might enhance serum secretory immunoglobulin A concentrations via promotion of the immunoglobulin production pathway in the ileum.

Keywords: paramylon; abdominal fat; DNA microarray; gene ontology; PPAR signaling

Citation: Aoe, S.; Yamanaka, C.; Mio, K. Microarray Analysis of Paramylon, Isolated from Euglena Gracilis EOD-1, and Its Effects on Lipid Metabolism in the Ileum and Liver in Diet-Induced Obese Mice. *Nutrients* **2021**, *13*, 3406. https://doi.org/10.3390/nu13103406

Academic Editor: Takao Kimura

Received: 30 July 2021
Accepted: 25 September 2021
Published: 27 September 2021

Publisher's Note: MDPI stays neutral with regard to jurisdictional claims in published maps and institutional affiliations.

1. Introduction

Obesity and overweight are worldwide problems caused by lifestyle and genetic backgrounds [1]. Abdominal fat obesity causes a pro-inflammatory status in several organs, such as the liver, adipose tissue, and the pancreas [2]. Chronic inflammation induces glucose intolerance, dyslipidemia, and an impaired immune defense system, which suggests that lifestyle changes, especially dietary habits, are very important. Dietary fiber is a beneficial food component, which is reported to improve obesity-related diseases [3]. Viscous or fermentable dietary fibers improve glucose intolerance [4], prevent body-weight gain [5], and improve dyslipidemia [6,7].

Beta-glucans are dietary fibers with very different physiological functions, depending on their origin and structure. Beta-1,3-1,4-glucan is a typical viscous and fermentable dietary fiber found in cereals; it is reported to play a crucial role in the modification of obesity-related disorders. Oat β-glucan reduces postprandial glucose increases [8,9] and serum LDL cholesterol levels in hypercholesterolemic subjects [9]. Barley β-glucan reduces the abdominal fat area [10] and improves glucose intolerance [11]. Beta-1,3-1,6-glucan, found in yeast and mushroom, is an almost insoluble dietary fiber, reported to enhance the immune response via dectin-1 [12].

Paramylon (PM) is a β-1,3-glucan, found in *Euglena gracilis* at levels of 20–70% (dry weight basis). It has been reported that PM has various effects, including an immunomodulating effect [13–15]. β-1,3-glucan has a triple helix structure and is classified as an insoluble dietary fiber [16]; therefore, it is considered that the physiological properties of PM differ from those of cereal β-glucan or yeast and fungal β-glucan.

There are very few reports on the effects of PM intake on metabolic disorders. Recent studies reported that PM did not have anti-obesity or anti-inflammatory effects in mice fed a high fat diet [17,18]; however, this may have been due to insufficient PM supplementation of the experimental diets [19]. We previously reported that PM had a beneficial effect in preventing obesity related parameters; in particular, abdominal fat accumulation and serum LDL cholesterol levels. We also observed an amelioration in postprandial glucose levels [20]. The mechanism underlying the anti-obesity effect of PM intake is unknown; the purpose of this study was to investigate the potential mechanism of PM modulation of lipid metabolism using DNA microarray with gene ontology (GO) classification and Kyoto Encyclopedia of Genes and Genome (KEGG) pathway analysis. It is reported that "GO characterizes the relationship between genes by specifically annotating and categorizing a gene product's molecular function and associated biological process" [21]. The KEGG pathway is a collection of pathway maps which includes molecular interactions, reaction and relationship networks for metabolism, genetic information processing, and cellular processes [22].

2. Materials and Methods

2.1. PM Isolation and Dietary Fiber Analysis

Kobelco Eco-Solutions Co., Ltd., (Kobe, Japan) provided PM powder isolated from *Euglena gracilis* EOD-1, which is known for its high yields of PM [23]. In accordance with previously reported culture conditions, glucose was used as the carbon source [20]. Total dietary fiber was 97.8 g/100 g, determined by the AOAC 991.43 method [24].

2.2. Animals and Study Design

Twenty C57BL/6J mice (4 week old, male) were purchased from Charles River Laboratories Japan (Yokohama, Japan). They were maintained on a 12 h light/dark cycle from 8:00 to 20:00. Mice were individually housed in plastic carbonate cages and given ad libitum access to water. This animal experiment was approved by the Otsuma Women's University Animal Research Committee (No. 19003, Tokyo, Japan) and was performed in accordance with the Regulation on Animal Experimentation at Otsuma Women's University. Mice were pre-reared for 7 d before being randomly divided into two groups (n = 10 per group) and fed the experimental diet containing 50 g/kg cellulose (control) or 51.2 g/kg PM (corresponding to 5% dietary fiber). Dietary compositions are shown in Table 1. The experimental diets were given to mice ad libitum for 12 weeks. The food intake and body weight of each mouse were measured three times per week during the experimental period. After eight hours of fasting, blood samples were withdrawn from the heart after isoflurane/CO_2 euthanasia. Serum was collected by centrifugation for lipid analysis. Liver, cecum with digesta, and abdominal fat (epididymal, retroperitoneal, and mesenteric fat) were dissected and weighed. Liver tissue and tissue of the ileum near the cecum (150 mg), were soaked in RNAprotect reagents (Qiagen, Hilden, Germany), and a freeze-dried residue of liver tissue was stored at −20 °C until cholesterol and triglyceride assays were performed.

2.3. Serum and Hepatic Biochemical Analysis

Hitachi 7180 auto-analyzers were used to analyze serum total, low-density lipoprotein (LDL), and high-density lipoprotein (HDL) cholesterols, triglycerides, non-esterified fatty acids (NEFA), and ketone body concentrations at Oriental Yeast Co., Ltd. (Shiga, Japan). Enzyme-linked immunosorbent assay (ELISA) was used to measure serum insulin (mouse insulin ELISA kit, Shibayagi Co., Ltd., Gunma, Japan) and sIgA concentrations (ELISA Kit for Secretory Immunoglobulin A, Cloud-Clone Corp., Katy, TX, USA). Hepatic cholesterol and triglyceride levels were analyzed according to a previous report [20,25].

Table 1. Test diet compositions (g/kg).

	Control	Paramylon
Corn starch	197.5	196.3
Dextrinized corn starch	132	132
Casein	200	200
Sucrose	100	100
Soybean oil	70	70
Lard	200	200
Cellulose	50	-
Paramylon	-	51.2
AIN-93G mineral mixture	35	35
AIN-93 vitamin mixture	10	10
L-cystine	3	3
Choline bitartrate	2.5	2.5
t-butylhydroquinone	0.014	0.014

2.4. Oral Glucose Tolerance Test

The experimental diets were given to mice for 12 weeks and an oral glucose tolerance test (OGTT) was performed, after 8 h fasting, during the final week. After oral glucose gavage (1.5 g/kg), a glucometer (Glutest Ace R, Sanwa Kagaku Kenkyusho Co., Ltd., Aichi, Japan) was used to measure blood glucose levels from the tail tip at 0, 15, 30, 60 and 120 min.

2.5. DNA Microarray Analysis

Total RNA in ileum and liver tissues were extracted using the RNeasy Mini kit (Qiagen, Hilden, Germany), according to the manufacturer's instructions. The RNA integrity number (RIN) of all total RNA samples was analyzed and samples with values greater than 6.5 were pooled by equal mixing for each group ($n = 10$ per group) before DNA microarray analysis [26]. Procedures for microarray analysis were performed according to the manufacturer's instructions and are described in the Supplementary File. The Gene Expression Omnibus (GEO) accession number is GSE181289.

Expression measurement annotations for upregulated and downregulated genes in each group probe were mapped to gene names using the Transcriptome Viewer (Kurabo Industries Ltd., Neyagawa, Osaka, Japan). Genes were regarded as differentially expressed when their fold change was higher than 1.3 or less than 0.77. All of the differentially expressed genes (DEGs) were characterized by their biological processes, cellular components, and molecular functions. GO enrichments were analyzed using clusterProfiler V3.6.0 (ver. 1.3.1093, R-Tools Technology Inc. (Richmond Hill, ON, Canada)). In the present study, the pathview package in R studio was used to identify and visualize the KEGG (Kyoto Encyclopedia of Genes and Genomes) pathways of the DEGs involved in lipid metabolism.

2.6. Expression Analysis of mRNAs Related to Lipid Metabolism in Liver and Ileum

Real-time PCR primer sequences are listed in Supplementary Table S1. mRNAs involved in liver and ileal metabolism were analyzed using SYBR® Green PCR master mix (Thermo Fisher Scientific, Waltham, MA, USA) and the Applied Biosystems Quant3 real-time polymerase chain reaction (PCR) system. Data analysis was performed using the $2^{-\Delta\Delta CT}$ method, where the threshold cycle (CT) indicates the fractional cycle number at which the amount of amplified target reaches a fixed threshold. The ΔCT was calculated from the difference of threshold cycles between reference gene (36B4) and target genes. The difference between the ΔCT for test diet and control diet determined the $\Delta\Delta CT$. Relative expression levels are expressed as fold changes from the control group (arbitrary unit/36B4).

2.7. Statistical Analysis

Based on our previous measurements [20], a total of 20 mice (10 mice per group) were determined to be required in this study. Statistical data were presented as mean ± SE (standard error of the mean). Significant differences ($p < 0.05$) between group means were determined by Student's *t*-test or Mann-Whitney U test. JMP (version 15.0, SAS Institute Inc., Cary, NC, USA) and R software were used for the statistical analyses.

3. Results

3.1. Food Intake, Body Weight and Organ Weight

Table 2 shows the growth parameters (food intake, body weight, and food efficiency ratio) in mice fed PM. The final weight and body weight gain were almost equal between the two groups. The organ weights in mice fed paramylon are shown in Table 3. Retroperitoneal and epididymal fat accumulation were lower in the PM group in comparison with the control group ($p < 0.05$). No significant differences in the weights of the liver and the cecum, and in their contents, were observed between the groups.

Table 2. Body weight gain, food intake, food efficiency ratio.

	Control	Paramylon
Initial weight (g)	20.7 ± 0.2	20.7 ± 0.2
Final weight (g)	43.4 ± 0.7	42.3 ± 0.8
Body weight gain(g/day)	0.27 ± 0.01	0.25 ± 0.01
Food intake (g/day)	2.83 ± 0.03	2.85 ± 0.03
Food efficiency ratio (%) *	9.42 ± 0.19	8.93 ± 0.28

Values are means ± SE; * food efficiency ratio = body weight gain/food intake × 100.

Table 3. Weight of organs.

	Control	Paramylon
Liver (g)	1.47 ± 0.05	1.42 ± 0.05
Cecum with digesta (g)	0.27 ± 0.02	0.30 ± 0.03
Retroperitoneal fat (g)	0.98 ± 0.07	0.77 ± 0.09 *
Epididymal fat (g)	2.49 ± 0.06	2.33 ± 0.06 *
Mesenteric fat (g)	0.83 ± 0.07	0.73 ± 0.07

Values are means ± SE; * significantly different from the control group ($p < 0.05$).

3.2. Oral Glucose Tolerance Test (OGTT)

Figure 1 shows the following results of the OGTT: blood glucose levels were significantly lower at 15 min and 120 min in the PM intake group in comparison with the control ($p < 0.05$). Blood glucose levels at the other time points or in the IAUC (data not shown) were not statistically significant between the groups.

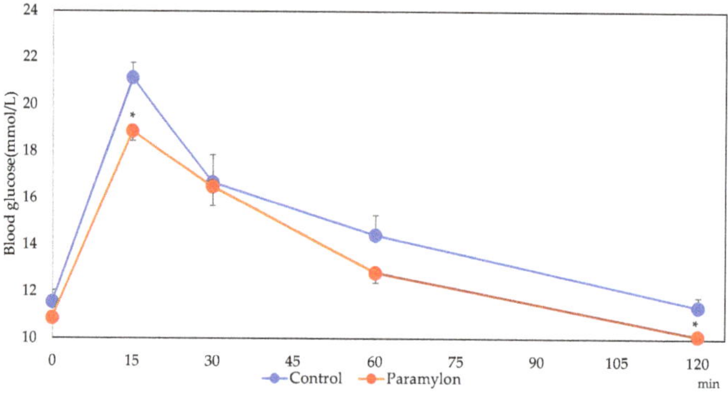

Figure 1. Blood glucose levels from the OGTT. Error bars represent standard error ($n = 10$). * Blood glucose levels were significantly lower at 15 min and 120 min in the PM group when compared with the control group ($p < 0.05$).

3.3. Biochemical Analysis of Serum and Liver Lipids

Table 4 shows the serum lipid concentrations. Significant reductions in the serum total and LDL cholesterol levels were observed in the PM group in comparison with the control group ($p < 0.05$). No significant differences in other serum lipids, ketone body, and insulin concentrations were observed between the two groups. The Hepatic lipid contents are shown in Supplementary Table S2; no significant differences in cholesterol and triglyceride contents were observed between the experimental groups. The Serum sIgA concentrations are shown in Figure 2. The Serum sIgA concentrations were significantly higher in the PM group when compared with the control group ($p < 0.05$).

Table 4. Serum lipid and insulin concentrations.

	Control	Paramylon
Total cholesterol (mmol/L)	5.06 ± 0.11	4.60 ± 0.16 *
LDL-cholesterol (mmol/L)	0.27 ± 0.02	0.21 ± 0.01 *
HDL-cholesterol (mmol/L)	2.17 ± 0.03	2.10 ± 0.04
Triglyceride (mmol/L)	0.76 ± 0.07	0.67 ± 0.04
NEFA (µmol/L)	536.9 ± 19.3	675.3 ± 26.2
Ketone body (µmol/L)	296.4 ± 44.6	330.6 ± 59.8
Insulin (ng/mL)	3.5 ± 0.6	2.2 ± 0.4

Values are means $\pm$ SE; * Significantly different from the control group ($p < 0.05$).

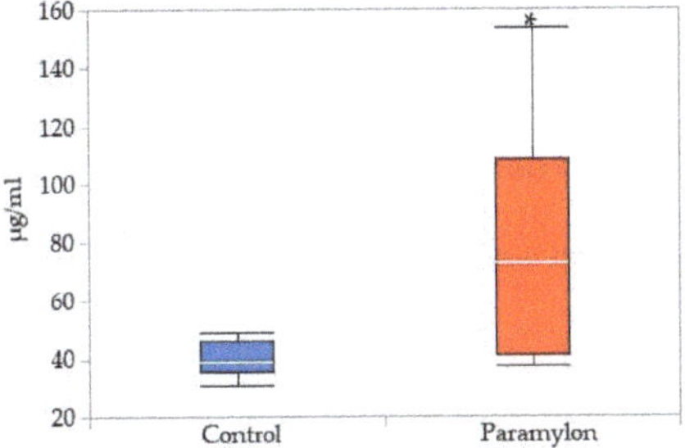

Figure 2. Serum sIgA concentrations. * Significantly different from the control group (Mann-Whitney test, $p < 0.05$).

3.4. Gene Ontology (GO) and KEGG Pathway Analysis of DEGs

A total of 41,345 mRNAs were screened. In the ileum, 1449 mRNAs were upregulated (>1.3 log-ratio), and 1614 mRNAs were downregulated (<0.77 log-ratio). In the liver, 1454 mRNAs were upregulated (>1.3 log-ratio), and 1452 mRNAs were downregulated (<0.77 log-ratio). GO enrichment analysis of the liver showed that the top 15 GO terms in DEGs with a greater than 1.3 (log-ratio) fold change included 'cholesterol metabolic process', 'secondary alcohol metabolic process', and 'sterol metabolic process' (Figure 3). No GO terms for DEGs with a less than 0.77 (log-ratio) fold change were detected in the liver.

GO enrichment analysis of the ileum showed that almost all of the top 15 GO terms in DEGs with a greater than 1.3 (log-ratio) fold change were involved in the immune response (Figure 4). GO terms for DEGs with a less than 0.77 (log-ratio) fold change were detected in the ileum and showed that the top 15 GO terms included 'lipid catabolic process', 'cellular lipid catabolic process', 'fatty acid metabolic process', 'digestion', 'positive regulation of triglyceride lipase activity', and 'positive regulation of lipid catabolic process' (Figure 5). Furthermore, KEGG pathway analysis indicated that the upregulated pathways in the liver included the 'PPAR signaling pathway', 'p53 signaling pathway', 'steroid biosynthesis pathway', and 'fatty acid degradation pathway' (Figure 6). The downregulated pathways in the ileum included 'fat digestion and absorption', 'protein digestion and absorption', 'PPAR signaling pathway', 'cholesterol metabolism', 'glycerolipid metabolism', 'steroid biosynthesis', and 'fatty acid degradation'.

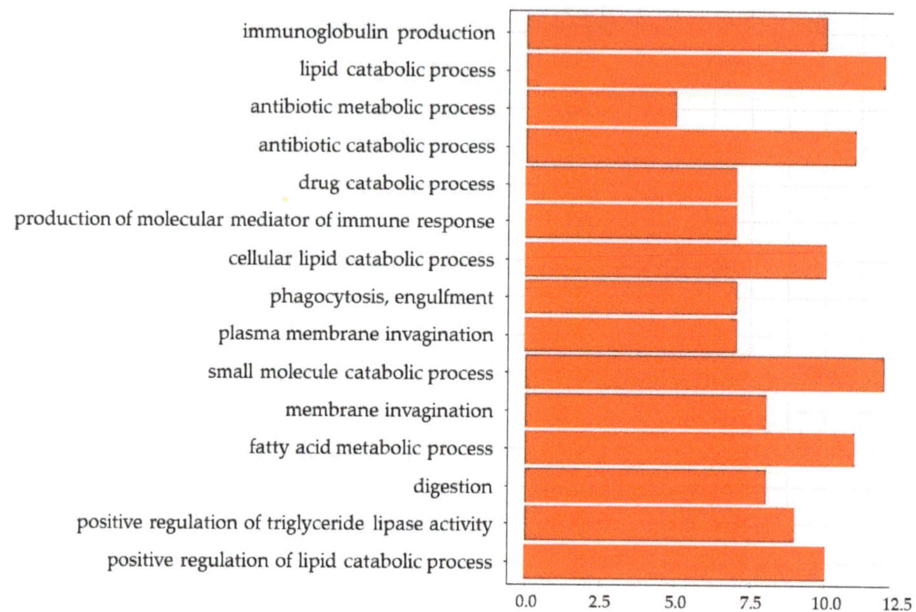

Figure 3. The top 15 GO terms from the Disease and Gene Annotations (DGAs) for upregulated genes in the liver.

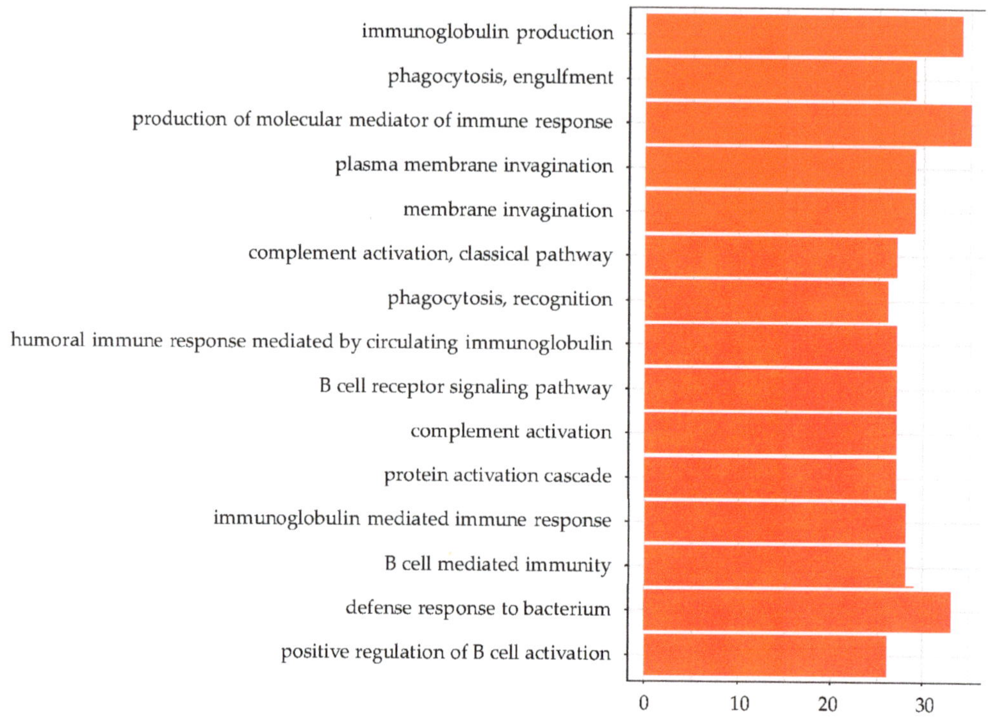

Figure 4. The top 15 GO terms from the DGAs for upregulated genes in the ileum.

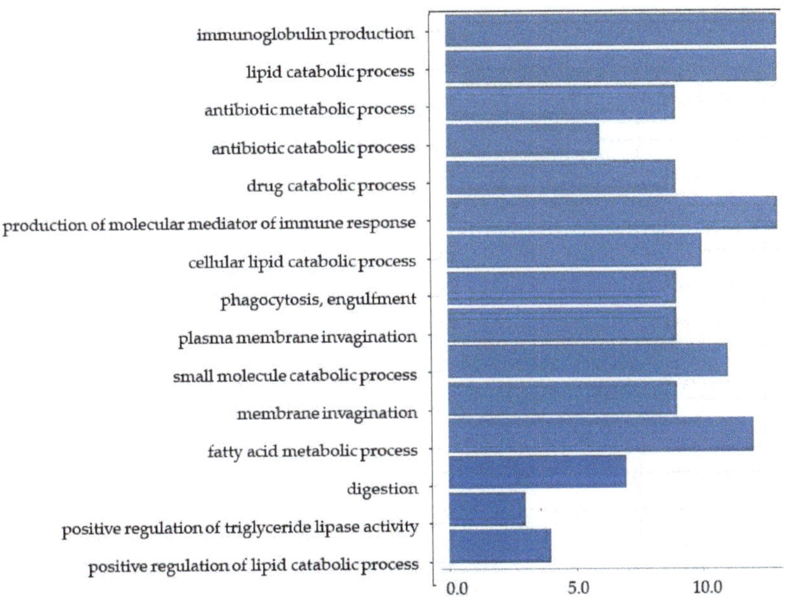

Figure 5. The top 15 GO terms from the DGAs for downregulated genes in the ileum.

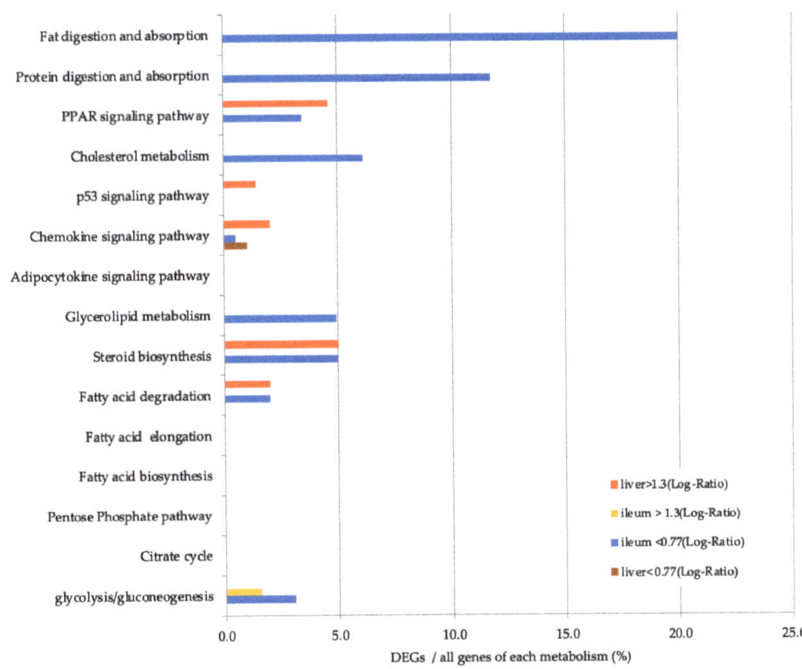

Figure 6. Percentage of upregulated (>1.3-fold difference to the control) and downregulated (<0.77-fold difference to the control group) differentially expressed genes (DEGs) in the liver and ileum. The y-axis shows the lipid metabolic pathway described in the Kyoto Encyclopedia of Gene and Genome (KEGG). The x-axis shows the percentage of DEGs per total genes involved in each pathway.

3.5. mRNA Expression Levels of Genes Involved in Lipid Metabolism in the Liver and Ileum

mRNA expression levels in the liver (a) and ileum (b) are shown in Figure 7. PPARα mRNA expression levels were significantly higher in the livers of mice in the PM group when compared with mice in the control group ($p < 0.05$; Figure 7a). SOD mRNA expression levels were significantly higher in the PM group in comparison with the control group ($p < 0.05$). No significant differences in the mRNA expression levels of the other genes analyzed were observed in the liver. PPARγ mRNA expression levels in the ileum were significantly lower in the PM group in comparison with the control group ($p < 0.05$; Figure 7b). The mRNA expression levels of AQP3 and AQP4 were significantly higher in the PM group, in comparison with the control group ($p < 0.05$). No significant differences in the mRNA expression levels of the other genes analyzed were observed in the ileum.

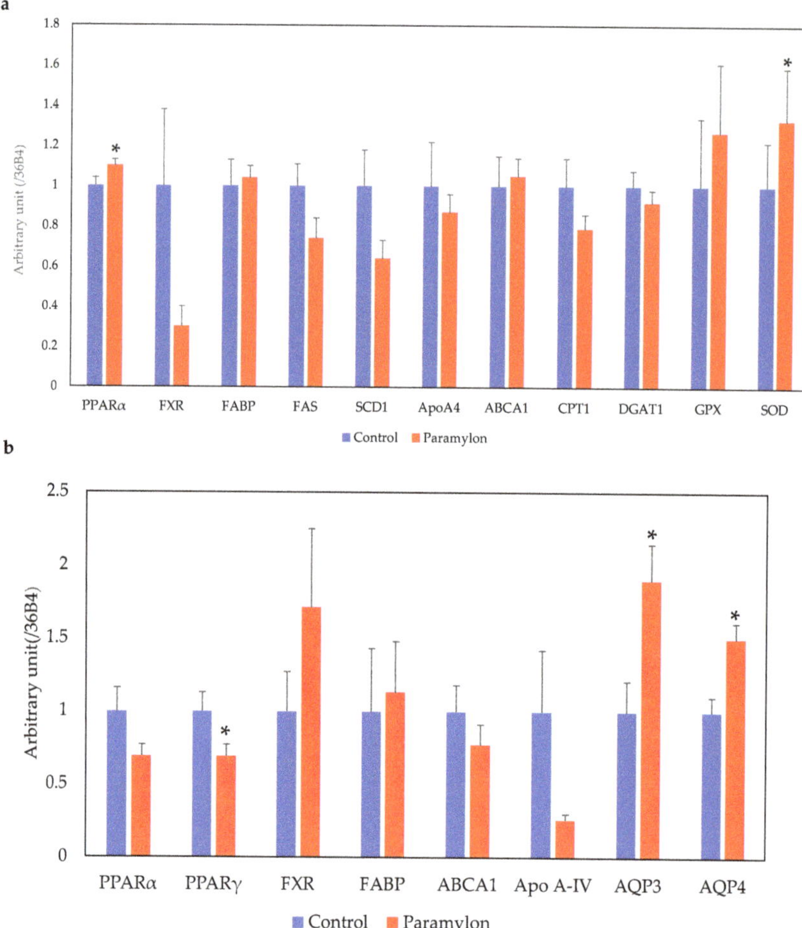

Figure 7. Relative expression levels of mRNAs involved in lipid metabolism in the liver (**a**) and ileum (**b**). Error bars represent standard error ($n = 10$). * Significantly different from the control group ($p < 0.05$). PPARα, peroxisome proliferator-activated receptor α; PPARγ, peroxisome proliferator-activated receptor γ; FXR, farnesoid X receptor; SCD1, stearoyl-CoA desaturase 1; Apo A-IV, apolipoprotein A-IV; ABCA1, ATP-binding cassette transporter A1; CPT1, carnitine palmitoyltransferase 1; DGAT1, diacylglycerol O-acyltransferase 1; GPX, glutathione peroxidase; SOD, superoxide dismutase; AQP 3,4, aquaporin 3,4.

4. Discussion

4.1. Fat Digestion and Absorption Pathway in the Ileum

The purpose of this study was to explore the potential mechanism of lipid metabolism modulation by PM, using GO classification and KEGG pathway analysis. Our previous report showed that PM intake reduced postprandial glucose levels and serum LDL cholesterol, and increased serum sIgA concentrations in mice, when compared with control mice [20]. The food efficiency ratio calculated from body-weight gain and food intake tended to be lower in the PM group (statistically not significant), which is similar to our previous report [20]. In addition, although no significant difference was only observed in mesenteric fat weight, the total abdominal fat weight (epididymal, retroperitoneal, and mesenteric fat) was significantly decreased in the PM group (data not shown). We confirmed that PM improved lipid metabolism in mice fed a high-fat diet, which was consistent with our previous report [20], before continuing with microarray analysis. This is the first report analyzing mRNA expression in the ileum, after PM intake. Interactions between indigestible substances, such as dietary fiber and intestinal cells, are more likely to occur in the ileum than the jejunum. In addition, the ileum contains both bile acid transporters and many L cells, which contain several receptors. The ileum is also affected by digesta. Fold changes of >1.3 and <0.77 were adopted in the microarray analysis, which is a range that has previously been used [27–29]. GO classification and KEGG pathway analysis consistently indicated that fat digestion and absorption was a key downregulated pathway in the ileum. Our previous report suggested that PM intake did not inhibit dietary fat absorption and did not affect microbiota constitution [20]; therefore, PM might modify the absorption process, and then inhibit chylomicron secretion. KEGG analysis of the fat digestion and absorption pathway suggested that PM suppressed the expression levels of scavenger receptor B1, FABP, ACAT, Apo A-IV, and ABCA1. A reduction in chylomicron secretion is a possible mechanism in reducing LDL cholesterol and abdominal fat accumulation. This is a new hypothesis, showing a reduction in the fat digestion and absorption pathway in the digestive tract after PM intake. However, serum lipid analysis in the chylomicron fraction was not determined. Further studies are needed to prove this hypothesis regarding chylomicron metabolism in the small intestine. We intend to investigate the influence of PM intake on chylomicron metabolism in future studies.

4.2. PPAR Signaling Pathway in the Ileum

KEGG pathway analysis showed downregulation of the PPAR signaling pathway in the ileum. A significant reduction in the expression of PPARγ mRNA in the ileum was observed in the PM group, in comparison with the control group ($p < 0.05$), and the same tendency was observed for PPARα (not statistically significant). PPARγ is a nuclear receptor that is involved in regulating lipid metabolism under high-fat diet conditions [30]. It is activated by fatty acids and their derivatives, and creates a lipid signaling network in the intestines, which are prone to inflammation when lipid metabolism is disturbed under high-fat diet conditions [31]. The expression of PPARs might be reduced through the reduction in fatty acid absorption; however, the expression levels of genes regulated by PPARs were not significantly altered. The data suggest that the contribution of PPARs on lipid metabolism in the ileum might be small.

4.3. Water Absorption Pathway in the Ileum

KEGG pathway analysis also showed upregulation of the water reabsorption pathway after PM intake. The expression levels of AQP3 and AQP4 mRNA were significantly increased in the PM group compared to the control group ($p < 0.05$), but that of APQ2 was not (data not shown). AQP3 and AQP4 have been shown to be expressed in the gastrointestinal tract [32], and preliminary data suggested that aquaporin has a role in dietary fat processing [32]. The expression of aquaporins in the intestinal tract suggests that they are involved in the transport of intestinal fluid; however, there are very few studies reporting on their functional significance. Further studies are needed to elucidate that PM may affect intestinal epithelial cells and modify the function of the gastrointestinal tract, through AQPs.

4.4. Immunoglobulin Production

GO classification analysis showed many enhanced GO terms for immunoglobulin production in the ileum and liver. KEGG pathway analysis highlighted 'TNF receptor superfamily member 13C (BAFFR)', 'major histocompatibility complex, class II (MHC class II)', and 'CC motif chemokine ligand 28 (CCL28)', suggesting an upregulated intestinal immune network for IgA production. Serum sIgA concentrations were elevated by PM intake, which was consistent with our previous report [20]. PM may have an activating effect on intestinal epithelial cells, promoting the secretion of sIgA [33]. A human study, involving the intake of *Euglena gracilis* EOD-1 biomass that was rich in PM, reported the production of PM-specific IgA antibodies and increased salivary IgA antibody titers [34]. Our data are in agreement with these previous reports. Further studies on the analysis of protein expression levels and intracellular signaling are needed to prove the activation of the IgA production pathway in the ileum.

4.5. PPAR Signaling Pathway in the Liver

GO classification analysis in the liver showed that the upregulated GO terms included 'cholesterol metabolic process', 'secondary alcohol metabolic process', and 'sterol metabolic process'. KEGG pathway analysis showed that the upregulated pathways included the 'PPAR signaling pathway', 'steroid biosynthesis pathway', and 'fatty acid degradation pathway'. Hepatic PPARα mRNA expression was significantly increased in the PM group, when compared with the control group ($p < 0.05$). It is speculated that the mRNA expression levels of ileal PPARs were decreased by a reduction in intestinal fatty acid absorption, whereas an increase in hepatic PPARα mRNA expression might be caused by an increase in abdominal fat degradation during the fasting period. The activation of PPARα might be a key pathway in hepatic metabolism, affecting downstream mRNA in the lipid catabolic process. We previously showed that PM intake increased hepatic PPARα mRNA expression and the subsequent induction of β-oxidation, through activation of ACOX, CPT and FATP2 mRNA expression [20]. We speculated that changes in fatty acid metabolism, through upregulation of the PPAR signaling pathway, may improve lipid and glucose metabolism. In this study, no significant differences in the expression levels of genes involved in lipid metabolism were detected between the PM and control groups; however, correlations were observed between PPARα mRNA and mRNA from genes involved in the pathway downstream of PPARα. It is well documented that PPARα exhibits significant anti-inflammatory properties [35]. The hepatic mRNA expression levels of SOD were significantly increased in the PM group, compared with the control group; GPX expression levels were also increased (not statistically significant). Therefore, upregulation of hepatic PPARα expression after PM intake plays a key role in anti-inflammation and modifying lipid metabolism.

4.6. Glucose Tolerance

PM reduced blood glucose levels at 15 min and 120 min in the OGTT, whereas PM reduced blood glucose levels at 60 min in the previous report [20]. The peak postprandial blood glucose rise was higher in this study than in the previous report, and the fasting serum insulin concentrations were also higher in this study than in the previous report. Therefore, it is possible that the experimental conditions of mice in this study were more sensitive to detecting a significant difference in the OGTT. Significant differences in serum insulin concentrations between groups were not detected in this study or the previous study. The previous study also showed that the mRNA expression of SREBP1c, controlled by insulin concentrations, was not affected by PM intake. Therefore, the contribution of serum insulin concentrations on lipid metabolism, by PM intake, may be small. However, since the insulin concentrations in the OGTT were not measured, and the insulin tolerance test was not performed, it was not possible to clarify how paramylon affected insulin sensitivity or resistance.

A major limitation of this study was that the possible mechanism for lipid metabolism was only speculated by mRNA expressions. A second limitation was that insulin concentrations were not measured during the OGTT test, to assess insulin resistance. Future studies are needed to elucidate the mechanism for glucose and lipid metabolism, by exploring the protein expression levels and insulin secretion.

5. Conclusions

In conclusion, PM intake reduced serum total and LDL cholesterol levels, and abdominal fat accumulation via suppression of the digestion and absorption pathway in the ileum, and activation of the hepatic PPAR signaling pathway. Changes in the glucose metabolism pathway, by PM intake, were not detected in the GO classification and KEGG pathway analysis. PM intake might enhance sIgA concentrations via promotion of the immunoglobulin production pathway in the ileum. However, key molecules, such as endogenous ligands, to upregulate hepatic PPARα expression and intracellular signaling related to immunoglobulin production, were not determined. Further investigation is required to elucidate the molecular mechanism behind the improvement in lipid metabolism by PM intake. The improvement in lipid metabolism and enhanced gastrointestinal immune function suggests that PM intake may be useful in people who are sensitive to intestinal fermentation, such as those with gut fermentation syndrome.

Supplementary Materials: The following are available online at https://www.mdpi.com/2072-6643/13/10/3406/s1. Supplementary Materials and Methods microarray analysis; Table S1: primers used for real-time reverse transcription polymerase chain reaction; Supplementary Table S2: liver lipid levels.

Author Contributions: Conceptualization, S.A.; data curation, S.A. and K.M.; formal analysis, C.Y.; investigation, S.A. and C.Y.; methodology, C.Y. and K.M.; project administration, S.A.; supervision, S.A.; writing—original draft, S.A. and K.M.; writing—review and editing, S.A. and K.M. All authors have read and agreed to the published version of the manuscript.

Funding: This study was financially supported by Kobelco Eco-Solutions Co., Ltd. (Kobe, Japan).

Institutional Review Board Statement: This animal experiment was approved by the Otsuma Women's University Animal Research Committee (No. 19003, Tokyo, Japan) and was performed in accordance with the Regulation on Animal Experimentation at Otsuma Women's University.

Conflicts of Interest: The authors (S.A., C.Y. and K.M.) have no conflict of interest to disclose. The PM used in this study was provided by Kobelco Eco-Solutions Co., Ltd. This company had no control over the interpretation, writing, or publication of this work.

References

1. Caballero, B. Humans against Obesity: Who Will Win? *Adv. Nutr.* **2019**, *10*, S4–S9. [CrossRef]
2. Vecchié, A.; Dallegri, F.; Carbone, F.; Bonaventura, A.; Liberale, L.; Portincasa, P.; Frühbeck, G.; Montecucco, F. Obesity phenotypes and their paradoxical association with cardiovascular diseases. *Eur. J. Intern. Med.* **2018**, *48*, 6–17. [CrossRef] [PubMed]
3. Chen, J.-P.; Chen, G.-C.; Wang, X.-P.; Qin, L.; Bai, Y. Dietary Fiber and Metabolic Syndrome: A Meta-Analysis and Review of Related Mechanisms. *Nutrients* **2018**, *10*, 24. [CrossRef] [PubMed]
4. Tucker, L.A.; Thomas, K.S. Increasing total fiber intake reduces risk of weight and fat gains in women. *J. Nutr.* **2009**, *139*, 576–581. [CrossRef] [PubMed]
5. Chandalia, M.; Garg, A.; Lutjohann, D.; von Bergmann, K.; Grundy, S.M.; Brinkley, L.J. Beneficial effects of high dietary fiber intake in patients with type 2 diabetes mellitus. *N. Engl. J. Med.* **2000**, *342*, 1392–1398. [CrossRef] [PubMed]
6. Jenkins, D.J.A.; Kendall, C.W.C.; Vuksan, V.; Vidgen, E.; Parker, T.; Faulkner, D.; Mehling, C.C.; Garsetti, M.; Testolin, G.; Cunnane, S.C.; et al. Soluble fiber intake at a dose approved by the US Food and Drug Administration for a claim of health benefits: Serum lipid risk factors for cardiovascular disease assessed in a randomized controlled crossover trial. *Am. J. Clin. Nutr.* **2002**, *75*, 834–839. [CrossRef] [PubMed]
7. Brown, L.; Rosner, B.; Willett, W.W.; Sacks, F.M. Cholesterol-lowering effects of dietary fiber: A meta-analysis. *Am. J. Clin. Nutr.* **1999**, *69*, 30–42. [CrossRef]

8. Zurbau, A.; Noronha, J.C.; Khan, T.A.; Sievenpiper, J.L.; Wolever, T.M.S. The effect of oat β-glucan on postprandial blood glucose and insulin responses: A systematic review and meta-analysis. *Eur. J. Clin. Nutr.* **2021**. Available online: https://www.nature.com/articles/s41430-021-00875-9 (accessed on 25 July 2021). [CrossRef]

9. Tosh, S.M.; Bordenave, N. Emerging science on benefits of whole grain oat and barley and their soluble dietary fibers for heart health, glycemic response, and gut microbiota. *Nutr. Rev.* **2020**, *78*, 13–20. [CrossRef]

10. Aoe, S.; Ichinose, Y.; Kohyama, N.; Komae, K.; Takahashi, A.; Abe, D.; Yoshioka, T.; Yanagisawa, T. Effects of high beta-glucan barley on visceral fat obesity in Japanese individuals: A randomized, double-blind study. *Nutritions* **2017**, *42*, 1–6.

11. Nilsson, A.C.; Ostman, E.M.; Holst, J.J.; Björck, I.M. Including indigestible carbohydrates in the evening meal of healthy subjects improves glucose tolerance, lowers inflammatory markers, and increases satiety after a subsequent standardized breakfast. *J. Nutr.* **2008**, *138*, 732–739. [CrossRef]

12. Goodridge, H.S.; Wolf, A.J.; Underhill, D.M. Beta-glucan recognition by the innate immune system. *Immunol. Rev.* **2009**, *230*, 38–50. [CrossRef]

13. Kondo, Y.; Kato, A.; Hojo, H.; Nozoe, S.; Takeuchi, M.; Ochi, K. Cytokine-related immunopotentiating activities of paramylon, a beta-(1→3)-D-glucan from *Euglena gracilis*. *J. Pharm.-Dyn.* **1992**, *15*, 617–621. [CrossRef] [PubMed]

14. Nakashima, A.; Sasaki, K.; Sasaki, D.; Yasuda, K.; Suzuki, K.; Kondo, A. The alga *Euglena gracilis* stimulates Faecalibacterium in the gut and contributes to increased defecation. *Sci. Rep.* **2021**, *11*, 1074. [CrossRef] [PubMed]

15. Taylor, H.B.; Gudi, R.; Brown, R.; Vasu, C. Dynamics of Structural and Functional Changes in Gut Microbiota during Treatment with a Microalgal β-Glucan, Paramylon and the Impact on Gut Inflammation. *Nutrients* **2020**, *12*, 2193. [CrossRef] [PubMed]

16. Marchessault, R.H.; Deslamdes, Y. Fine structure of (1→3)-β-d-glucans: Curdlan and paramylon. *Carbohydr. Res.* **1979**, *75*, 231–242. [CrossRef]

17. Sakanoi, Y.; Yamamoto, K.; Ota, T.; Seki, K.; Imai, M.; Ota, R.; Asayama, Y.; Nakashima, A.; Suzuki, K.; Tsuduki, T. Simultaneous Intake of *Euglena Gracilis* and Vegetables Synergistically Exerts an Anti-Inflammatory Effect and Attenuates Visceral Fat Accumulation by Affecting Gut Microbiota in Mice. *Nutrients* **2018**, *10*, 1417. [CrossRef]

18. Okouchi, R.; Yamamoto, K.; Ota, T.; Seki, K.; Imai, M.; Ota, R.; Asayama, Y.; Nakashima, A.; Suzuki, K.; Tsuduki, T. Simultaneous Intake of *Euglena gracilis* and Vegetables Exerts Synergistic Anti-Obesity and Anti-Inflammatory Effects by Modulating the Gut Microbiota in Diet-Induced Obese Mice. *Nutrients* **2019**, *11*, 204. [CrossRef]

19. Nakashima, A.; Sugimoto, R.; Suzuki, K.; Shirakata, Y.; Hashiguchi, T.; Yoshida, C.; Nakano, Y. Anti-fibrotic activity of Euglena gracilis and paramylon in a mouse model of non-alcoholic steatohepatitis. *Food Sci. Nutr.* **2018**, *7*, 139–147. [CrossRef]

20. Aoe, S.; Yamanaka, C.; Koketsu, K.; Nishioka, M.; Onaka, N.; Nishida, N.; Takahashi, M. Effects of Paramylon Extracted from *Euglena gracilis* EOD-1 on Parameters Related to Metabolic Syndrome in Diet-Induced Obese Mice. *Nutrients* **2019**, *11*, 1674. [CrossRef]

21. Ashburner, M.; Ball, C.A.; Blake, J.A.; Botstein, D.; Butler, H.; Cherry, J.M.; Davis, A.P.; Dolinski, K.; Dwight, S.S.; Eppig, J.T.; et al. Gene ontology: Tool for the unification of biology. The Gene Ontology Consortium. *Nat. Genet.* **2000**, *25*, 25–29. [CrossRef]

22. KEGG. Kyoto Encyclopedia of Genes and Genomes. Available online: https://www.genome.jp/kegg/pathway.html (accessed on 25 July 2021).

23. Takahashi, M.; Kawashima, J.; Nishida, N.; Onaka, N. Beta-Glucan Derived from *Euglena* (Paramylon). In *Trends in Basic Research and Applied Sciences of β-glucan*; Fukui, Y., Tameda, N., Eds.; CMC Publishing Co., Ltd.: Tokyo, Japan, 2018; pp. 174–182.

24. Lee, S.C.; Rodriguez, F.; Storey, M.; Farmakalidis, E.; Prosky, L. Determination of soluble and insoluble dietary fiber in psyllium containing cereal products. *J. AOAC Int.* **1995**, *78*, 724–729. [CrossRef] [PubMed]

25. Folch, J.; Lees, M.; Sloane-Stanley, G.H. A simple method for the isolation and purification of total lipids from animal tissues. *J. Biol. Chem.* **1962**, *226*, 497–509. [CrossRef]

26. Steffen, M. *Application Note: Optimizing Real-Time Quantitative PCR Experiments with the Agilent 2100 Bioanalyzer*; Agilent Technologies, Inc.: Santa Clara, CA, USA, 2008.

27. Mio, K.; Yamanaka, C.; Matsuoka, T.; Kobayashi, T.; Aoe, S. Effects of beta-glucan Rich Barley Flour on Glucose and Lipid Metabolism in the Ileum, Liver, and Adipose Tissues of High-Fat Diet Induced-Obesity Model Male Mice Analyzed by DNA Microarray. *Nutrients* **2020**, *12*, 3546. [CrossRef] [PubMed]

28. Huang, B.B.; Liu, X.C.; Qin, X.Y.; Chen, J.; Ren, P.G.; Deng, W.F.; Zhang, J. Effect of High-Fat Diet on Immature Female Mice and Messenger and Noncoding RNA Expression Profiling in Ovary and White Adipose Tissue. *Reprod. Sci.* **2019**, *26*, 1360–1372. [CrossRef]

29. Park, E.J.; Jung, H.J.; Choi, H.J.; Cho, J.I.; Park, H.J.; Kwon, T.H. miR-34c-5p and CaMKII are involved in aldosterone-induced fibrosis in kidney collecting duct cells. *Am. J. Physiol.-Renal Physiol.* **2018**, *314*, F329–F342. [CrossRef] [PubMed]

30. Wahli, W.; Michalik, L. PPARs at the crossroads of lipid signaling and inflammation. *Trends Endocrinol. Metab.* **2012**, *23*, 351–363. [CrossRef] [PubMed]

31. Tomas, J.; Mulet, C.; Saffarian, A.; Cavin, J.B.; Ducroc, R.; Regnault, B.; Kun Tan, C.; Duszka, K.; Burcelin, R.; Wahli, W.; et al. High-fat diet modifies the PPAR-gamma pathway leading to disruption of microbial and physiological ecosystem in murine small intestine. *Proc. Natl. Acad. Sci. USA* **2016**, *113*, E5934–E5943. [CrossRef] [PubMed]

32. Ma, T.; Verkman, A.S. Aquaporin water channels in gastrointestinal physiology. *J. Physiol.* **1999**, *517*, 317–326. [CrossRef]

33. Spaeth, G.; Gottwald, T.; Specian, R.D.; Mainous, M.R.; Berg, R.D.; Deitch, E.A. Secretory immunoglobulin A, intestinal mucin, and mucosal permeability in nutritionally induced bacterial translocation in rats. *Ann. Surg.* **1994**, *220*, 798–808. [CrossRef]

34. Ishibashi, K.; Nishioka, M.; Onaka, N.; Takahashi, M.; Yamanaka, D.; Adachi, Y.; Ohno, N. Effects of *Euglena* gracilis EOD-1 Ingestion on Salivary IgA Reactivity and Health-Related Quality of Life in Humans. *Nutrients* **2019**, *11*, 1144. [CrossRef] [PubMed]
35. Bougarne, N.; Weyers, B.; Desmet, S.J.; Deckers, J.; Ray, D.W.; Staels, B.; De Bosscher, K. Molecular Actions of PPARalpha in Lipid Metabolism and Inflammation. *Endocr. Rev.* **2018**, *39*, 760–802. [CrossRef] [PubMed]

MDPI
St. Alban-Anlage 66
4052 Basel
Switzerland
Tel. +41 61 683 77 34
Fax +41 61 302 89 18
www.mdpi.com

Nutrients Editorial Office
E-mail: nutrients@mdpi.com
www.mdpi.com/journal/nutrients